Continuously Variable Transmissions for Passenger Vehicles

(Selected papers through 1986)

Prepared under the auspices of the Transmission and Driveline Committees
Passenger Car Activity and Farm, Construction and Industrial Machinery Activity

SAE GLOBAL MOBILITY DATABASE

The papers included in this volume are abstracted and indexed in the SAE Global Mobility Database

Published by:
Society of Automotive Engineers, Inc.
400 Commonwealth Drive
Warrendale, PA 15096

ISBN 0-89883-118-0
SAE/PT-87/30

Library of Congress Catalog Card Number: 86-63691
Printed in U.S.A.

PREFACE

A continuously variable transmission (CVT) has long been a goal of transmission designers to provide an efficient stepless drive. A CVT provides infinitely variable ratio change within a finite range and thus has potential to allow the engine to operate in its most efficient range or at maximum performance.

This is the first publication to bring together the available SAE technical presentations dealing with CVT related topics applicable to motor vehicles of the passenger car type. These presentations have been sponsored by the Transmission and Driveline Committees of the Passenger Car Activity and Farm, Construction and Industrial Machinery Activity.

While some concepts of the Continuously Variable Transmission (CVT) date back to very early motor vehicles, the technology has been emerging more rapidly in recent years. No single design or product has yet fully met the operational, reliability, driveability, and manufacturability requirements of the modern passenger car well enough to be accepted as a full competitor to the conventional automatic transmission. Analysis and actual design implementations have demonstrated some of the problems and complexity of even the deceivingly simple concepts.

The purpose of this publication is to aid further developments, stimulate thinking, and encourage future presentations. A successful commercial product will demonstrate the integration of solutions to design problems into a total system concept. The committees thank the authors and all others who have contributed to this publication and invite paper submissions from any authors who have contributions to make to this body of knowledge. Such papers will be considered for future SAE presentations.

SAE makes no endorsement of any of the material nor establishes any position relative to patents. Authors and readers are left to discharge their own responsibilities in these areas.

Ernest W. Upton
for the
Transmission & Driveline Committee
of the Passenger Car Activity

Editorial Advisory Committee Preparing PT-30

TABLE OF CONTENTS

Efficiency and Performance

System Control and Design

Metal Belt or Chain Drives

Non-Metallic Belt Drives

Traction Drives

Efficiency and Performance

Comparison of CVT Engine Operating Schedules**

David Ganoung
Consultant

**Paper 830574 presented at the International Congress and Exposition, Detroit, Michigan, February, 1983.

ABSTRACT

Maximum fuel economy and low exhaust emissions can exist together if a predominantly wide-open-throttle engine operating schedule is used to complement a continuously variable transmission. Moreover, the concurrently required engine re-calibration often entails less effort than the more usual fuel consumption and emission mapping procedure.

IN 1952, two G.M. research engineers named Caris and Richardson pointed the way toward extensive wide-open-throttle operation of engines powering passenger cars through wide-ratio continuously variable transmissions (1)*. While explicity endorsing lean WOT operation in the case of an "experimental Research engine," these engineers apparently did not translate their endorsement into hardware. Moreover, they reverted to the description, "near full-throttle," after broadening their discussion to include production engines as well as their mysterious research engine. Also, exhaust emission levels were not exactly an overriding concern in 1952. For these, or whatever reasons, the "near full-throttle" approach is what we have seen translated into research hardware.

To follow the lead of Caris and Richardson, exhaust emissions will be relegated to secondary importance, but only for the moment. As a preliminary concession to reduced emissions, however, recirculated exhaust gas will replace the excess air of lean combustion as the diluting constituent of the intake charge.

Extensive WOT engine operation will now be compared to the "near full-throttle" engine operating schedule most typically chosen for use with a wide ratio continuously variable transmission. In the usual approach, a dynamometer study maps the efficiency of an entirely standard passenger car engine against crankshaft speed and brake torque. On this engine map, an infinite number of combinations of RPM and torque will produce a given fixed horsepower, but only one combination will do so with optimum efficiency. If power output is now varied throughout the engine's horsepower range, the point of optimum efficiency will trace a line across the engine map. A system for controlling the engine throttle and the transmission ratio is then programmed to restrict engine operation to this schedule whenever driveability and available transmission ratios permit.

It is only logical to select one of the more efficient passenger car engines before applying the above strategy. As verified in SAE paper #810010 (2), high efficiency certainly characterizes the Nissan NAPS-Z engine. SAE paper #810010 also indicates that the optimum efficiency schedule for this Datsun engine would include much operation at the high torque values where EGR rates are high relative to many production engines. In contrast to these other production engines, the NAPS-Z engine benefits significantly

*Numbers in parentheses designate references at end of paper.

from high EGR because it employs fast-burn design features which counteract the EGR-caused decrease in burn rate. SAE paper #800920 (3), by G.M. Research Laboratories, further documents the advantages of fast-burn engines.

Summarizing to this point, the conventional approach accepts the completely stock engine calibration as sacred and varies throttle opening in search of optimum efficiency. In contrast, the wide open position of the engine throttle may take precedence, with the stock EGR calibration being abandoned so that EGR rate can be varied to find the minimum brake specific fuel consumption at each fixed crankshaft speed. Additionally, the WOT approach requires that both a stoichiometric air-fuel ratio and minimum spark advance for best torque be maintained as the EGR rate is varied in search of the optimum value at each of a progression of representative engine speeds. Which, if either, of these two procedures for determining a CVT engine operating schedule is truly optimum from the standpoint of efficiency?

We can assume for now that each of the two procedures would generate the same indicated torque curve as a function of the crankshaft speed of the NAPS-Z engine. Then, to use a somewhat simplified viewpoint in comparing the two operating schedules, the two average temperatures of the combustion products will be in inverse proportion to the two densities of the intake charge because the indicated mean value of cylinder pressure has just been assumed to be the same for both cases. This is simply an application of the perfect gas equation at equal pressures. Thus, the higher charge density of WOT engine operation would secure reduced combustion temperatures in comparison to the conventional CVT engine operating schedule.

It might at first seem that WOT engine operation would have to produce more torque than throttled operation at the same RPM. On the contrary, extra EGR or other increased dilution of the air-fuel charge can displace fuel-containing mixture in an amount more than offsetting the increased throttle opening. If indicated efficiency is improved, this must in fact be the case.

THERMAL EFFICIENCY

Three classically recognized mechanisms improve efficiency at the lower combustion temperatures available using the WOT schedule. Most obvious is the reduction in the temperature difference which drives heat to the cylinder walls and other surfaces that contain the hot, expanding combustion gases. Second, the ratio of the constant pressure specific heat to the constant volume specific heat remains more favorable at lower temperature, and third, molecular dissociation of the combustion gases is reduced.

On the negative side, the extra EGR dilution of the air-fuel charge required for the WOT schedule would reduce the burn rate of the charge, owing mainly to a decrease in the rate at which expansion of the heated combustion products advances the flame front (4). Even so, engines such as the NAPS-Z are explicitly designed to offset decreased expansion velocity through enhancement of other components of the burn rate. The net result is shown qualitatively in Figure 1.

In illustration of indicated efficiency, which is plotted on the vertical axis in Figure 1, chemical energy in the fuel is converted into a pressure force on the top of the piston with greater and greater efficiency as increasing EGR reduces the three temperature-related losses. Nevertheless, a concurrent deterioration in burn rate allows less and less of the charge to contribute its heat energy close to the top dead center piston position where the heated gases may undergo a full expansion nearly equal to the mechanical compression ratio of the engine. As shown in Figure 1, the deterioration in burn rate quickly predominates in engine designs dating to the pre-emissions era. In contrast, reduced burn rate does not strongly affect the NAPS-Z engine until the three temperature-related factors have already improved thermal efficiency significantly.

But how do we know that the extra increment of EGR required for the WOT schedule is on a generally upward sloping curve as shown in Figure 1? Even the curve for a fast-burn engine will eventually turn downward, if for no reason other than misfire due to excessive EGR.

First, as confirmed by Figures 2 and 6 in SAE paper #810010, the NAPS-Z brake efficiency curve at 2000 RPM and only 15% of available brake torque does not turn downward until EGR reaches fully 20%. Because throttling sufficient to limit brake torque to 15% would also severely curtail the contribution of turbulent intensity to the burn rate, one would expect an equal burn rate at WOT and roughly 35% EGR. So the fast-burn curve of Figure 1 should not turn downward until about 35% EGR. The corresponding WOT brake torque would lie right in the area typical of conventionally-chosen CVT engine operating schedules. (The effect of pumping loss does cause an indicated efficiency curve such as in Figure 1 to turn downward somewhat sooner than the constant 15% brake torque curve, but minimized pumping loss is itself an, until now, unmentioned advantage of the WOT schedule.)

As perhaps a more familiar example of the strong dependence of burn rate on turbulent intensity, vacuum diaphragms have long been used in passenger cars to advance spark timing

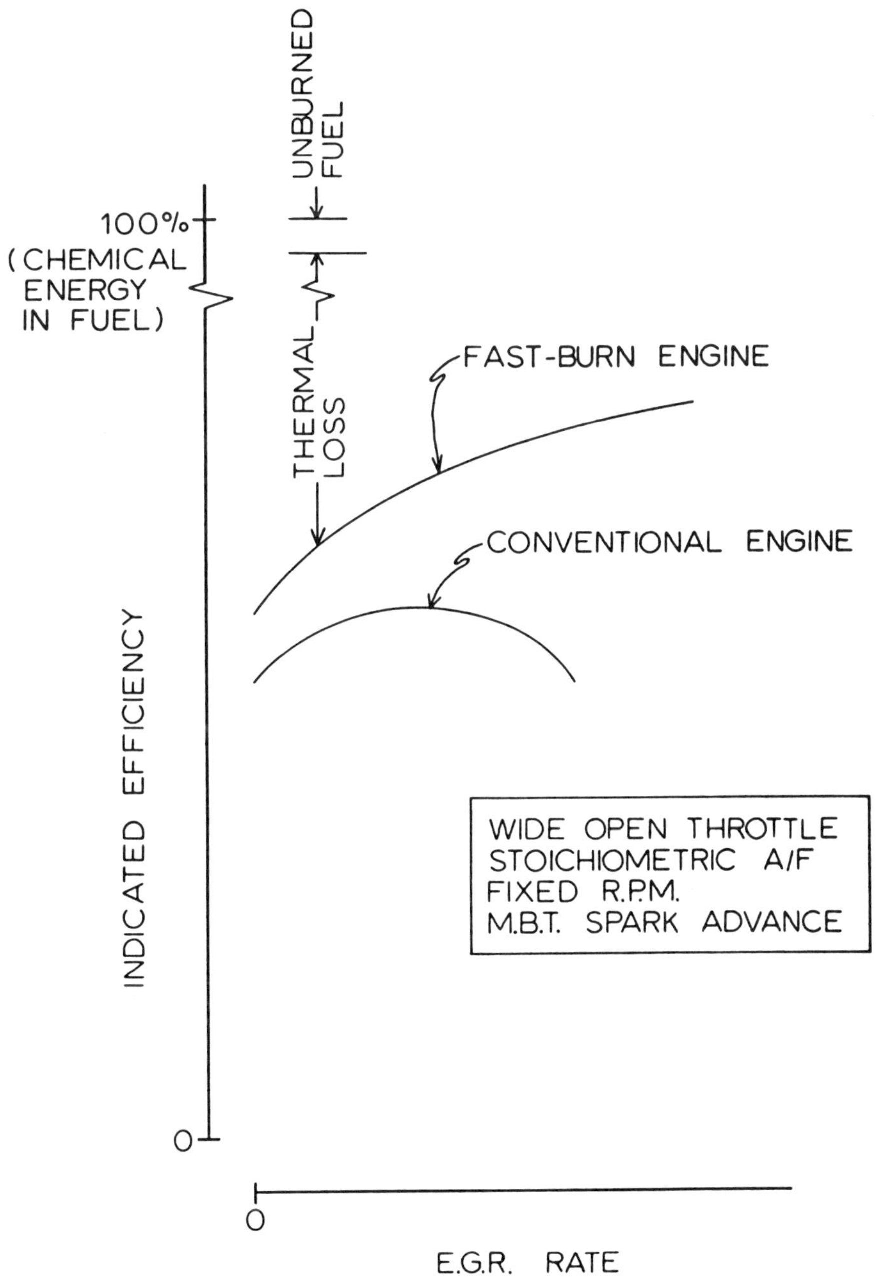

Fig. 1 The efficiency advantage of Fast-Burn engine design features versus EGR rate.

during low torque engine operation, such as the 15% torque condition being considered. Extra spark advance of course means that the burn period is increased after top dead center as well as before top dead center. To anyone familiar with typical spark advance curves, it is certainly plausible that the burn rate in the NAPS-Z engine at WOT and 35% EGR would at least equal that at the same RPM and air-fuel ratio, but 20% EGR and only 15% of available torque.

Although WOT operation is almost certain to improve efficiency because of the reasons already given, even better WOT efficiency will result if the torque curve of the conventional CVT engine operating schedule is abandoned. As suggested before, the EGR rate should be varied during WOT dynamometer testing to locate the minimum brake specific fuel consumption at each RPM, rather than to duplicate a torque curve that is essentially arbitrary. The predominant effect here is to find the best compromise between losses to engine friction and losses to the three temperature-related causes previously described.

MECHANICAL EFFICIENCY

In more detail, Figure 2 assumes the same operating conditions as Figure 1 and qualitatively illustrates the importance of engine friction. Since engine friction remains relatively independent of the useful torque actually delivered to the dynamometer brake at constant RPM, it is shown as a flat line in Figure 2. Pumping loss also has a constant but smaller value, because Figure 2 assumes constant WOT in addition to constant RPM.

If, as shown exaggerated on the

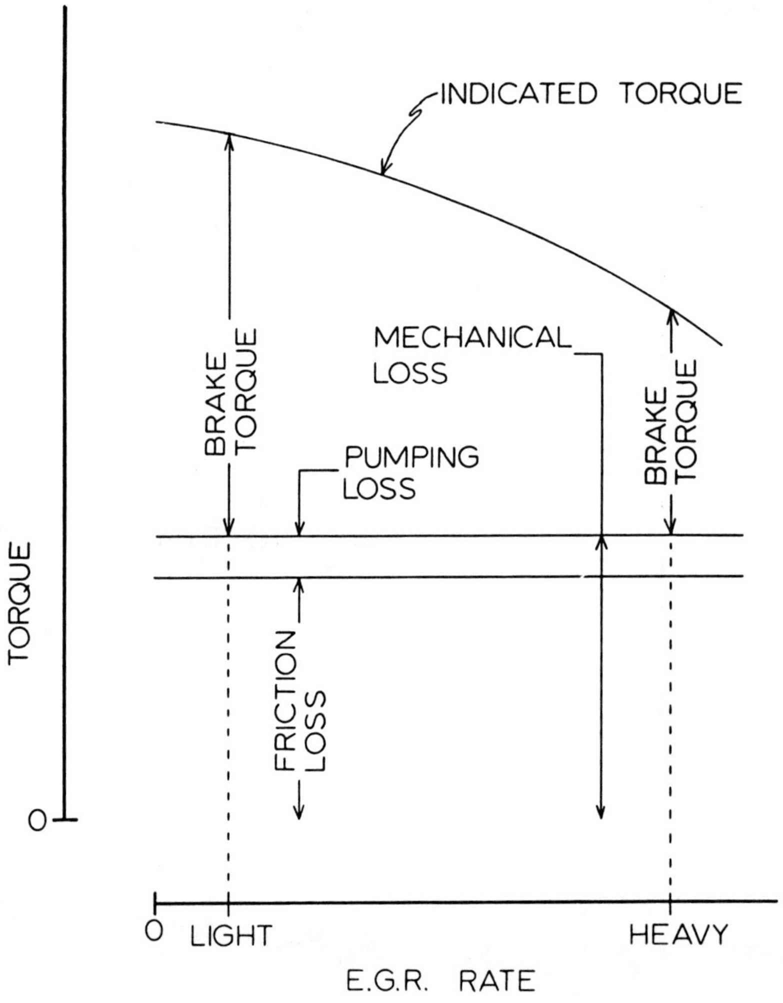

Fig. 2 Engine mechanical loss versus EGR rate.

right hand side of Figure 2, a high rate of EGR reduces torque by displacing a large amount of air-fuel mixture, then engine friction and pumping loss will together absorb a very significant portion of the power delivered to the piston head by the heat of combustion. On the other hand, a much more moderate EGR rate causes mechanical loss to consume a much less significant portion, as on the left hand side of Figure 2.

OVERALL EFFICIENCY

Combining Figures 1 and 2 reveals what is left as useful brake work after all losses are accounted for. But first, Figure 2 must be presented on the same basis as Figure 1. That is, Figure 1 already shows that, for a given amount of fuel consumed, the amount of the thermal loss varies with the EGR rate. Figure 2 must now be adjusted to show that the amount of the mechanical loss also varies with EGR rate when the same given amount of fuel is consumed. In other words, the given amount of fuel will power the engine through more crankshaft revolutions when EGR is increased to displace more of the stoichiometric air-fuel mixture, WOT and constant RPM still being assumed. Since each crankshaft revolution extracts about the same friction and pumping work, regardless of EGR rate, the mechanical loss will increase with EGR. In contrast, thermal loss decreases with EGR.

Figure 3 illustrates the best

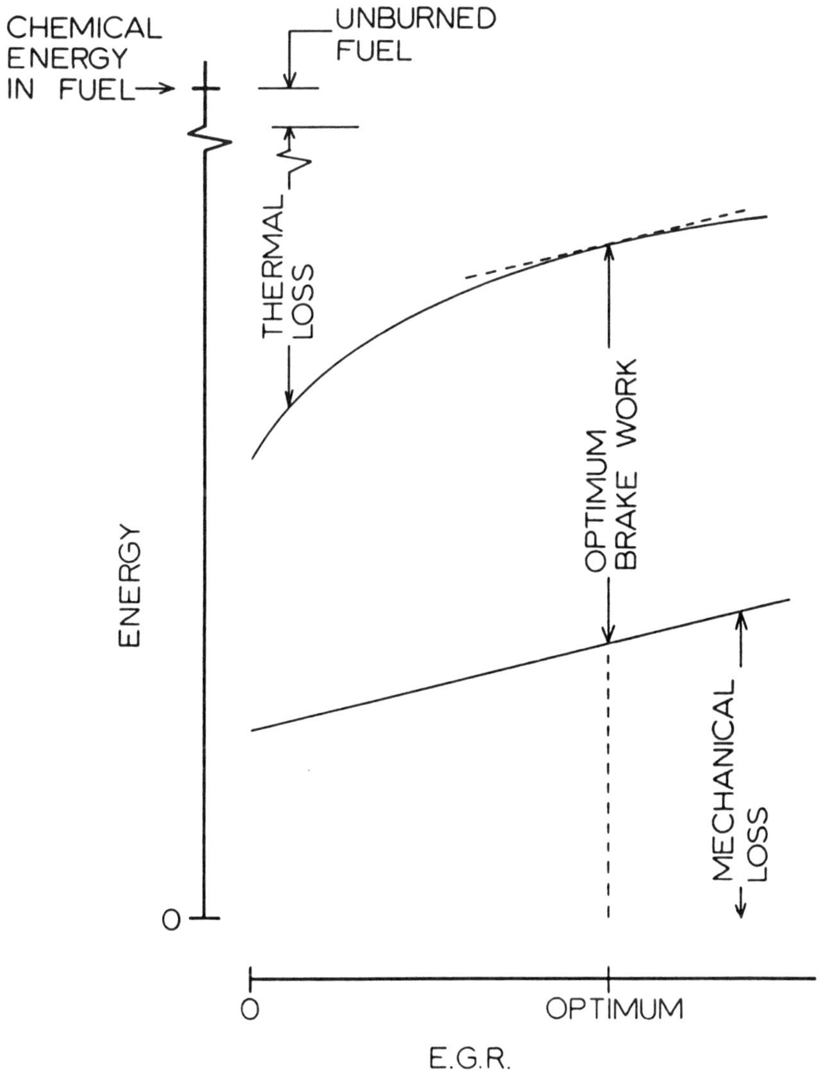

Fig. 3 The overall effect of EGR rate on efficiency.

compromise between mechanical and thermal losses, or equivalently, the EGR rate achieving the lowest brake specific fuel consuption for the engine speed in question. As shown in Figure 3, the EGR scale can be adjusted so that mechanical loss has a constant slope. Then the optimum EGR rate is conveniently found where the tangent to the indicated efficiency curve parallels the line of mechanical losses.

But again, what evidence is there to confirm the qualitative features of Figure 3, and what is the significance of these features? The answer to both questions relates to exhaust emissions.

EXHAUST EMISSIONS

Figure 3 can be used to explore the effect of the conventional NAPS-Z engine calibration feature of rapidly curtailing EGR as the throttle approaches the fully open position. Until now, constant WOT has been implied for Figure 3 because determination of the WOT engine operating schedule requires the EGR rate to be varied at constant WOT to locate its optimum value. Constant RPM also characterizes Figure 3, regardless of whether the conventional or the WOT schedule is being considered. For the conventionally-chosen CVT engine operating schedule, WOT would exist only at the left hand side of Figure 3 where EGR is zero. At the right hand side of Figure 3, slight, but nevertheless significant throttling would accompany the heaviest EGR rate.

As already explained, the mechanical loss line has an upward slope in Figure 3 because EGR displaces fuel-containing mixture, thus causing more crankshaft revolutions to be associated with the consumption of a given amount of fuel. With the conventional EGR calibration, slight throttling would accompany EGR to further restrict the fuel consumption per crankshaft revolution. The increased pumping loss which accompanies throttling also acts to increase the slope of the mechanical loss line in comparison to the case of constant WOT.

Going all the way back to Figure 1 now provides an example of how decreased burn rate causes an indicated efficiency curve, such as the upper curve in Figure 3, to turn downward sooner. The reduced turbulent intensity due to throttling precipitates this earlier downturn in the present case. The combined result in Figure 3 of the increased slope of the lower, mechanical loss boundary, and the decreased slope of the upper, thermal boundary, is of course a much lower optimum EGR rate. This is clear in Figure 3 using the technique of placing a line tangent to the indicated efficiency curve and parallel to the mechanical loss line. As a direct result, the EGR rate of the conventionally-chosen CVT engine operating schedule is unnecessarily biased toward levels too low for effective control of nitrogen oxides emissions. The common practice of enriching the air-fuel ratio past stoichiometric as WOT is approached further reinforces this effect, usually with even more disastrous consequences for carbon monoxide emissions.

Using a slightly different viewpoint of the conventional schedule, some loss in thermal efficiency is more than offset by the higher torque and thereby reduced mechanical losses of enrichment and lower EGR. However, the benefit of extra torque can be gained by fully opening the throttle, rather than by making any sacrifice in thermal efficiency, and especially, in exhaust emission control.

All this is not to say that enrichment of air-fuel ratio and deletion of EGR cannot be used to enhance all-out acceleration of the associated passenger car. Such engine operation would only accompany full depression of the accelerator pedal, but the control system interposed between the pedal and the throttle would institute WOT, together with heavy EGR and no enrichment, throughout a wide range of pedal positions.

CONTROL SYSTEM

Other advantages of the WOT engine operating schedule have yet to be considered. These advantages concern the control system which adjusts both transmission ratio and throttle position to follow the chosen CVT engine operating schedule. Since the control system responds to error, any particular engine operating condition is very rarely obtained exactly, especially under dynamic conditions. When constant WOT is being maintained, however, control system error displaces the engine operating condition along the chosen schedule, rather than causing any reduction in engine efficiency due to displacements away from the schedule.

These displacements can reach

surprisingly high levels because relatively small throttle movements have a pronounced effect on intake manifold vacuum at low crankshaft speed. Since conventional engine controls commonly utilize manifold vacuum to actuate both fuel-air enrichment and EGR systems, slight errors in throttle position often precipitate large excursions in exhaust emission levels.

Such excursions may prompt some intentional deviation of the conventionally-chosen engine operating schedule away from minimum BSFC. In fact, low RPM, high torque engine operation might have to be eliminated entirely from the engine operating schedule just to guarantee that control system error would not cause manifold vacuum to temporarily drop below the value originally selected for minimum BSFC.

RATIO RANGE

Simulation studies based on the performance maps of completely conventional engines often conclude that an essentially negligible increment in fuel economy is afforded by extension of the CVT ratio range beyond about 5 to 1. This is not surprising because, as we have just seen, completely conventional engines will almost always have difficulty when

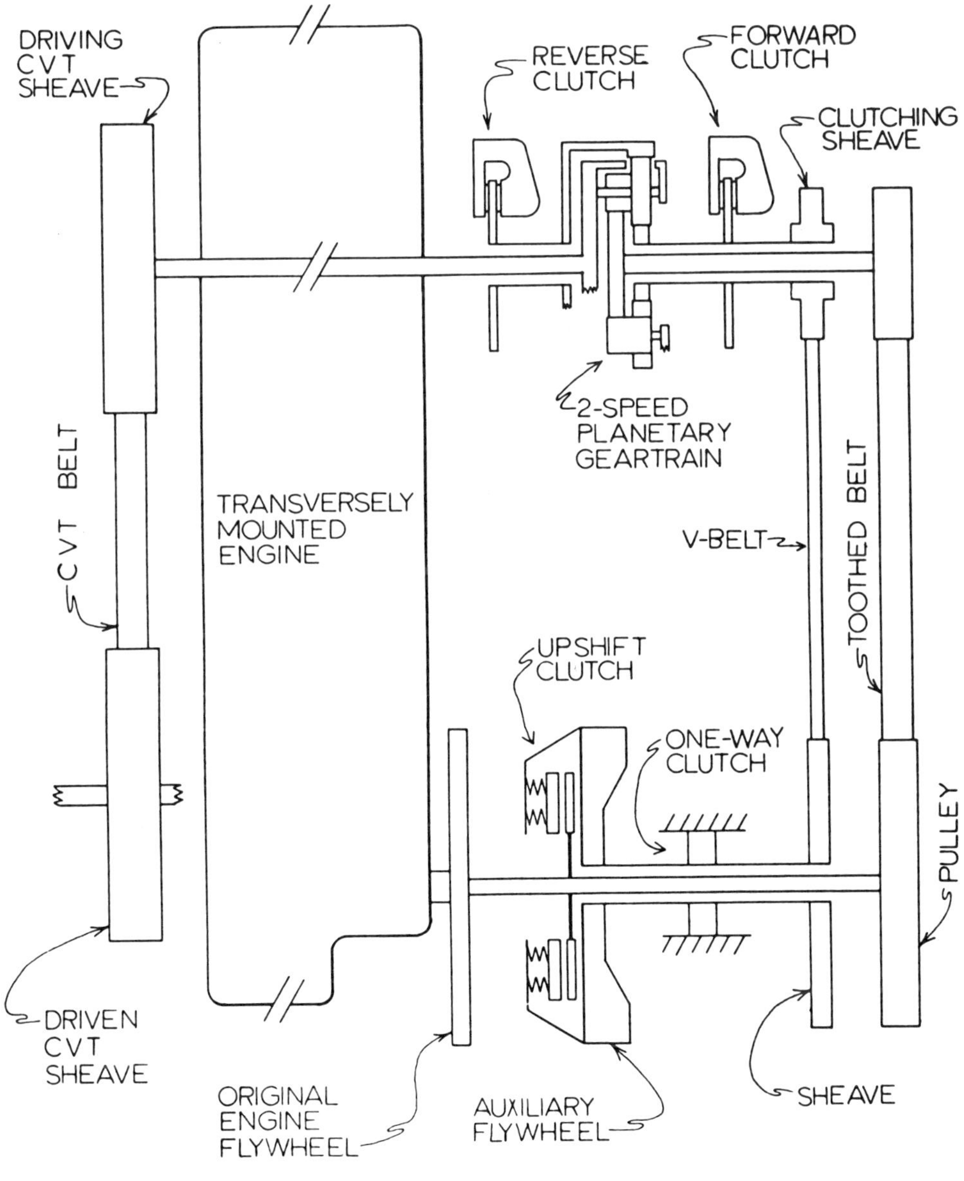

Fig. 4 A hybrid transmission.

operating in the low RPM, high torque regime made consistently accessible by extreme overall gear ratios in the overdrive direction. As another example of engine operating difficulty, conventional carburetion usually leaves much to be desired at low air flow and heavy throttle.

High torque operation at mid-range engine speed poses much less of a problem, but the high vehicle speeds associated with extreme overdrive cruising at this engine operating condition are rarely included in normal driving or in the driving cycles used in simulation studies. Also rarely included in driving cycles are common real world conditions such as extended highway cruising with tailwinds and slight downgrades. These are the conditions which benefit the most from extreme overdrive.

In summary of ratio range, only rather minor engine improvements stand between an extended ratio range and a really significant increment in CVT fuel economy.

DRIVEABILITY

As just explained, extension of the CVT ratio range well beyond 5 to 1 will offer a worthwhile increment in fuel economy. Nevertheless, inexpensive and practical belt-drive CV transmissions do not provide quick downshift response, and an extension of ratio range only compounds this problem. Futhermore, the smaller passenger cars most likely to use rubber belt drives will have engines with four cylinders at the most. Smooth WOT operation of such engines in extreme overdrive ratios will require increased flywheel damping. This damping puts an even bigger hole in downshift response.

If, as shown in Figure 4, a two-speed transmission is placed in series with a belt-drive CVT to expand the ratio range, then both transmissions may downshift simultaneously. In this way, extension of the ratio range improves, rather than degrades available downshift response. However, the very quick downshift possible with the added Ravigneau geartrain precludes the use of a large engine flywheel.

The auxiliary flywheel of Figure 4 gives the required variable flywheel effect and utilizes existing clutch components because the reaction sun gear in the Ravigneau geartrain is kept stationary in Ravigneau low range, but coupled to the crankshaft in high range where the extreme ratios in the overdrive direction would be available.

The Ravigneau geartrain adds relatively little extra hardware when it replaces the somewhat simpler planetary geartrain more typically used to add reverse to belt-drive and other CV transmission types. In this same vein, the Ravigneau geartrain allows simple disc brakes to be used as clutches for initiating vehicle motion, and no fluid coupling or similar drive element is necessary for cushioning upshifts of the Ravigneau gearset. In fact, using conservation of angular momentum to size the auxiliary flywheel will preserve the smoothness of a pure CVT.

In final consideration of Figure 4, the Ravigneau gearset, CVT driver sheave and other transmission components can occupy space high in the engine compartment of a front wheel drive car. This space above the conventional transmission is often either wasted or partially occupied by components that may be relocated. Much of the extra volume of a rubber belt transmission might be accomodated in this way.

An alternative approach is not fully automatic but gives even better response to abruptly increased depression of the accelerator pedal, as well as allowing the use of an engine flywheel having a very large moment of inertia. Briefly, downshifting at a rather leisurely rate may precede the actual demand for increased power if the driver can anticipate a possible need for sudden acceleration. As discussed in the Appendix, only a minor addition to a quite simple CVT control system is necessary to offset prematurely elevated crankshaft speed with throttling. Thus, response depends primarily on how quickly the engine accepts a suddenly re-opended throttle, rather than on how quickly the CVT can downshift. The undesirable inertia torque of a large flywheel is also thereby largely avoided.

SUMMARY AND CONCLUSIONS

Conventional engine calibration features seriously compromise the performance available when transmission ration range is extended and made continuous. Furthermore, appropriate engine re-calibration usually entails less effort than the dynamometer mapping procedure required before a CVT engine operating schedule can be chosen in a conventional manner. In the case of WOT operation, the basic operating schedule is already known prior to any dynamometer testing. In essence, an EGR operating schedule replaces a throttle operating schedule for determination on the engine dynamometer.

When both the WOT schedule and a very wide ratio range are chosen, part-throttle operation is used predominantly to join engine idle to the minimum practical power level along the WOT line. Consequently, good results may be achieved using extrapolation in lieu of extensive dynamometer testing for hot cycle calibration of all part-throttle conditions above minimal engine speed.

With respect to the driveability of CVT-equipped cars, the WOT engine operating schedule also offers a uniquely simple approach. And last, but certainly not least, fast-burn engines with highly desirable features such as multi-point fuel injection are already available in production passenger cars.

APPENDIX

When a wide-ratio continuously variable transmission is used to improve passenger car fuel economy, some type of feedback control system usually replaces the simple mechanical linkage normally used between the accelerator pedal and the engine throttle valve. This is so because the efficiency of the gasoline engine can be improved by using the CVT to expand the role engine speed plays in determining power output. With the importance of crankshaft speed thus increased, the effect of the engine throttle is correspondingly decreased, and the control system then often becomes necessary to re-establish a responsive feel at the accelerator pedal.

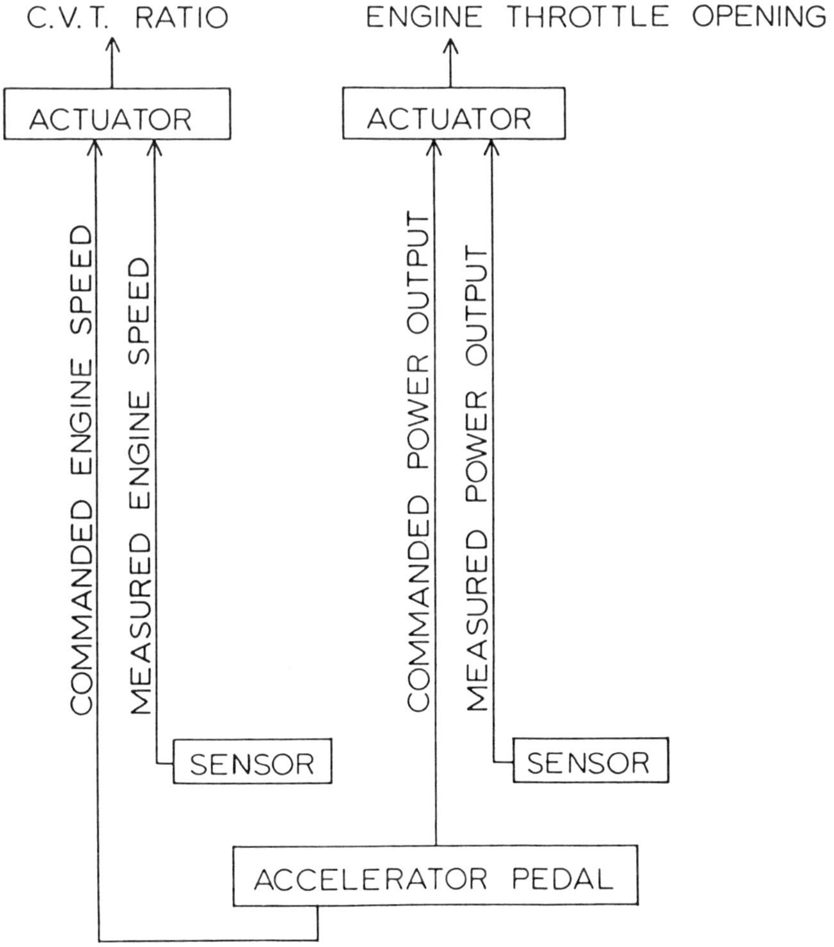

Fig. A1 CVT control system diagram.

Since engine speed thus becomes quite important in determining power output, the control system ties transmission ratio, as well as throttle opening, directly to accelerator pedal position. Figure A1 reveals one basic way of doing this by using two actuators along with engine sensors for crankshaft speed and power output.

At each engine speed, only one throttle setting will produce the optimum compromise between efficiency and reduced exhaust emissions. The corresponding engine operating schedule of power output versus crankshaft speed can be stored in a simple memory device for simultaneously generating the two command signals as a function of accelerator pedal position. Nevertheless, one might wonder why the throttle actuator should not respond directly to commanded and sensed values of throttle setting, rather than to values of power output.

To begin with, the more practical types of CVT have a finite range of ratio in the overdrive direction. Consequently, engine speed will often reach equilibrium at values distinctly exceeding the commanded value, as when the associated passenger car experiences tailwinds or slight downgrades at highway speed. Without a compensating factor, this range of equilibrium crankshaft speed at a single accelerator pedal position leads to an indefinite feel and can require bothersome and unnecessary manipulation of the pedal.

Using power output, rather than throttle position, as the commanded and measured variable inherently provides the needed compensation. For example, even though engine speed may deviate widely from the commanded value, appropriate changes in throttle opening will still regulate power output to a single equilibrium value at each accelerator pedal position. Briefly stated, throttle control is inherently resurrected as the main influence over power output when transmission ratio is locked in the most extreme overdrive value and is therefore not available for effective control of power output.

A possibly more important aspect of driveability again relates both to the characteristics of practical CVT's and to deviation in crankshaft speed from the value chosen purely in considertion of efficiency and acceptable exhaust emissions. In this case, however, the error in crankshaft speed is intentional.

In a conventional passenger car, the engine usually operates at higher speed and with more throttling than in a CVT-equipped vehicle. Thus, sudden acceleration in the CVT vehicle depends more heavily on downshift response, which often leaves something to be desired, especially for belt-drive CVT's. If, however, the driver can momentarily elevate engine speed, but without disturbing power output, then quick response again becomes available.

In more detail, the driver would press and hold a button or other added control device when he anticipates extra power might soon be needed. The button would cause commanded engine speed to be multiplied by some conveniently variable factor, but no corresponding change would occur in the commanded power level. As a result, downshifting and concurrent closing of the throttle would precede the actual demand for increased power. Because this action on the part of the driver is in anticipation, the unchanged power level avoids an unnecessary and possibly dangerous situation where compensation with the accelerator pedal would be required. When the driver does demand extra power a short time later, a substantial portion of that power can be had as quickly as the throttle reopens.

Another situation where extra engine speed and additional throttling might be desirable is during cold engine operation. Here again it is advantageous to maintain the normal functional relationship between power output and accelerator pedal position in spite of the temporary changes in crankshaft speed and throttling.

At this point, one could reasonably ask if the foregoing advantages are worth the cost of a sensor for measuring engine power output. In light of the precision normally required to keep the engine operating acceptably close to the chosen schedule, the answer might be negatve. (Digital electronic technology does offer the following approach. First, a microprocessor would store an engine data map of actual power output as a function of accurately sensed values of crankshaft speed and throttle angle. When a difference were to exist between commanded and actual values of power output, the data map would further provide the new throttle angle needed at the existing engine speed.)

Nevertheless, a predominantly wide open throttle (WOT) engine operating schedule breaks from convention in several important ways, including its control system requirements. Most importantly though, the WOT schedule and its concomitant changes in engine calibration provide the best available compromise between improved fuel economy and acceptable exhaust emissions. The reasons for this superior compromise are explained in the body of this paper.

To return to control system considerations, Figure A2 shows a basic arrangement for the WOT engine operating schedule. In Figure A2, a torque potentiometer extracts a power output signal from the existing crankshaft speed signal. This is often possible using an inexpensive vacuum-actuated diaphragm because intake manifold pressure usually gives a reasonable approximation of the torque output of a gasoline engine. In addition then, the crankshaft speed sensor should generate its signal in direct proportion to the actual speed.

Also notable in Figure A2 is the lack of any memory element for coordinating the command signals for output power and crankshaft speed. A simple accelerator pedal potentiometer supplies the same voltage for both command signals.

In operation, the ratio control actuator will act to match crankshaft speed to the commanded value. As long as the actuator can accomplish this, the voltage applied to the top of the resistive element in the torque potentiometer will remain equal to the corresponding command signal supplied to the throttle actuator. The top of the resistive element is also the position occupied by the wiper in the potentiometer when the throttle is wide open. Consequently, the throttle will remain wide open.

If, on the other hand, the maximum overdrive ratio can only limit engine speed to, say 125% of the commanded value, then the voltage at the top of the resistive element will similarly be 125% of the command signal supplied to the throttle actuator. Accordingly, the actuator will close the throttle until the vacuum diaphragm pulls the attached potentiometer wiper down to sample 80% (1/125%) of the applied voltage. The resistive element can usually employ a calibration where the manifold vacuum at 80% voltage gives about 80% of WOT torque throughout a broad crankshaft speed range. (Other inexpensive alternatives are often available, such as a torque signal derived from the pulse width of a fuel injection system.)

Although the "power signal"

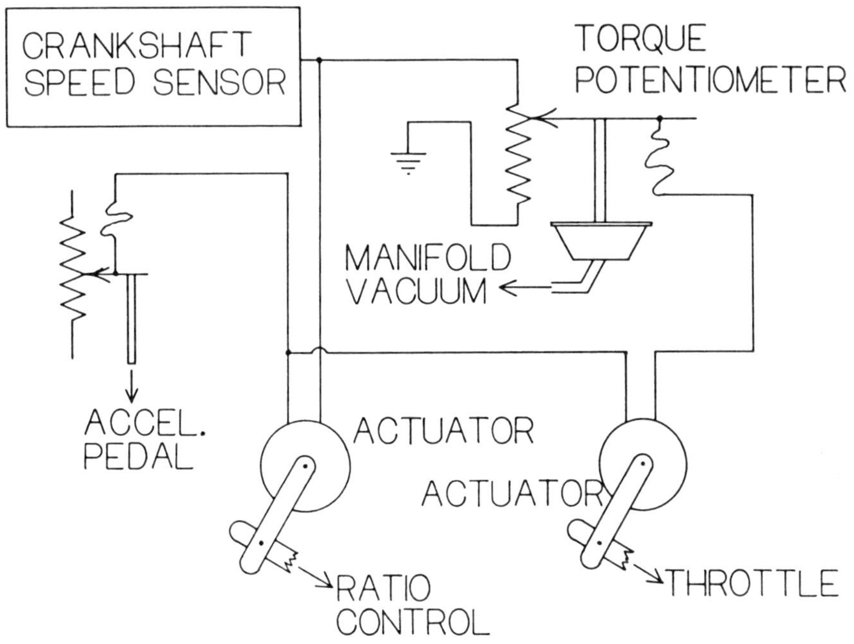

Fig. A2 Basic WOT control system.

generated at the wiper of the torque potentiometer is basically a signal approximating engine intake flow, and a rough approximation at that, control system errors exist almost exclusively as displacements along the efficient WOT schedule. In contrast, more conventional engine operating schedules cause excursions into less efficient operating areas even when the control system is much more precise than the configuration just described. Using an intake flow signal also allows heavy exhaust gas recirculation to enhance WOT operation, but without upsetting control system operation when the EGR is abruptly eliminated to thereby maximize power output.

A simple operational amplifier may be added to temporarily increase the command signal for crankshaft speed as supplied to only the ratio actuator. Such an amplifier provides one method of elevating engine speed for anticipated acceleration or cold engine operation, or both. A similar technique avoids WOT operation at inappropriately low crankshaft speeds and is mandatory.

In conclusion, the WOT engine operating schedule enjoys several advantages. Sophisticated control system performance with only simple and inexpensive hardware is one of these advantages, though by no means the most important one.

REFERENCES

1. CARIS, D.F. and RICHARDSON, R.A., "Engine-Transmission Relationship for Higher Efficiency," SAE Summer Meeting, Atlantic City, June 6, 1952.

2. MASANORI, H., KADOTA, T. and SUGIYAMA, Y., "Nissan NAPS-Z Engine Realizes Better Fuel Economy and Low NO_x Emission," SAE Paper 810010, 1981.

3. MATTAVI, J.N., "The Attributes of Fast Burning Rates in Engines," SAE Paper 800920, 1980.

4. MATTAVI, J.N., GROFF, E.G. and MATEKUNAS, F.A., "Turbulence, Flame Motion and Combustion Chamber Geometry - Their Interactions in a Lean-Combustion Engine," Paper C100/79, presented at I. Mech. E. Conference on Fuel Economy and Emissions of Lean Burn Engines, London, June 1979.

CVT-Efficiency Measured under Dynamic Running Conditions*

Anders Bonthron
Volvo Components Corp.
Transmission Division
Sweden

*Paper 850569 presented at the International Congress and Exposition, Detroit, Michigan, February, 1985.

ABSTRACT

The CVT, designed by Volvo Components Corporation for the Volvo LCP 2000, did not give the expected acceleration to the cars in which it was tested. That led us to suspect that a CVT of the chain or belt type has a poor efficiency during ratio change.

Our measurements prove that the changing of ratio only is not a problem. The measurements illustrate the importance of controlling the rate of change of the engine speed. How this control is carried out influences the response when going from low power at low engine speed to high power at high engine speed.

BACKGROUND

For the light weight, energy efficient car project Volvo LCP 2000 a CVT was developed by Volvo Components Corporation. This was designed for a maximum engine torque of 120 Nm and had a ratio span of about 6:1. The highest ratio is equivalent to 49 km/h at 1 000 rpm engine speed.

The control system was designed and developed by Kongsberg Vaapenfabrikk AS, Norway, and incorporated a micro-processor.

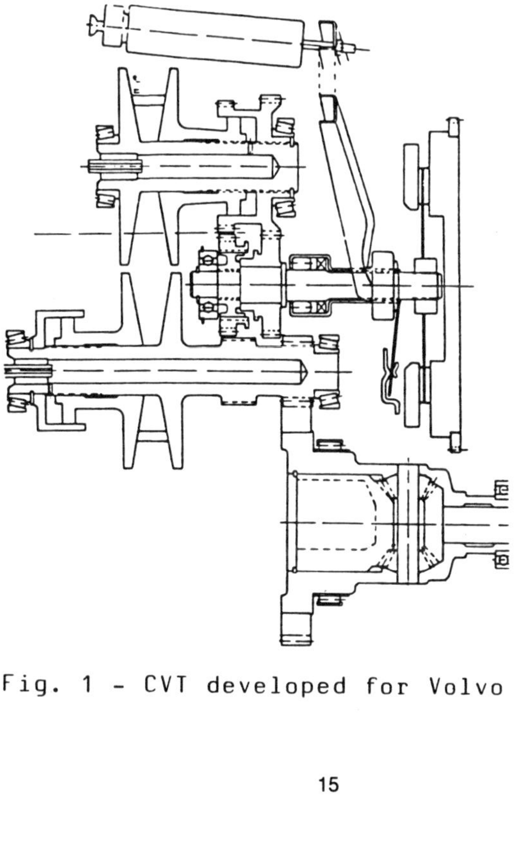

Fig. 1 - CVT developed for Volvo LCP 2000

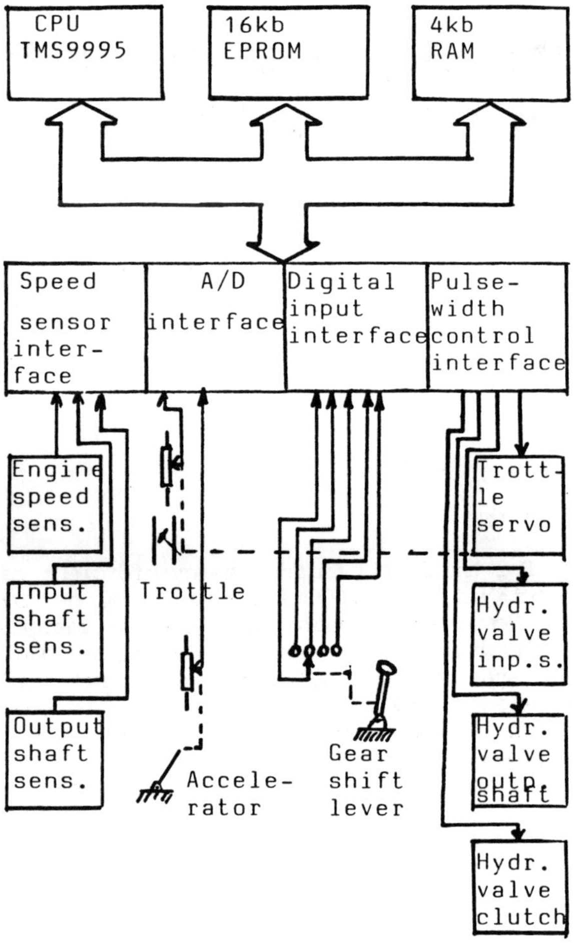

Fig. 2 - CVT - control system schematic diagram

Input information to the microprocessor includes speeds, throttle opening, accelerator position and gear shift lever-position (D, N, R, P). The output consists of drive currents to three pressure control valves and one servo engine. The valves control the hydraulic pressure for the two disc actuators and the clutch actuator. The servo engine controls the throttle opening.

With this system it is possible to control the chain clamping loads according to the torque, ratio and speed with which the CVT is running. The torque is estimated from the engine speed and the throttle opening position. Our own experiments and co-operation with PIV in Germany lead us to a mathematical model for suitable clamping loads for trade off between bad mechanical efficiency and low risk of slip. We think, however, that this model is not yet optimized because it gives us bad mechanical efficiency at low power (road load at constant speed) - 80-85%, hydraulic losses not taken into account.

When testing our CVT in test cars, we therefore knew there was a potential for better fuel efficiency than measured. Also an improvement in acceleration was desirable, but how was that to be achieved?

Our results gave rise to the opinion that the mechanical losses in our CVT were fairly high in constant running conditions and even higher in dynamic running conditions. We had studied the losses in constant running conditions and knew that with the axial loads we were using, we could expect a mechanical efficiency of 80-85%. With a smaller safety margin against slip, we knew it was possible to improve this. We were also aware of some methods of improving the efficiency by changing the design.

We knew nothing about CVT-efficiency during ratio change and our discussion partners gave us different answers, so we measured it.

This work has been carried out within Volvo Components Corporation under supervision and support from Volvo Car Corporation, Power-train Engineering.

MEASURING METHOD

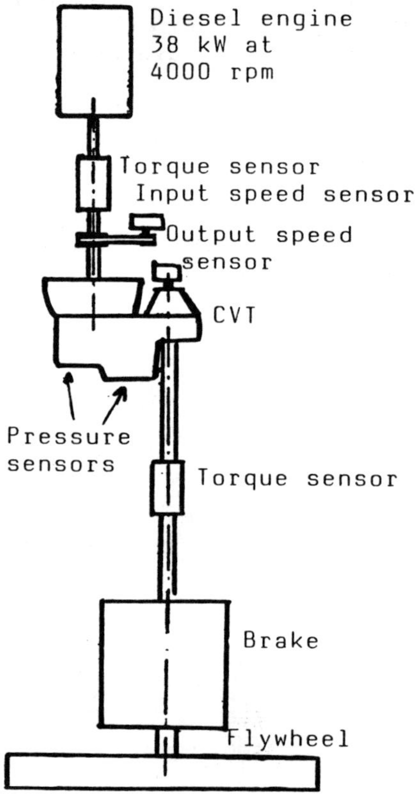

Fig. 3 - Test-bench for study of CVT-efficiency

The CVT was run in a test bench with a 1 600 cm^3 diesel engine, a hydraulic brake and a fly wheel, see figure 3. There is one torque sensor at the CVT input shaft and one at the CVT output shaft. Speed transducers with a high number of pulses per turn were used - 2 500 at the output and 500 at the input. Those are the parameters needed to be measured for calculating the mechanical efficiency.

Furthermore we used pressure transducers for measuring the CVT control pressures and the pump output pressure. From the CVT control pressures we can easily derive the axial forces acting from the pulleys on the chain. The pump output pressure gives the power consumption of the hydraulic system. (The pump is run by an electric engine at constant speed giving constant flow.)

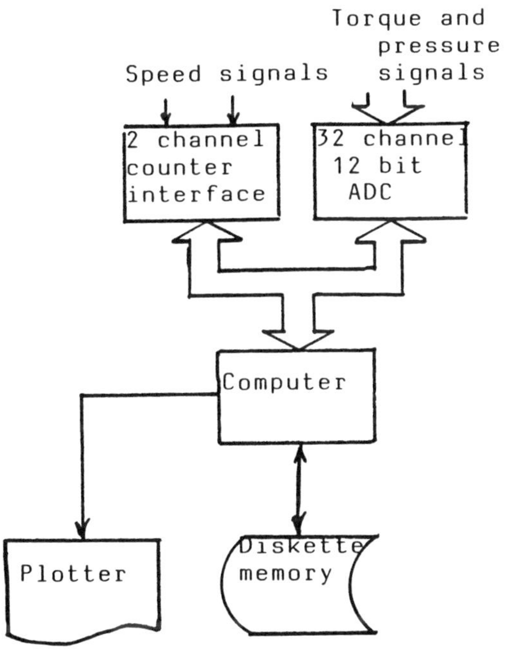

Fig. 4 - Data collecting system schematic diagram

The data collecting system consists of a rather small personal computer (32 Kbyte RAM) based upon Z-80 micro-processor equipped with one 32 channel 12 bit ADC and one two channel 12 bit counter interface for the speeds. See figure 4.

A fast data collecting program (written in Assembler) just reads data and stores it in the internal RAM. When the available space is full it is dumped on a diskette. Then the evaluation program, written in high level language, reads the values from the diskette, calculates the actual values in the wanted dimensions and calculates the values of such parameters as the efficiency. The result is then presented on a plotter.

A typical sample period consists of 16 ms counting time for the speeds during which the 6 analogous channels used are read 26 times, the values being added to one sum per channel. During about 5 ms the 6 sums are divided by the number of readings to form average values and stored away together with the number of counted pulses from the speed transducers, and the counter interface is started up again. (See figure 5.)

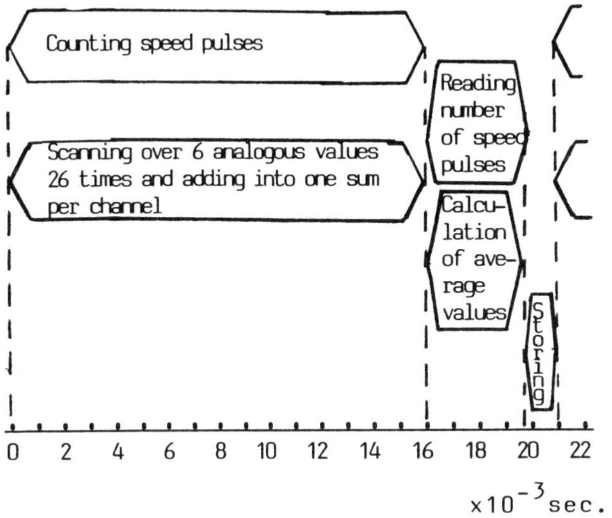

Fig. 5 - A typical sample period

While running, the CVT is controlled by a special control system, where the required engine speed is needed as an input signal and the input torque also is fed to the control system. The required engine speed is set by the operator, but he can set two different values and the ramp time for changing from one to the other.

RESULTS

A number of tests were made with the initial condition constant speed with road load and relatively low engine speed. Then the engine was given full injection at the same time as the required engine speed was changed to a value corresponding with maximum power.

The results from these tests are here presented in diagrams showing input speed, output power, input torque and efficiency as function of time. The efficiency shown in the diagrams is calculated as output torque x output speed divided by input torque x input speed. The following notations are used:

T_i = input torque P_{out} = output power

n_i = input speed η = mechanical efficiency

The hydraulic losses are not shown in the diagrams but we have estimated them to be around 250 W while running at constant speed, 300 W while accelerating with constant input speed and 420 W while changing input speed.

These figures are highly dependent on the layout of the components in the hydraulic system.

Figures 6 and 7 show the results from tests made from initial conditions corresponding with a steady 35 km/h resp. a steady 90 km/h and a comparison between them says something about the influence of different initial speed when starting to accelerate.

The dominating difference between the two diagrams lies in the efficiency while still in the initial state.

With the very low load at 35 km/h (c:a 2 kW) the CVT efficiency is as bad as 74% while with the higher load of 90 km/h (c:a 7 kW) it reaches 85%. When the input torque and speed have started to increase both diagrams show almost the same thing:

- The torque rise is limited due to the acceleration of engine inertia.
- The output power has a rather small first increase step and then gradually increases with the input speed.
- The efficiency goes almost immediately to the final value, c:a 88-90%.

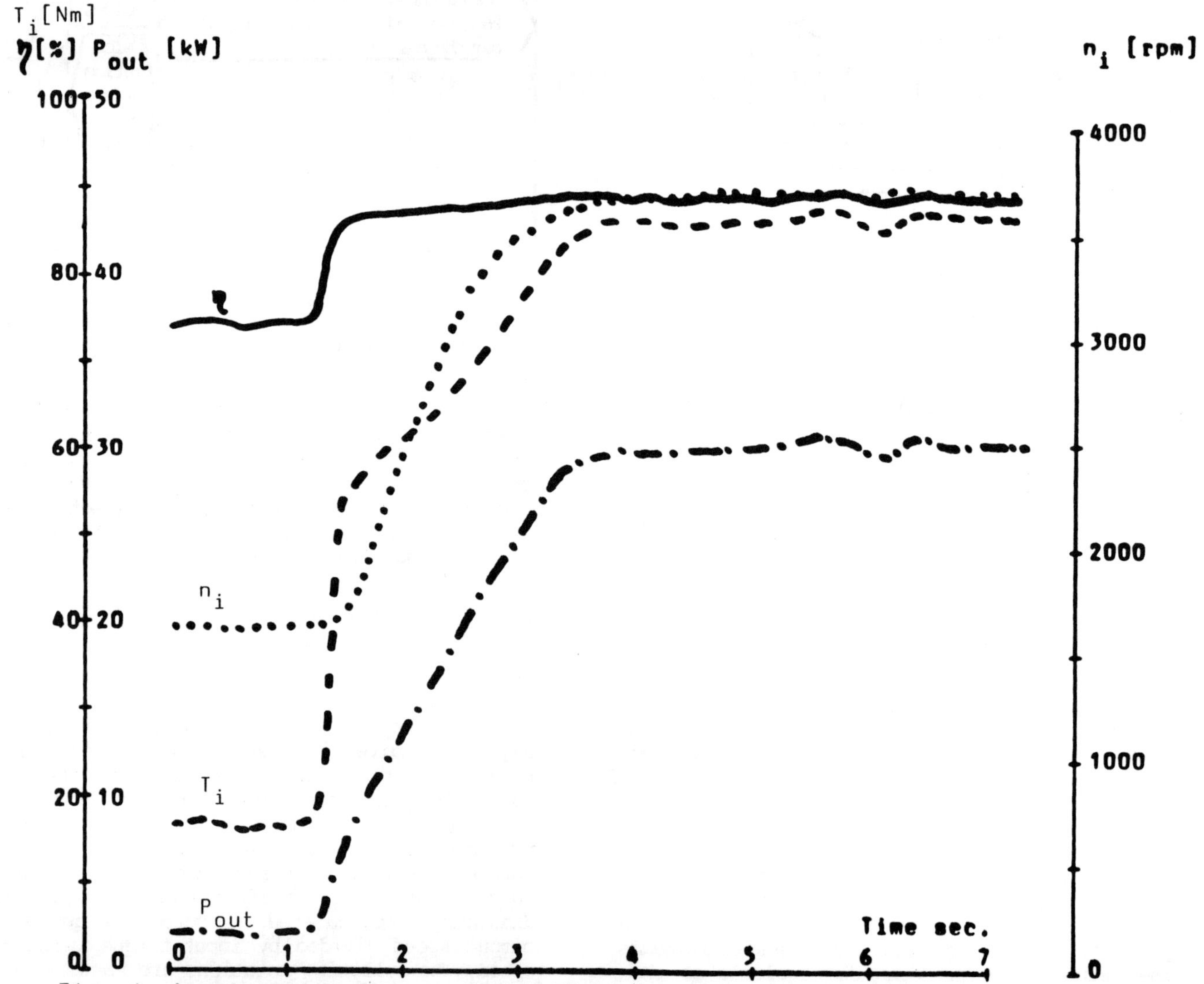

Fig. 6 -Accelerating from constant 35 km/h speed

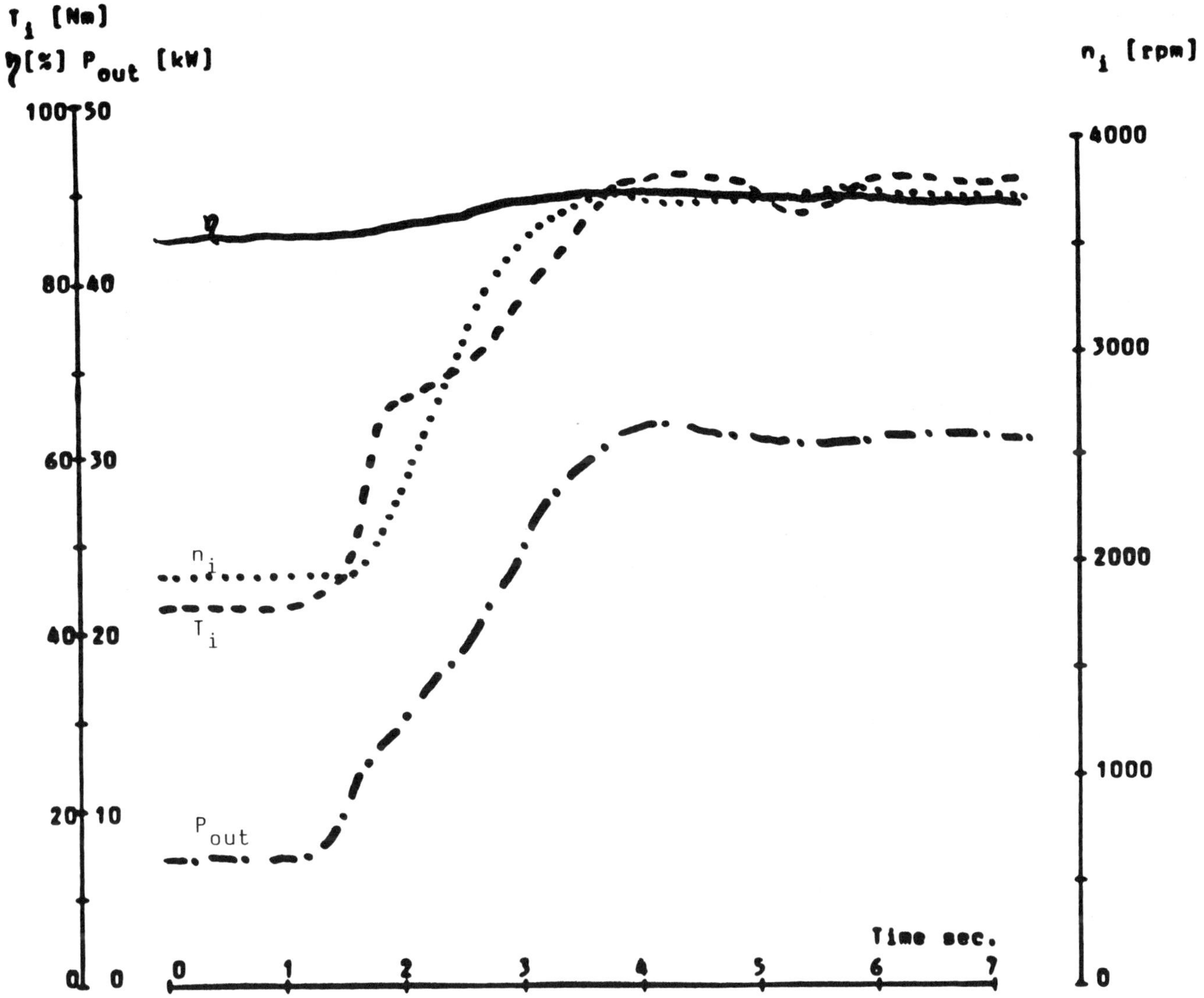

Fig. 7 - Accelerating from constant 90 km/h speed

Figures 8 and 9 show the results from tests made in the same initial conditions, corresponding with a steady 50 km/h, but the rate of change of the input speed is lower in figure 9 than in figure 8.

The difference between those two diagrams is the obvious result from the mechanical laws: Due to the smaller acceleration of engine inertia in figure 9 the immediate increase in input torque and of course in output power is greater.

The increase of efficiency is rather immediate in both cases.

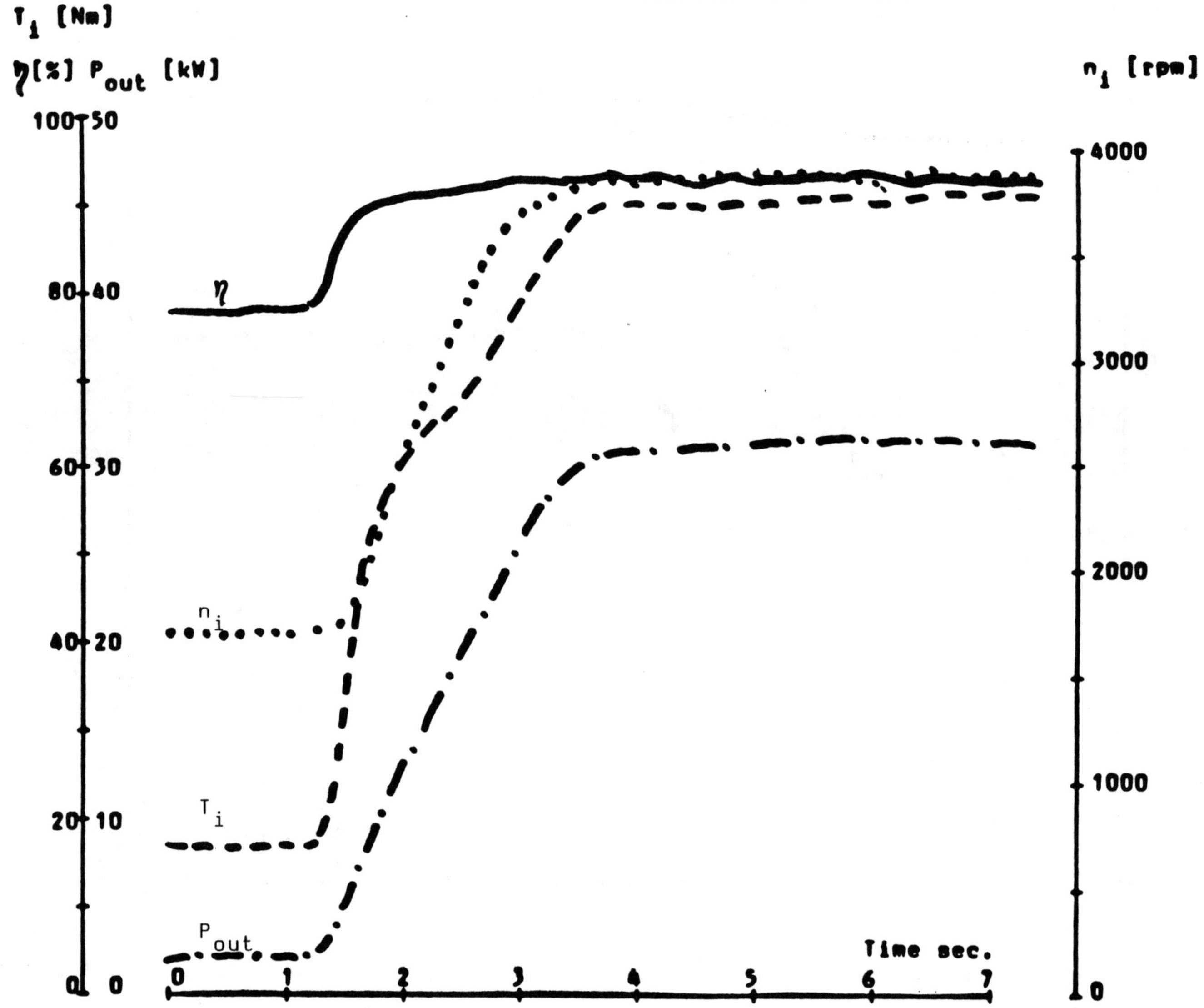

Fig. 8 - Accelerating from constant 50 km/h speed, normal engine acceleration

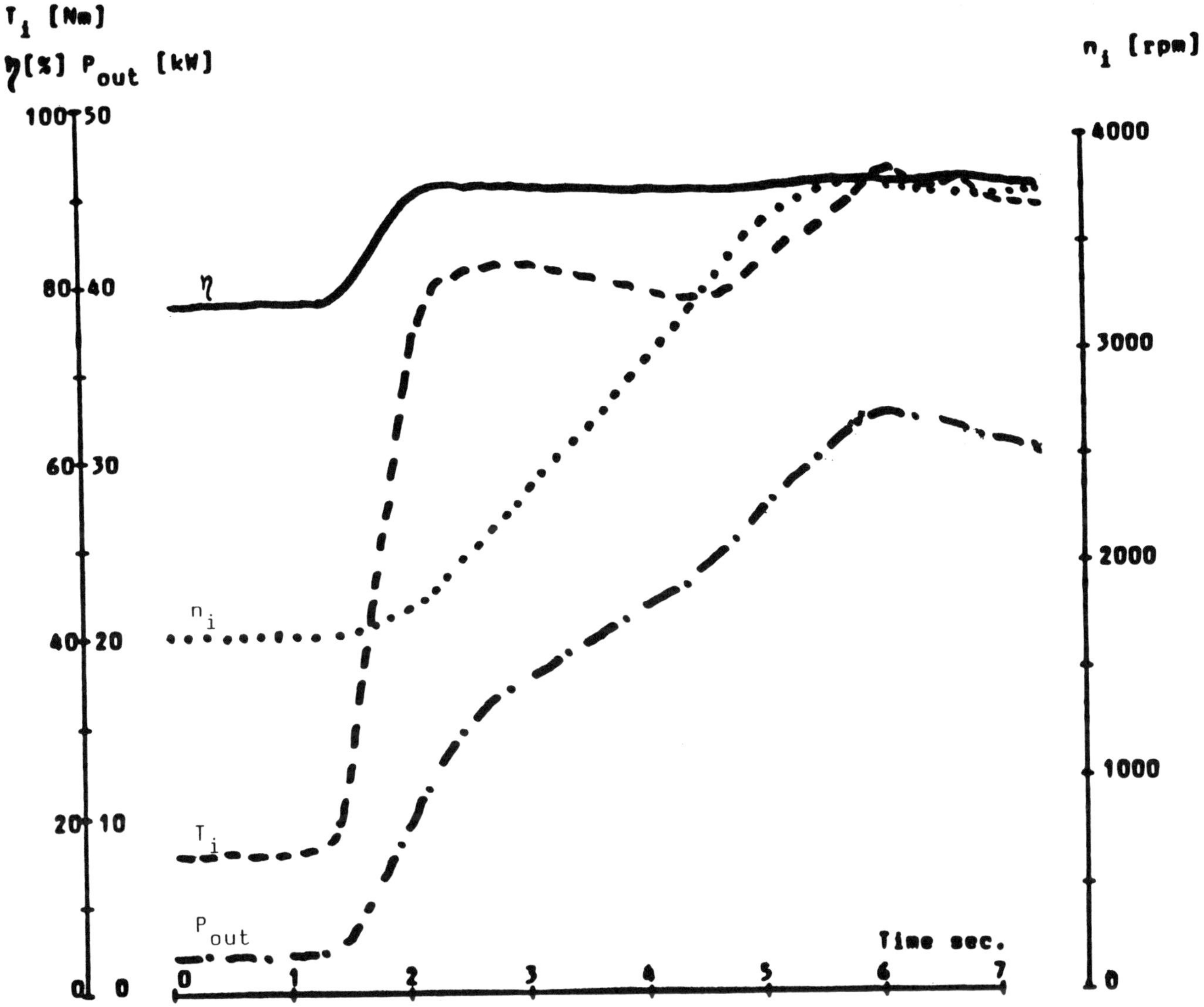

Fig. 9 - Accelerating from constant 50 km/h speed, low engine acceleration

Figure 10 shows the result from a test which was run in the same way as the that shown in figure 8, but the axial forces from the pulleys on the chain are at an average 10% lower than in the test shown in figure 8. There is no significant difference in the efficiency when accelerating with constant input speed and 2-3% unit at constant 50 km/h.

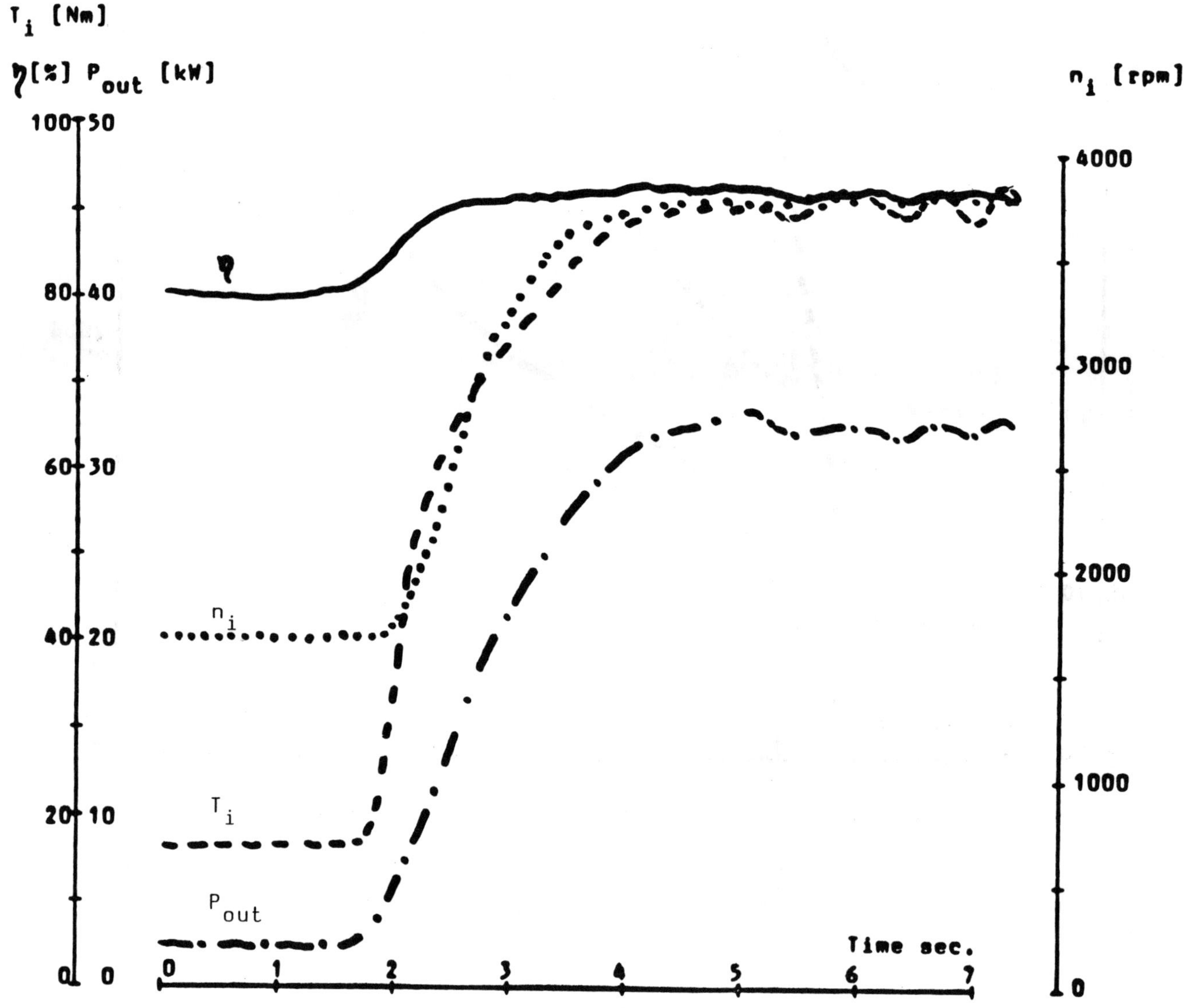

Fig. 10 - Accelerating from constant 50 km/h speed, low axial forces

Figure 11 shows the result from a test where the initial condition was acceleration with input speed at 3 700 rpm. Then, when output speed corresponding to 90 km/h was reached, changing input speed to 1 900 rpm and torque to the level of road load at that speed. A phenomenon which is almost the inverse to that shown in figure 7 can be seen here.

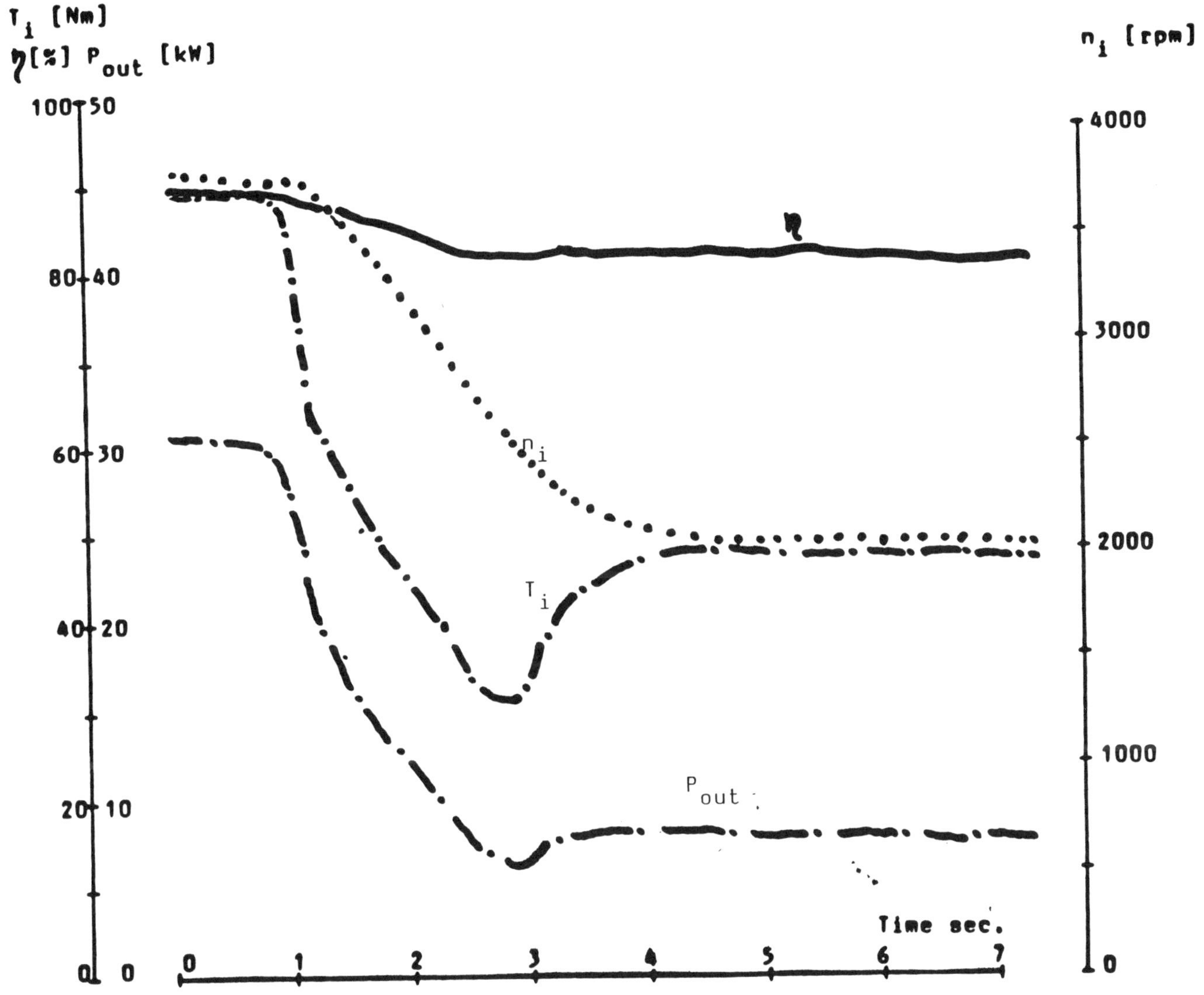

Fig. 11 - Change from acceleration to constant 90 km/h speed

Figure 12 shows efficiency as a function of input torque at constant ratio 1:1 between the CVT-pulleys and at input speeds 1 700, 2 560 and 3 300 rpm. In the diagram points marked with asterisks represent the efficiency and the torque when the ratio was allowed to change through ratio 1:1 under dynamic tests. The input speed when the ratio passed through 1:1 is written beside the different points.

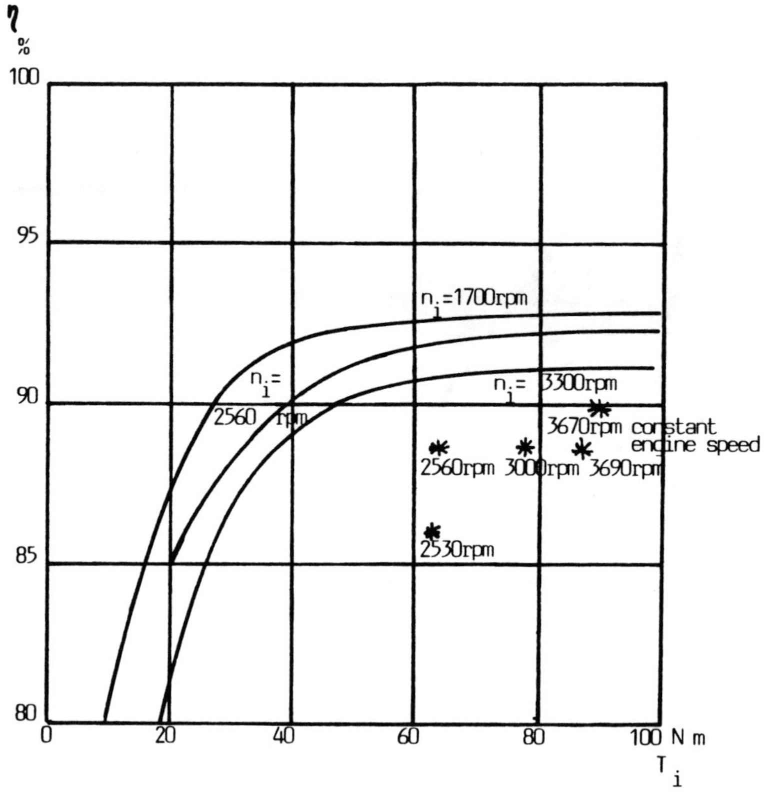

Fig. 12 - Efficiency as a function of input torque at constant ratio 1.0

Figure 13 shows efficiency as a function of input torque at constant ratio 1,4 and at input speeds 1 700 and 3 300 rpm. Points marked with asterisks represent results from dynamic tests in the same way as described for figure 12 above.

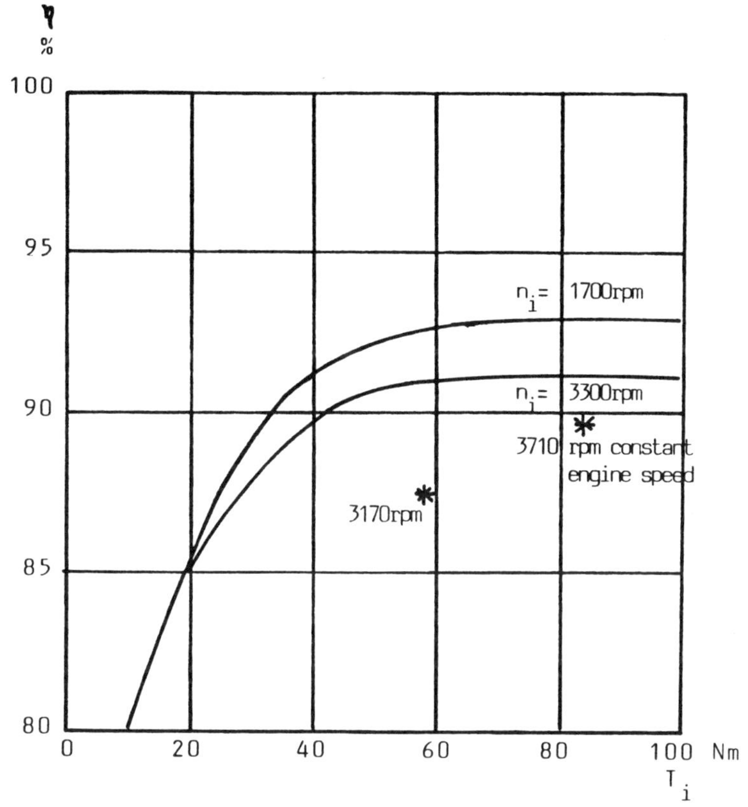

Fig. 13 - Efficiency as a function of input torque at constant ratio 1.4

Figure 14 shows efficiency as a function of input torque at constant ratio 0,7 and at input speeds 1 700 and 3 200 rpm. Points marked with asterisks represent results from dynamic tests in the same way as in figures 12 and 13.

The fact that the points passed through when the ratio was allowed to change dynamically lie slightly under the curve obtained when the ratio was held constant, shows that there is some influence on the efficiency from the changing of ratio.

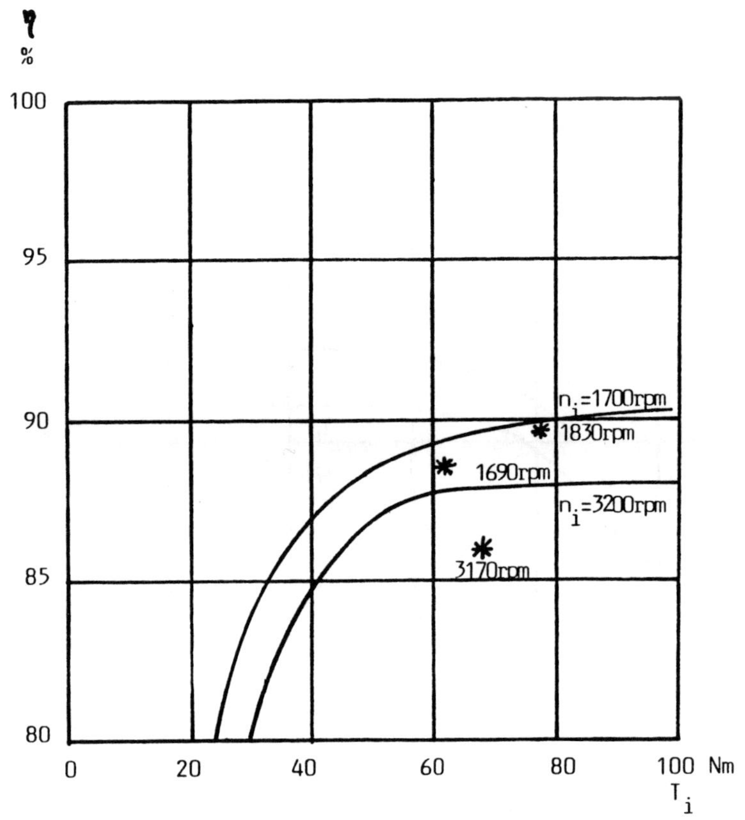

Fig. 14 - Efficiency as a function of input torque at constant ratio 0.7

CONCLUSION

The main object of our measurements was to show if there was any major decrease of efficiency due to changing ratio. The answer is "no". The main influence on efficiency seems to be the amount of power through the CVT.

Hydraulic losses do not change this situation because they are not affected by the power through the CVT and only slightly influenced by the rate of change of the ratio.

In our CVT cars we had to set limits on rate of change of engine speed and throttle opening speed for system stability reasons. Together with extremely low engine speed and high torque at road load this gave us a slow response when a higher output power was requested.

As a spin off from the measurements we can study the mechanical system in which the CVT is one part, running under dynamic and transient conditions. Engine inertia represents an energy storage that has to be filled before a higher power at higher engine speed is available at the output. Such limitations can be studied from the measurements made. The resulting response using different control strategies can be measured in order to develop an optimized system.

Performance of CVT Transmissions*

Howard E. Chana
Advanced Engineering Staff
General Motors Corp.

*Paper 860637 presented at the International Congress and Exposition, Detroit, Michigan, February, 1986.

ABSTRACT

Continuously Variable Transmissions (CVT's) have become a part of many discussions in Automotive circles today while on the horizon there appears to be units that are approaching production status. An analytical method has been used to determine Fuel Economy - Performance trade-offs on drivetrain component features including, ratio range, number of ratio steps, clutched torque converters, and Continuously Variable Transmissions. Indications from both the analytical approach as well as vehicle tests are that todays conventional transmission concepts will still be competitive with Continuously Variable Units.

OVER THE PAST DECADE, there has been an increased number of changes in the Automotive Powertrain Systems which have been brought about by the fuel economy requirements (CAFE), emmission standards, and customer requirements. During this period, many manual transmissions have changed from 3 and 4 speeds to 5 speeds and have increased their overall ratio coverage. Just a few years back it was common to find manual transmissions having 3 to 1 overall ratio range; however, today many have ratio coverage of 5 to 1 and some are even higher. In the automatic transmission field, wide ratio four speeds with torque converter clutches are rapidly becoming the majority - while only a few years ago the 3-speed with an open converter was the common transmission.

There are now reports that claim that the Continuously Variable ratio Transmission (CVT) is on the horizon and will soon be available in some passenger vehicles. In the initial introduction of the gasoline engine passenger cars during the late 1890's and early 1900's, most of them were powered by friction drives of variable ratio (Figure 1). These drives in later years were replaced with units similar to what we know today; however, they have been developed into very sophisticated forms over the years.

Figure 1

While many different types of CVT drives have been experimented with during this interval of time, none have apparently shown the promise to the Automotive Engineer than those that use the steel push type belt or tension chain. A picture of the VanDoorne push type belt is shown in (Figure 2). With the recent efficiency and improvements made in conventional transmission systems, the new family of CVT's will have stiff competition. We will now look at what our analytical analysis of conventional and CVT type transmissions indicates for comparative performance and economy.

Figure 2

First, it is important to understand how this data was derived. A sophisticated vehicle analytical program called GPSIM was used. This program is a complex general purpose vehicle simulation and if given the proper inputs can accurately predict vehicle performance and economy. When looking at any vehicle modification, it is important to insure that the trade off between performance and economy is properly made. To accomplish this, we have developed what is called the optimum performance "vs" economy line. A very detailed discussion of this method, how it is developed, and how it is used has been covered in a previous SAE Paper No. 770418.

Briefly it consists of developing an axle final drive ratio hook for the vehicle being considered (Figure 3). This chart shows such an axle hook for effective final drive ratios between 1.9 and 3.6 to 1. Improving vehicle performance is plotted on the vertical axis while increasing economy is shown on the horizontal axis.

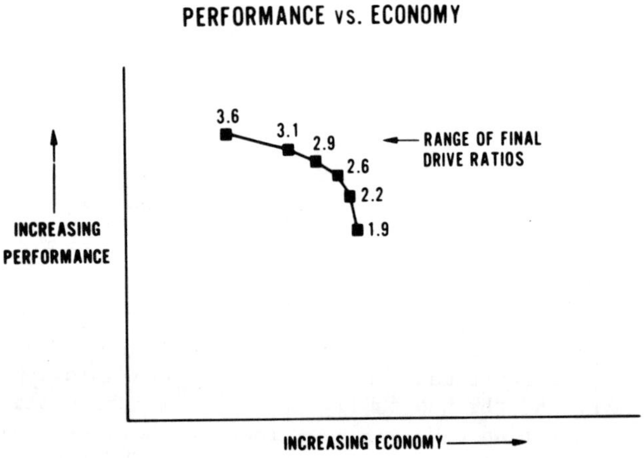

Figure 3

In (Figure 4) the graph has been expanded to show change with engine size. A line has been drawn tangent to the axle hooks and is labeled the, "Optimum Performance vs Economy line". It is important to note that it is impossible to build a vehicle with the parameters selected that would have performance and economy in the region above and rightward of this line - they can only be built either on the line or below to the left side of it.

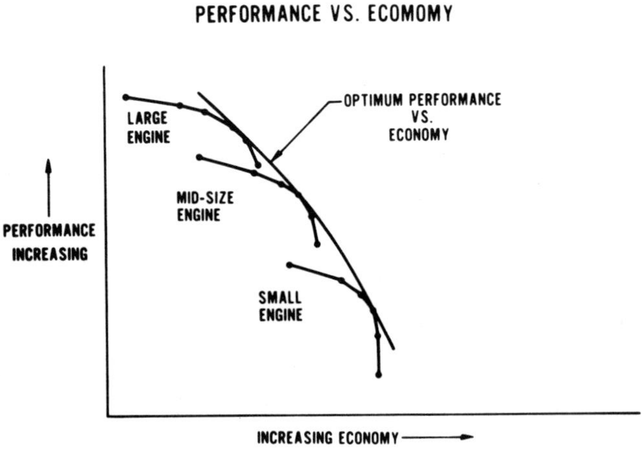

Figure 4

If we now want to analyze any modification to this vehicle whether it be Aero, Torque Converters, Shift Schedules, Transmission Gear Box Modification, etc., it becomes an easy task.

(Figure 5) shows the development of such an optimum line for this vehicle with another type transmission. It can be seen how easy it is to determine the optimum economy at constant performance. Any trade-off between performance and economy can easily be determined. This pro-

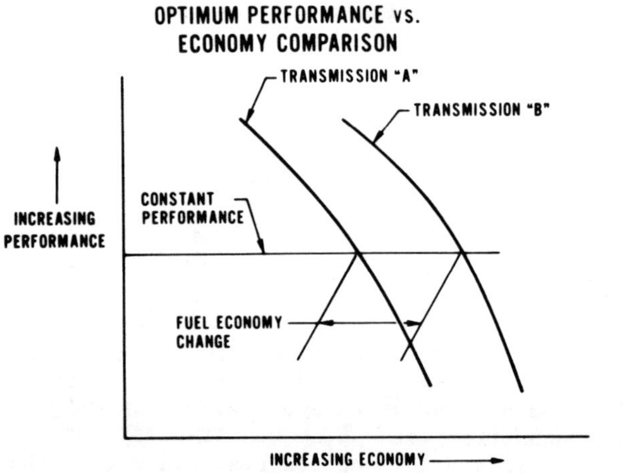

Figure 5

cess was used for all analytical study data that is discussed here. For these comparisons, constant 0-60 mph performance was maintained and composite fuel economy determined.

(Figure 6) compares manual transmissions when operated on the EPA test using the specified EPA shift speeds. It becomes apparent why overall ratio range has been increasing over the past years in manual transmissions. It also would indicate that the number of speeds was increased because of customer demands and driver pleasibility and not specifically for fuel economy reasons.

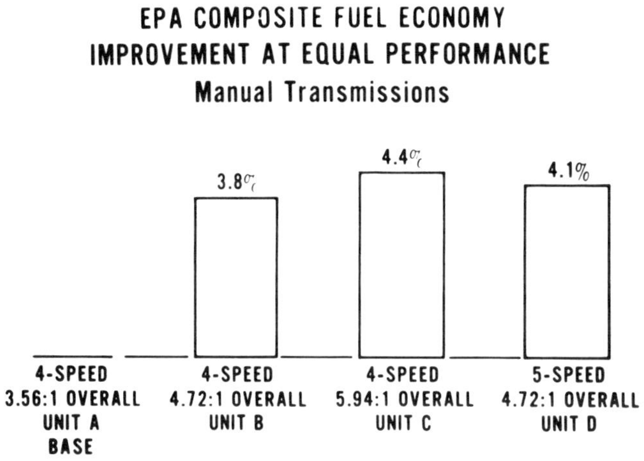

Figure 6

Likewise, in (Figure 7) this same type information for various automatic transmissions having converters with and without converter clutches, various converter clutch engagement schedules, several ratio ranges, and also three simulations of CVT's is shown. It should be noted that the lower overall ratio range 4 speed and CVT gave similar results while at the wider ratio range

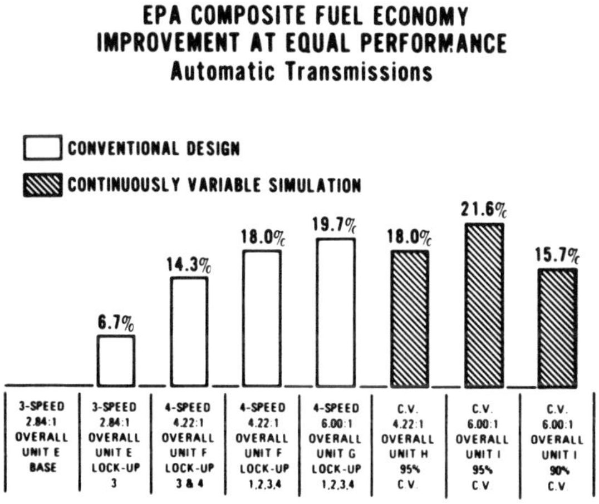

Figure 7

the CVT indicated an improvement of a couple percent - this advantage was lost; however, if the CVT component efficiency drops 5%; that is, to change from 95 to 90%.

(Figure 8) has been developed for automatic transmissions and shows percent change in composite fuel economy on the vertical axis against overall ratio spread on the horizontal axis. Break-off points for 2-speeds, 3-speeds, 4 speeds, and 12 and 20 speed transmissions are also shown. the 12 and 20 speed units have been used to represent CVT's. The lower portion of this curve has been substantiated with many car tests. All simulations used torque converters with torque converter clutches released in the lowest underdrive gear until it became more efficient to be engaged: it then continued to be engaged in all the other gears. It was also assumed that spin losses were the same for all units. The two speed had a Ravigneaux gearset, the three speed a Simpson, the four speed was of integral design, and the 12 and 20 speeds used a flat 95% efficiency for their gear train simulation of the CVT.

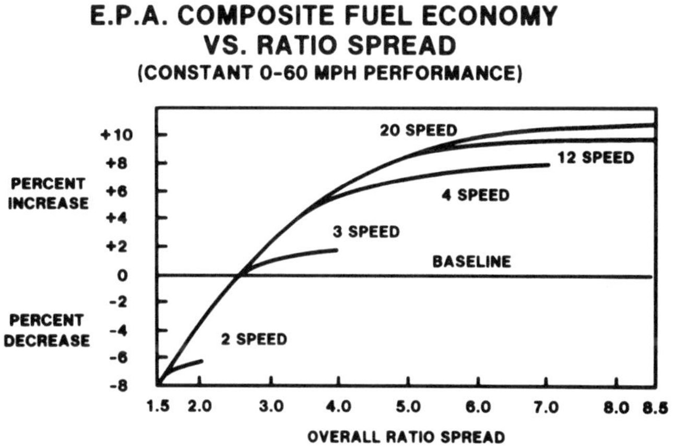

Figure 8

I would like to briefly touch on some actual vehicle testing between a 4-speed manual having an overall ratio range of 3.8/1 and a CVT having an overall ratio range of 3.9/1. All testing was done in the same vehicle and the manual transmission was driven using EPA shift point rules for the economy tests. The CVT for these tests was a design which would have minimum parasitic type losses. It can be noted that in (Figure 9) for this CVT, fuel economy benefits were minimal and a performance loss was substantial. The majority of the performance loss occurring below 30 mph.

FUEL ECONOMY COMPARISON

	CITY	HWY	COMP
4-SPEED MANUAL	28.6	40.1	32.8
CVT	29.2	39.3	33.0
	+ 2.0%	- 2.0%	+ .6%

PERFORMANCE COMPARISON

	0-30 MPH	0-60 MPH
4-SPEED MANUAL	4.5 SEC.	13.2 SEC.
CVT	4.9 SEC.	13.7 SEC.
	- 8.9%	- 3.8%

Figure 9

Steady state fuel economy was also measured on these transmissions at various operating N/V's. This is shown in (Figure 10). In general at constant N/V's, the CVT transmission produced fuel economys approximately 1 to 2 mpg lower than the conventional manual transmission.

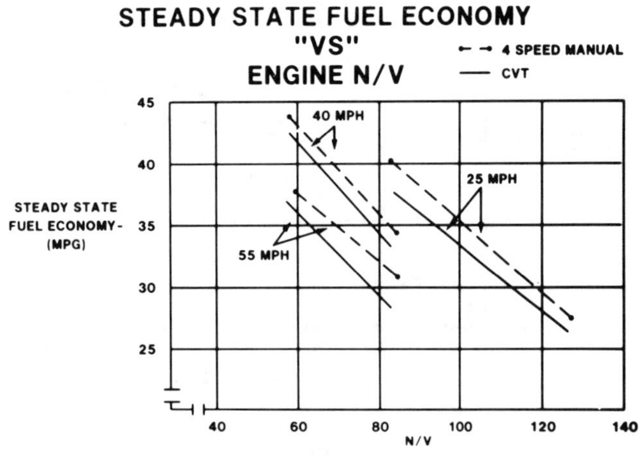

Figure 10

In conclusion, it is very important that drivetrains be evaluated at their optimum performance and fuel economy capability since it is by this means that the true potential of a transmission can be determined. It is also important that during analysis and car tests that oranges be compared to oranges and not to lemons. For instance, since ratio range is so important, when comparing different concepts similar ratio ranges should be considered so that one concept does not have a definite advantage over the other when it is probably possible that this feature could be common to both. With the improvements that have been made in todays conventional transmissions - that is, increased number of speeds, increased ratio coverage, and converter clutches or other methods of reducing torque converter losses; the demands placed on the future CVT's will be just that much more challenging.

While it is not indicated that the CVT transmission will offer any substantial improvments in the optimum trade-off of fuel economy and performance, it may have other merits; that is, stepless ratio changing, customer pleaseability, and sales appeal.

ACKNOWLEDGMENT

The author wishes to thank the General Motors Corporation for their permission to publish this information.

System Control and Design

Torque Sensing Variable Speed V-Belt Drive*

Larry R. Oliver and Dewey D. Henderson
Dayco Corp.

*Paper 720708 presented at the Combined National Farm, Construction, Industrial Machinery and Powerplant Meetings, Milwaukee, Wisconsin, September, 1972.

THE PURPOSE of this paper is to present design criteria for and information on the application of a torque-controlled tensioning mechanism for variable speed V-belt drives. The tensioning mechanism (Fig. 1) consists of a simple helical cam which is incorporated into the sliding flange of a variable speed V-belt sheave and a cam follower which is attached to the other flange. This latter flange is rigidly connected to the loaded shaft. When torque loading is increased, a belt tends to rotate the sliding flange relative to the loaded flange and, as a consequence of the helical cam and follower, the sliding flange is forced toward the fixed flange. This decrease in the sheave's groove width creates a slightly larger belt diameter, which causes an elongation in the belt since the diameter of the companion sheave is fixed. Such elongation increases belt tension which, in turn, enables the drive to transmit higher loads without slippage.

Increasing tension is the sole purpose of the cam angle for the applications discussed in this paper. Specifically, it is not considered as a means of changing speed ratio, which is accomplished by some other operator-controlled shifting mechanism incorporated into the companion sheave or into an adjustable idler.

Conventional agricultural and industrial variable speed drives usually consist of one spring-loaded variable sheave, which is connected through a V-belt to another mechanically or hydraulically-actuated variable sheave that is adjusted to create whatever speed ratio an operator wants from the drive. V-belt drives of this type must be tensioned to a level which will prevent slippage for any load that the drive might be required to transmit. Otherwise, peak loads which might cause slippage only for a short period of time could generate enough heat to destroy the belt. If the belt is tensioned adequately for peak loads, its fatigue life is greatly diminished because of the high tension which it carries continuously even though most loads require less than half that tension. Over-tensioning a drive should be avoided if possible since high tension, with the exception of small diameters, is the most important factor in reducing belt life.

A solution to the problem of having to tension a belt for peak loads to avoid slippage and then having to accept shorter belt life because of it lies in the use of a torque-controlled tensioning mechanism in conjunction with a relatively light compression spring. When loads are high the device increases belt tension and when they are low it reduces belt tension. This variation in tension with load, if the tension is forced to the most efficient levels, will maximize belt life. In fact, it

ABSTRACT

Equations have been derived for designing automatic tensioning variable speed V-belt drives. Such drives employ a screw principle to increase tension in the belt as loading of the drive increases. Adjustable tensioning makes it possible for variable speed drives to transmit significantly larger loads without overly penalizing belt life. This feature is possible because high tensions required by heavy loads are not retained during lighter loads experienced under normal operations.

Drives have been designed by these equations and have been proven successful both in the field and laboratory. Experience in designing the drives is discussed and general recommendations are provided for the design of new torque sensing drives. Other types of applications are described that would also benefit from torque-actuated tensioning.

Fig. 1 - Torque sensing variable speed sheave

will make successful field drives out of ones which otherwise would have never gotten off the drawing board because of the unacceptable belt life that would result from having to tension for high peak loads.

DESIGN OF TORQUE SENSING CAM

The design of a torque actuated cam can be approached in two ways depending upon whether the cam angle (β) is allowed to vary or is restricted to a constant. In either case the first step would be to establish the maximum tight strand (T_1) to slack strand (T_2) belt tension ratio permitted at the peak operating load (recommended values for such tension ratios will be discussed in a later section). Also, approximate tension ratios desired for all lower loads should be estimated. The peak-load tension ratio and torque ($D/2\ [T_1 - T_2]$) together with the belt pitch diameter (D), coefficient of friction (μ), wrap angle (θ), and included groove angle (α) are used to calculate equilibrium axial forces for the driver (F_r) and driven sheaves (F_n). Depending upon whether the torque sensing feature is employed on the driver or driven sheave, the appropriate axial force and peak-load torque along with a specified value for the cam diameter (D_a) and the coefficient of friction for the cam and cam follower interface (μ_a) are used to determine the helical angle which will create that axial force. This cam angle would then govern the axial force at all lower loads. The tension ratios which would exist at these lesser but more typical operating loads are computed and studied. If they are too tight or too loose, the cam angle is adjusted and the corresponding new tension ratios are examined. This process is continued until the best compromise for a helical cam angle is reached.

If the helical angle is allowed to vary along the cam's surface, the above process is repeated for incremental speed ratios ranging from low to high speed adjustments on the mechanically or hydraulically-actuated sheave. If the cam angle is restricted to a constant because of length of cam follower contact, manufacturing, or other reasons, the selected angle should be based on a compromise between tension ratios calculated for all speed and load conditions.

K. G. Hornung (1)* has developed the most rigorous and accurate method for relating tension to axial force in a V-belt drive. However, this method requires a computerized numerical solution to five simultaneous differential equations. Such a procedure is thought to be too involved for general application usage so a simpler, yet relatively accurate, set of equations presented by W. S. Worley (2) are employed in this paper. These equations in their published form are:

$$F_n = \frac{T_1 - T_2}{2k}\left(\frac{1 - \mu_R \tan\frac{\alpha}{2}}{\mu_R + \tan\frac{\alpha}{2}}\right) \tag{1}$$

$$k = \frac{\ln T_1/T_2}{\theta_n} \tag{2}$$

$$\mu = \sqrt{\mu_R^2 + \mu_T^2} \tag{3}$$

$$\mu_T = k\left(\sin\frac{\alpha}{2} + \mu_R \cos\frac{\alpha}{2}\right) \tag{4}$$

$$F_r = \frac{T_1\theta_r}{2}\left(\frac{1 - \mu\tan\frac{\alpha}{2}}{\mu + \tan\frac{\alpha}{2}}\right) \tag{5}$$

where the subscripts r and n represent driver and driven sheaves and R and T indicate radial and tangential directions. In this paper these equations are manipulated into various forms which eliminate the variables k, μ_R, and μ_T. However, they include an additional variable ϕ which indicates the direction of the frictional force acting at the belt-sheave interface. This new variable is related to the eliminated ones by

$$\tan\phi = \frac{\mu_R}{\mu_T} \tag{6}$$

An equation for the axial force produced by a torque-sensing tensioning mechanism can be obtained from an equilibrium analysis of forces acting on the sliding flange. Such axial force depends on torque, helical cam diameter and angle, compression spring pre-load (F_o) and rate (K), and belt diameter. Fig. 2 depicts for impending sheave closing a free body diagram of the cam where R_a denotes the normal force on its surface and F_a represents the reaction axial force. Equilibrium in axial and tangential directions can be expressed in terms of the following two respective force balances:

$$-R_a\cos\beta + \mu_a R_a \sin\beta + F_a = 0 \tag{7}$$

*Numbers in parentheses designate References at end of paper.

Fig. 2 - Free body diagram of torque actuated cam

$$R_a \sin\beta + \mu_a R_a \cos\beta - \frac{D}{D_a}\left(\frac{T_1 - T_2}{2}\right) = 0 \tag{8}$$

In Eq. 8 the last term represents a torque induced tangential force which presses cam and cam follower together. Solving these two equations simultaneously yields in terms of known variables the axial force produced by a helical cam in response to an applied torque load.

$$F_a = \frac{D}{D_a}\left(\frac{T_1 - T_2}{2}\right)\left(\frac{\cos\beta - \mu_a \sin\beta}{\sin\beta + \mu_a \cos\beta}\right) \tag{9}$$

Impending sheave opening is also a possibility since a tensioned belt is continually trying to force sheave flanges apart. An equation for this situation is obtained simply by reversing the sign for the cam's coefficient of friction,

$$F_a = \frac{D}{D_a}\left(\frac{T_1 - T_2}{2}\right)\left(\frac{\cos\beta + \mu_a \sin\beta}{\sin\beta - \mu_a \cos\beta}\right) \tag{10}$$

This impending opening equation will be used throughout the remainder of the paper for calculating axial force produced by a cam. Any further discussions about cam coefficients of friction pertain to Eq. 10 and, therefore, positive and negative values indicate, respectively, impending opening and closing.

Total axial force on a sliding flange and hence on the side of a belt can be expressed as a summation of axial forces produced by the cam and compression spring. Spring force is a combination of installation pre-load, measured when the sheave is closed (D_o), and an additional force that results from compressing the spring by an amount equal to the increase in groove width (ΔG). This increase depends on groove angle and change in belt diameter.

$$\Delta G = \Delta D \tan\frac{\alpha}{2} \tag{11}$$

The final expression for belt axial force resulting from a cam and spring tensioning system now becomes

$$F = \frac{D}{D_a}\left(\frac{T_1 - T_2}{2}\right)\left(\frac{\cos\beta + \mu_a \sin\beta}{\sin\beta - \mu_a \cos\beta}\right) + F_o + K\Delta D \tan\frac{\alpha}{2} \tag{12}$$

CAM ON DRIVEN–If the helical angle is allowed to vary along a cam's length, a range of cam angles can be established that correspond to the range of drive speed ratios which extend from low to high-speed operating condition. In other words, there is a one-to-one correspondence between operating speed ratios and positions of follower contact along the cam's surface. A variable angled cam is theoretically desirable because the helical angle required to create optimum tension can be matched to the variation in torque and spring compression associated with each speed ratio.

For each speed range and associated position of follower contact on the cam's surface the corresponding peak-load tension ratio is substituted into the following friction angle equation:

$$\sin\phi = \frac{-2\mu \cos\frac{\alpha}{2}\sin\frac{\alpha}{2}\left(\ln\frac{T_1}{T_2}\right)^2 + \sqrt{\left[2\mu \cos\frac{\alpha}{2}\sin\frac{\alpha}{2}\left(\ln\frac{T_1}{T_2}\right)^2\right]^2 - 4\left[\left(\mu\cos\frac{\alpha}{2}\ln\frac{T_1}{T_2}\right)^2 + (\mu\theta_n)^2\right]\left[\left(\ln\frac{T_1}{T_2}\sin\frac{\alpha}{2}\right)^2 - (\mu\theta_n)^2\right]}}{2\left[\left(\mu\cos\frac{\alpha}{2}\ln\frac{T_1}{T_2}\right)^2 + (\mu\theta_n)^2\right]} \tag{13}$$

This friction angle is then used to determine the required axial force on a driven sheave.

$$F_n = (T_1 - T_2)\left(\frac{\cos\frac{\alpha}{2} - \mu\sin\frac{\alpha}{2}\sin\phi}{2\mu\cos\phi}\right) \tag{14}$$

This value for belt axial force is inserted into the following inverse form of Eq. 12 to find the cam angle.

$$\beta = \tan^{-1}\left\{\frac{1 + \mu_a \frac{D_a}{D_n}\left[\frac{2}{T_1 - T_2}\right]\left[F_n - F_{no} - Kn\,(D_{no} - D_n)\tan\frac{\alpha}{2}\right]}{\frac{D_a}{D_n}\left[\frac{2}{T_1 - T_2}\right]\left[F_n - F_{no} - K_n\,(D_{no} - D_n)\tan\frac{\alpha}{2}\right] - \mu_a}\right\} \tag{15}$$

These values for the helical angle and cam diameter specify a tensioning mechanism which will develop the desired tension ratio at a specified load.

The angle is applicable to all loading conditions at a given operating speed ratio. Tension ratios which will be created at loads other than the peak value must be investigated. These tensions can be calculated by inversely solving the above equations. Axial forces produced by a cam with this angle for other values of torque are calculated from Eq. 12. The friction angles corresponding to these torques and axial forces can be computed from

$$\sin\phi = \frac{2\mu\cos\frac{\alpha}{2}\sin\frac{\alpha}{2} + \sqrt{\left(2\mu\cos\frac{\alpha}{2}\sin\frac{\alpha}{2}\right)^2 - 4\left[4\mu^2\left(\frac{F_n}{T_1 - T_2}\right)^2 + \left(\mu\sin\frac{\alpha}{2}\right)^2\right]\left[\cos^2\frac{\alpha}{2} - 4\mu^2\left(\frac{F_n}{T_1 - T_2}\right)^2\right]}}{2\left[4\mu^2\left(\frac{F_n}{T_1 - T_2}\right)^2 + \left(\mu\sin\frac{\alpha}{2}\right)^2\right]} \tag{16}$$

Tension ratios for the new lesser loads are finally found from the following exponential expression:

$$\frac{T_1}{T_2} = \exp\left(\frac{\mu\theta_n\cos\phi}{\mu\sin\phi\cos\frac{\alpha}{2} + \sin\frac{\alpha}{2}}\right) \tag{17}$$

If tensions at all loads meet the recommendations presented in this paper, the peak-load angle should be used. However, if tension ratios are either too high or too low for primary operating loads, this angle should be adjusted until the best compromise is found. When a helical angle for the specified speed ratio has been determined, a new speed condition and corresponding peak-load tension ratio should be selected. This tension ratio is again used to find an initial trial value for a cam angle at the new speed range. The process is continued until a map relating cam angle to speed ratio has been established throughout the range from low to high.

If a helical angle is restricted to a constant value along the cam surface, the procedure for determining that angle is the same as above with the exception that it is not changed from one speed condition to another. In this case, the single value selected for a cam angle must be the best compromise between tension ratios for all speed ratios as well as for all loading conditions within each speed range.

No matter what cam angle is selected, the same procedure is used to calculate the driver axial force that is necessary to keep a belt in equilibrium at conditions assumed in calculating that angle. The difference between tight and slack strand tensions is known from torque and belt diameter inputs. Tension ratio is either an input or is calculated from Eq. 17. Tension difference and ratio expressions can then be solved simultaneously for tight and slack strand tensions. The value for tight strand tension and the belt wrap angle on the driver sheave are substituted into Eq. 5 to find the required driver axial force for a given cam, load, and speed condition.

CAM ON DRIVER—If the torque sensing mechanism is placed on a driver sheave, the procedure for selecting a cam angle is somewhat simpler because friction angles do not have to be calculated. The peak load and corresponding tension ratio fix the tight strand tension which is used in Eq. 5 to find the equilibrium driver axial force. This axial force is substituted into Eq. 15 (where all n subscripts should be changed to r) in order to find the initial trial cam angle. Eq. 12 is then employed to calculate the axial force produced at that angle for smaller loads. This axial force is substituted back into Eq. 5 to compute the associated tight strand tension, which in conjunction with the known tension difference establishes the resulting tension ratio. The remainder of the procedure is the same as that described in the "cam on driven" section.

Once a cam angle has been chosen, the axial force which must be exerted by the driven sheave can be found by solving Eq. 13 for the friction angle in terms of tension ratio and then computing the axial force from Eq. 14.

The location of the cam and sliding flange on the driver as on the driven sheave should be such that when the sliding flange rotates in the direction of increasing belt tension around the sheave, the sheave will begin to close.

SELECTION OF COEFFICIENT OF FRICTION FOR BELT

In order to predict tension ratios and driver and driven axial forces, values for the coefficient of friction between the belt and sheave and for the cam surface must be known.

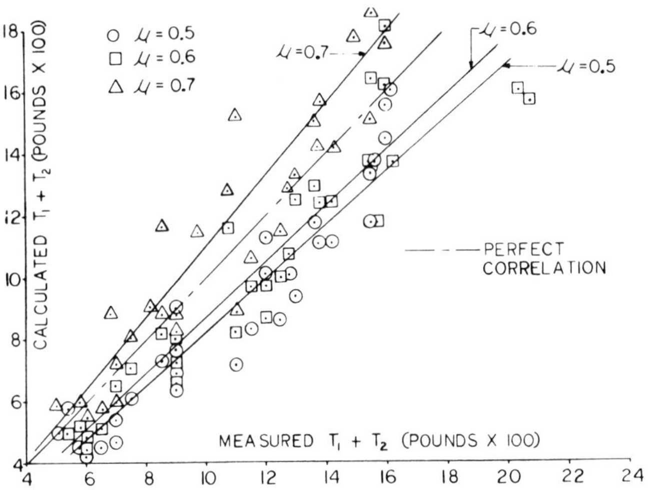

Fig. 3 - Correlation of theoretical and measured total tensions for different values of μ

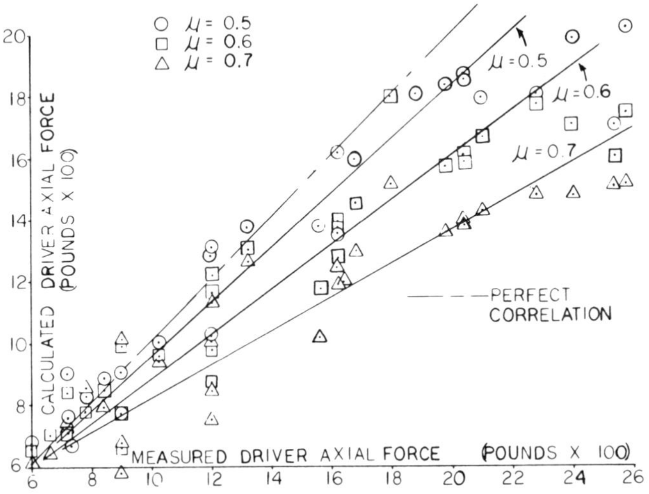

Fig. 4 - Correlation of theoretical and measured axial force for the driver sheave for different values of μ

An analysis of 36 tests conducted by Allis-Chalmers Corp. in their laboratory using an actual combine variable speed header drive, which utilized a hydraulically-controlled driver sheave and a torque sensing driven sheave, shows the effect of μ on total tension and driver axial force.

Fig. 3 shows the calculated $T_1 + T_2$ plotted against the measured $T_1 + T_2$ using the measured driven axial force to calculate T_1 and T_2 from Eq. 16 and 17 for various values of μ. Cam angles of 55, 60, and 65 deg and horsepower loads of 8-48 hp were involved. The plots show the best agreement between calculated and measured $T_1 + T_2$ for a coefficient of friction of the belt at the driven sheave to be between 0.6 and 0.7.

Similarly, the measured driver axial force is plotted against the calculated driver axial force (from Eq. 5) for various values of μ in Fig. 4, using the T_1 computed at the driven sheave. These plots show the best agreement between calculated and

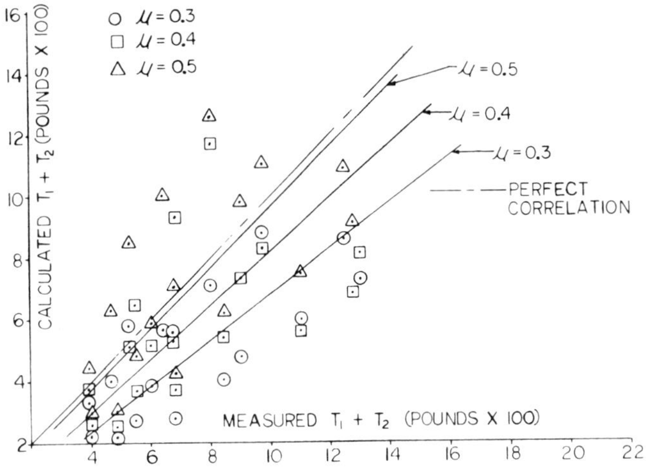

Fig. 5 - Correlation of theoretical and measured total tension for different values of μ (field belt)

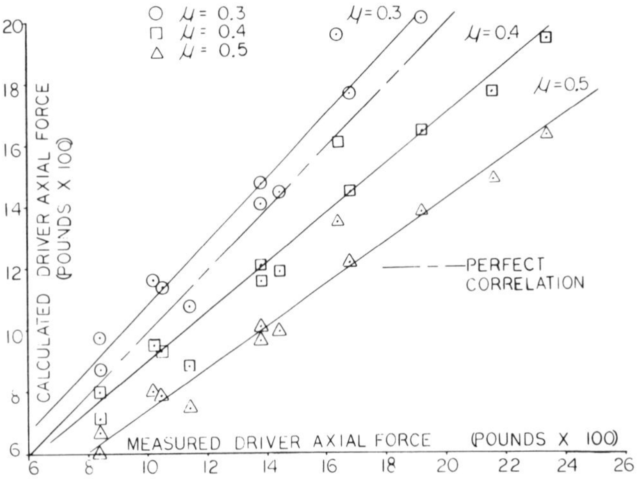

Fig. 6 - Correlation of theoretical and measured axial force for driver sheave for different values of μ (field belt)

measured driver axial force for a coefficient of friction at the driver sheave to be between 0.4 and 0.5.

Since the driver and driven sheaves were made of the same materials and had the same surface finish, a single value of μ must be used. This would place the value of μ for a molded cog belt on worn-in machined gray iron castings at 0.55. The use of $\mu = 0.55$ will result in calculated total tensions and driver axial force being approximately 10% below the actual. It should be noted that this value was obtained under laboratory conditions and will be less under field conditions. It is recommended that the μ of 0.55 be used only for designing drives that are to be operated in a clean environment, such as a protected stationary industrial application.

Keller (3) reports a 48% reduction in μ for molded cog belts after being run under field conditions for 250 h. Fig. 5 shows calculated $T_1 + T_2$ versus measured $T_1 + T_2$ and Fig. 6 shows calculated driver axial force versus measured driver axial force for Keller's tests of field belts. Fig. 5 shows the μ for the best agreement between the calculated $T_1 + T_2$ and the measured

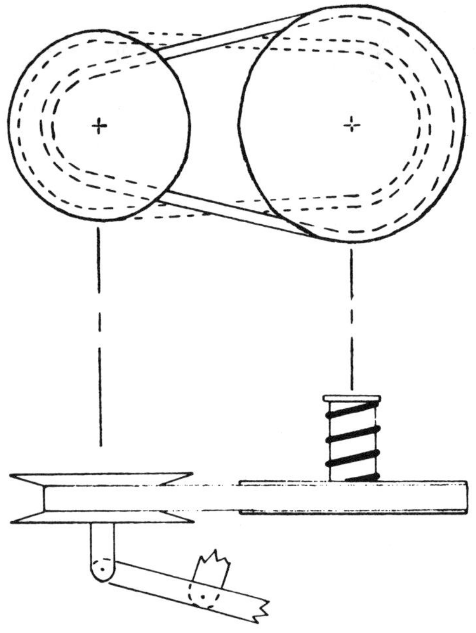

Fig. 7 - Conventional spring loaded variable speed drive

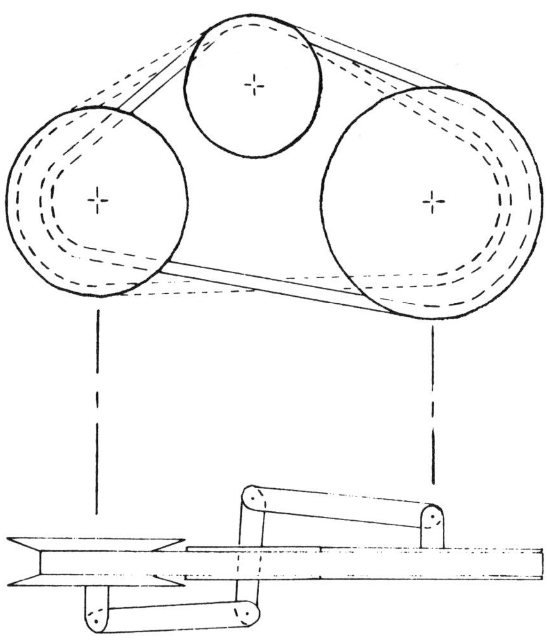

Fig. 8 - Conventional cross linked variable speed drive

$T_1 + T_2$ to be slightly greater than 0.5. Fig. 6 shows a value of μ for the best agreement between calculated driver axial force and measured driver axial force to be slightly greater than 0.3. In this case, a value of μ which is common to the driver and driven sheaves is found to be 0.4. It is recommended that the μ of 0.4 be used for designing drives that are to operate under field conditions. The use of 0.4 will result in calculated total tensions and driver axial forces being approximately 15% below the actual. The reduction in μ for field operation would then be 27%.

COEFFICIENT OF FRICTION OF CAM SURFACE

Keller (3) reports values of frictional coefficients for a cast gray iron cam and follower surface of 0.22 when greased and 0.26 when dry. Swenson (4) found a value of 0.11 for an aluminum cam with a nylon follower. These values should be used for cams made of similar materials. For other cam materials, coefficients should be determined experimentally, if possible; and, if not, they should be selected from the range of handbook listings or from material manufacturer's recommendations.

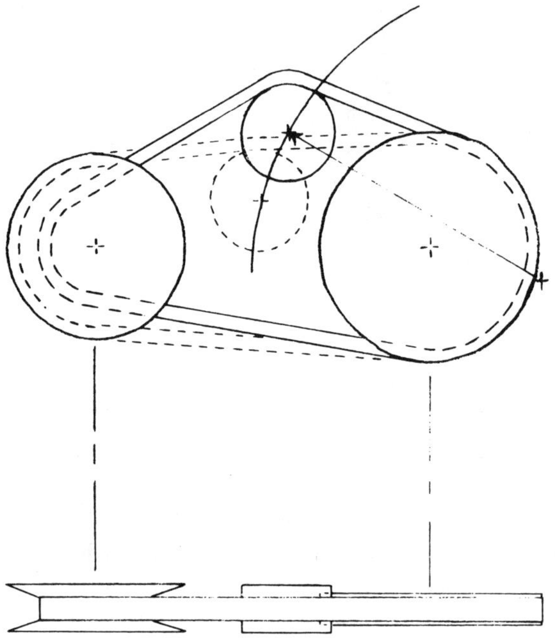

Fig. 9 - Conventional single adjustable sheave variable speed drive

CONVENTIONAL AGRICULTURAL AND INDUSTRIAL VARIABLE SPEED DRIVES

The most commonly used variable speed or adjustable speed drives are, first, two adjustable sheaves where either the driver or driven sheave is spring loaded (Fig. 7) or where the moveable flanges are cross linked together (Fig. 8) with some means of compensating for the difference in belt length between the 1:1 ratio and the maximum and minimum ratios. Second, one fixed diameter sheave and one adjustable sheave (Fig. 9) with an idler employed to take up the excess belt length when the adjustable sheave is moved to smaller diameters. Third, a two-belt variable (Fig. 10) with a fixed driver and driven and a floating double groove adjustable sheave with a common moveable center disk.

TORQUE SENSING VARIABLE SPEED DRIVES–The torque sensing system is most suitable for use on the first three of the conventional drives outlined above with the following advantages being realized. The belt tension is regulated by the drive load rather than being maintained at a level sufficient to transmit the maximum peak load. The over-tensioning result-

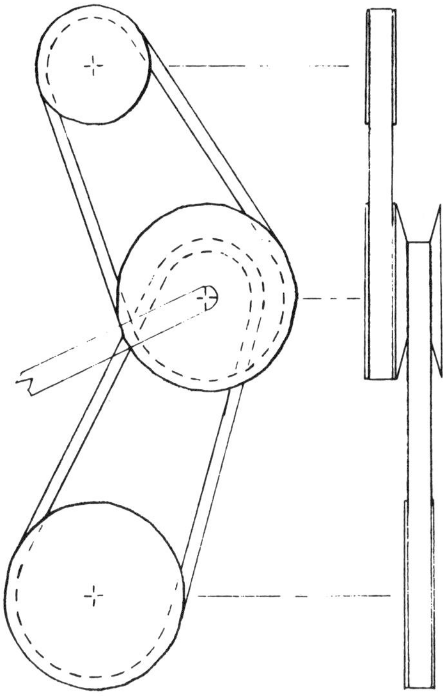

Fig. 10 - Conventional two-belt, floating center sheave variable speed drive

ing from the conventional drives results in lower belt and hardware life.

DESIGN PARAMETERS FOR TORQUE SENSING DRIVE

Having selected the appropriate coefficient of friction for the cam material to be used and for the belt, the next step is to determine belt section and sheave diameter that will yield the desired speed range. The cam actuated sheave must be designed to operate under normal load with a nominal gap of 0.1 or greater in its closed position. The sheave will then be sensitive to torque changes while operating in the closed position. In the case of the torque sensing driven sheave, the low range PD should be chosen to yield the sheave gap for normal loads and resulting in a slight loss of driven rpm under the peak operating loads. The drive should incorporate a belt length take-up device which will allow the sheave to be adjusted to the correct gap. The mean cam diameter is selected by the designer according to the drive space limitations. The spring preload should only be enough to prevent excessive belt whip at the no-load condition. Experience has shown the minimum preload requirement for HL and HM section belts to be approximately 200 lb, with the spring rate held to a minimum.

ALLOWABLE TENSION RATIOS

Cam angles should be selected to yield a tension ratio of approximately 3:1 for the primary operating load and allowing the tension ratio to increase to 5:1 for the maximum sustained operating load. If the peak loads are approximately three to four times the primary operating load, the maximum tension ratio for peak-load condition should not exceed 12:1. If the primary operating load results in a tension ratio lower than 3:1 and the peak load ratio is above the maximum 12:1, laboratory tests at the selected cam angle should be conducted.

FIXED SPEED TORQUE TENSIONED DRIVES

The torque sensing tensioning system can be applied to fixed speed drives by incorporating a split sheave with a cam on the moveable flange. The same procedure used in variable speed drives for selecting the cam angle is used for the fixed speed drive. In the case of the fixed speed drive there is only one speed range to consider when selecting the cam angle.

SUMMARY

The problem of having to tension conventional variable speed V-belt drives tightly enough to transmit peak loads and of having to accept the associated reduction in belt life can be solved through the use of a torque controlled tensioning mechanism. The mechanism, which consists of a helical cam attached to a sliding sheave flange, converts high torques into tight belt tensions and low torques into reduced tensions. Load dependent tensioning of this nature is essential for long belt life on drives which transmit a wide range of torques including some that are quite high but of short duration.

Design equations for this tensioning device have been developed and experimentally verified. The equations can be used to establish proper values for the helical cam angle and diameter, for the compression spring pre-load and rate, and for the required axial force on the companion sheave. These design parameters are finalized through a balancing procedure aimed at establishing a set of values which will result in tension ratios greater than 3:1 for primary operating loads, less than 12:1 for peak loads, and equal to 5:1 for the maximum sustained operating load. Inputs to the equations are application torques, belt diameters and wrap angles, and belt and cam coefficients of friction.

Experimental results indicate that the coefficient of friction for a molded cog belt running on a cast gray iron sheave in a clean environment is probably about 0.55 although a value of 0.4 should be used for designing actual field drives. A frictional coefficient of 0.26 was found for the cast gray iron cam and follower which were used in this experiment.

Although this tensioning mechanism is not suited for the two-belt, floating-center-sheave type of variable drive, it can be successfully incorporated into any conventional agricultural variable speed drive of the single or double adjustable sheave types. The device can also be used on a constant speed drive if one of the drive's sheaves is split into fixed and moveable flanges.

REFERENCES

1. K. G. Hornung, "Factors Influencing the Fatigue Characteristics of Rubber-Textile Machine Elements." Doctor of Philosophy Dissertation, The Ohio State University, 1959.
2. W. S. Worley, "Designing Adjustable-Speed V-Belt Drives for Farm Implements." SAE Transactions, Vol. 63 (1955).
3. D. L. Keller and R. E. Wilson, "Design & Development of a High Horsepower Torque Sensing Variable Speed Drive." Paper 72079 presented at SAE Farm, Construction and Industrial Machinery Meeting, Milwaukee, September 1972.
4. J. E. Swenson, "Performance of Torque and Speed Controlled Variable Speed V-Belt Drives." M.Sc. Thesis, Mechanical Engineering Department, The Ohio State University, 1972.

APPENDIX

Subscripts r and n appended in the text to variables in the following list of nomenclature refer to driver and driven sheaves, respectively.

D = belt pitch diameter (in)
D_a = diameter of torque sensing cam measured at center of cam wall (in)
D_o = maximum belt pitch diameter (in)
F = axial force (lb)
F_o = pre-load on compression spring (lb)
K = rate of compression spring (lb/in)
T_1 = tight strand belt tension (lb)
T_2 = slack strand belt tension (lb)
α = included groove angle (deg)
β = angle between cam surface and a perpendicular to shaft of sheave (deg)
θ = arc of belt-sheave contact (rad)
μ = coefficient of friction between belt and sheave surface
μ_a = coefficient of friction on cam of torque sensing sheave
ϕ = friction angle, direction opposite to relative motion of belt and sheave surfaces measured from tangent of belt strand which is entering sheave (deg)

Design and Development of a High Horsepower Torque Sensing Variable Speed Drive**

Duane L. Keller and Ronald E. Wilson
Allis-Chalmers Corp.

**Paper 720709 presented at the Combined National Farm, Construction, Industrial Machinery and Powerplant Meetings, Milwaukee, Wisconsin, September, 1972.

THE PURPOSE OF this investigation was to determine certain design parameters necessary in developing a high horsepower, torque-sensing, variable speed drive. The advantage of a torque-sensing drive is that it will allow the belt to be run at low tensions when the load is low, but will increase belt tension applied to the drive under peak loads. This will result in longer belt life, increased drive hardware life, and a more aggressive drive when field conditions demand high loads to be transmitted.

The objectives were to determine:

1. The effect of the independent variables (cam surface, cam angle, and belt construction) on the drive design parameters (tight and slack side tension, driven and driver axial forces).
2. Verification of Dayco's equations in predicting the design parameters of a torque-sensitive design.
3. Coefficients of friction for various cam surfaces, and various belt construction to cast-iron sheaves.

TORQUE SENSING DRIVE EQUATIONS

Oliver (1)* developed equations which relate belt axial and tensile forces and drive rpm to output torque, belt pitch diameters, and torque-sensing sheave design. Derivations are quite concise, and no attempt will be made to derive the equations. For this application, the generalized equations may be simplified for a torque-sensing sheave design with only driven cam actuators and a compression spring.

The axial force on the belt due to torque sensing (driven axial) is:

$$F_N = \frac{I}{D_a}\left[\left(\frac{T_1 - T_2}{2}\right) D_N\right]\left[\frac{\cos\beta + \mu_c \sin\beta}{\sin\beta - \mu_c \cos\beta}\right] + F_{NO} + K_N (D_{NO} - D_N) \tan\frac{\alpha}{2} \qquad (1)$$

*Number in parentheses designates Reference at end of paper.

ABSTRACT

The application of torque-sensitive drives for combine belt drives affords the advantage of allowing the drive to be normally tensioned at lower tensions but still maintain the aggressiveness to transmit high peak horsepower loads. Dayco's equations will predict drive design (driven and driver axial forces, tight and slack side tensions) for various output loads. Varying the cam actuators surface and ramp angle will increase or decrease the effect of torque sensing. Belt-sheave and actuator surface coefficients of friction were experimentally determined.

By knowing the driven axial force and strand tension difference (effective pull), the friction angle is solved:

$$\sin \phi = \left\{ 2\mu_\beta \cos\frac{\alpha}{2} \sin\frac{\alpha}{2} + \frac{\sqrt{4\mu_B^2 \cos^2\frac{\alpha}{2} \sin^2\frac{\alpha}{2} - 4\left[4\mu_B^2\left(\frac{F_N}{T_1 - T_2}\right)^2 + \mu_B^2 \sin^2\frac{\alpha}{2}\right]\left[\cos^2\frac{\alpha}{2} - 4\mu_B^2\left(\frac{F_N}{T_1 - T_2}\right)^2\right]}}{2\left[4\mu_B^2\left(\frac{F_N}{T_1 - T_2}\right)^2 + \mu_B^2 \sin^2\frac{\alpha}{2}\right]} \right\} \tag{2}$$

$$\cos \phi = \sqrt{I - \sin^2 \phi}$$

Utilizing $\sin \phi$ and $\cos \phi$, the ratio of tight side to slack side tension is:

$$\frac{T_1}{T_2} = e^{\frac{\mu_B \theta_N \cos \phi}{\mu_B \sin \phi \cos\frac{\alpha}{2} + \sin\frac{\alpha}{2}}} \tag{3}$$

Values for the tight side and slack side tension forces are then solved from:

$$T_1 = \frac{T_1 - T_2}{\left(I - \frac{I}{T_1/T_2}\right)} = \frac{\text{effective pull}}{\left(I - \frac{I}{T_1/T_2}\right)} \tag{4}$$

$$T_2 = T_1 - (T_1 - T_2) = T_1 - \text{effective pull} \tag{5}$$

The algebraic sum of T_1 and T_2 yields the total tension for the drive. The driver axial force on the belt is determined as a function of T_1.

$$F_R = \frac{T_1 \theta_R}{2}\left[\frac{I - \mu_B \epsilon \tan\frac{\alpha}{2}}{\mu_B + \epsilon \tan\frac{\alpha}{2}}\right] \tag{6}$$

To apply these equations, it is necessary to determine appropriate values for cam surfaces (μ_C) and belt-sheaves (μ_B) coefficient of frictions.

TORQUE SENSING DRIVE DESIGN

In conducting the investigation of the characteristics of torque-sensing drives, a variable speed corn head drive was selected for testing. The driver was a hydraulic positioning, variable speed sheave design with a fixed inner sheave and a movable outer sheave. The torque-sensing driven unit consisted of a fixed inner sheave, movable outer sheave, two cam actuators, and a compression spring, as shown in Fig. 1. A fixed inside idler was used for belt take-up. All sheaves were machined from gray iron castings with a machine finish of 63 μ in.

The cam actuators were castings that were used in the "as-cast" conditions, with the cam actuator surfaces being typical of the casting process. The cam actuators were designed to provide three surface areas of contact. Three different cam actuator surfaces: "as-cast" with greased actuator surfaces, "as-cast" without greased actuator surfaces, and "as-cast" hardened cams with greased actuator surfaces, were tested. Four different cam angles: 48, 55, 60, and 65 deg, were studied. The measured angle is between the actuator surface and plane of the belt's path.

Adjustable speed belts with an HL section were used with four different constructions. Those tested were: wrapped cover, wrapped cover-notched, molded Cog-Belt with fiberglass load cords, molded Cog-Belt with dacron load cords. Belts with various amounts of laboratory running time and field operating conditions were tested. The belt described as "field belt" was run for 250 h in actual field conditions.

EXPERIMENTAL PROCEDURE

To evaluate the drive design, it was necessary to measure the following parameters experimentally:

1. Driver speed.
2. Driven speed.
3. Driver sheave axial force.
4. Driven sheave axial force.
5. Total tension.
6. Slack side tension.
7. Driven shaft output torque.
8. Driver pitch diameter.
9. Driven pitch diameter.

Fig. 1 - Photograph of torque sensing sheaves with cam actuators and compression spring revealed

Fig. 2 shows the experimental test setup with the necessary electronic instrumentation that was employed. A Brush "Mark 200" oscillograph was used to provide a continuous trace of the measured parameters. The driver and driven speeds were measured by means of a magnetic pickup, located on the respective shafts, and fed to a "Flow Tech" flowmeter to the oscillograph input. The driver axial force was measured by the use of a pressure transducer, installed in the hydraulic line from the hydraulic-positioned sheaves. Driven axial force on the belt was the sum of the preload compression on the driven spring, plus a measured force due to the cam actuators, which was measured by means of a strain gage located on an extension of the driven shaft. Total tension was determined as the resultant bearing load on the driven shaft, measured by means of a strain gage cluster fed through a slip ring. The total tension was actually the vector sum of slack side tension and tight side tension, which is slightly less than the algebraic sum of slack side and tight side tension. Slack side tension was determined by measuring the force on the idler due to the belt tension in the slack side span and trigonometrically calculating the tangential component (slack side tension). The idler force was measured by means of a load cell.

A Taylor dynamometer was used to load the drive, with a load cell installed to measure driven shaft torque. Driver and driven pitch diameters were measured by means of a hand scale.

For a series of tests, the belt was installed and properly positioned, by means of the idler, to provide a certain initial belt tension. Loads were then applied to the drive, and, after stabilizing, readings were recorded at various intervals for a range from 0 to 45 hp.

RESULTS

BELT CONSTRUCTION - Belt-sheave coefficients of friction, for the various belt constructions, were determined by solving the driver axial force equation, Eq. 6, using the measured driver axial force for each individual test. A mean value with its variance was calculated for each belt construction, with the results listed in Table 1.

The belt-sheave coefficient of friction increased slightly from the lower loads to peak values in the 24-32 hp range, and then decreased at severely high loads for the drive. The decreasing μ_B at the higher loads would indicate a change from a static coefficient to that of a dynamic coefficient of friction.

The difference in μ_B for dacron and fiberglass was not as expected, due to the belts' similar physical appearance, with the stronger tensile cords of the fiberglass belt being the only internal difference. The fiberglass cords reduce the amount of elongation in the belt under heavy loads, resulting in a more constant output speed. The "field belt," which was

Table 1 - Belt-Sheave Coefficients of Frictions for Different Belt Constructions

Belt Construction	Mean μ_B	Standard Deviation
Molded cog-belt with dacron cords	0.622	±0.122
Molded cog-belt with fiberglass cords	0.534	±0.066
Wrapped cover	0.378	±0.076
Wrapped cover, notched	0.466	±0.090
Molded cog-belt, field belt	0.327	±0.036

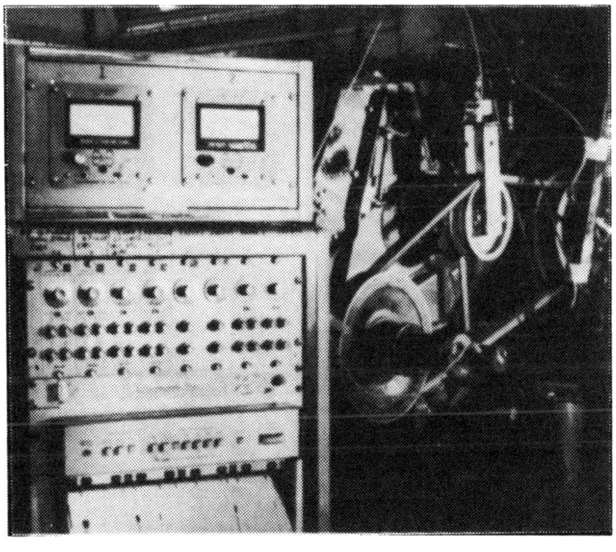

Fig. 2 - Photograph of header drive with testing instrumentation installed

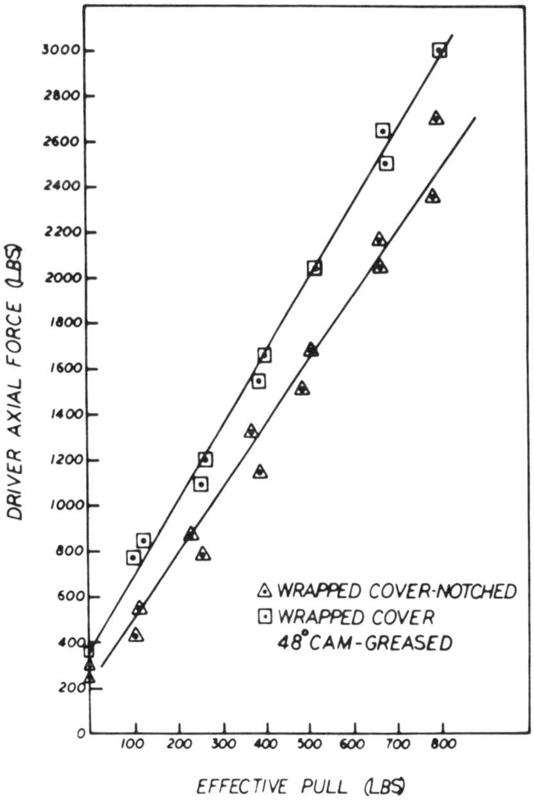

Fig. 3 - Driver axial force versus effective pull for wrapped cover-notched and wrapped cover belt with 48 deg greased cams

embedded with sand and dirt particles from field operating conditions, produced a lower coefficient of friction than laboratory tested molded cog-belts.

The wrapped cover-notched belt was a cured, wrapped cover belt that was mechanically notched. The notching of the belt increased the μ_B, but was still lower than the molded cog-belts. The primary reason for the wrapped covered belt having a lower μ_B than the molded Cog-Belt was the fabric in the belt cover.

The main objective of testing the wrapped cover-notched and wrapped cover belt was to determine how belts with different moduli would affect driver axial forces. Early test results showed the wrapped cover belt required greater driver axial force than the molded Cog-Belt. By notching a wrapped cover belt, it was possible to isolate and identify the effect of the modulus of stiffness. For the limited number of tests run,

Table 2 - Effect of Cam Angle on Driven Axial Force Due to Torque Sensing

Cam Angle, deg	Torque Sensing Axial Force Multiplier	
	Calculated	% Increase
65	0.758	–
60	0.906	19.5
55	1.079	19.1

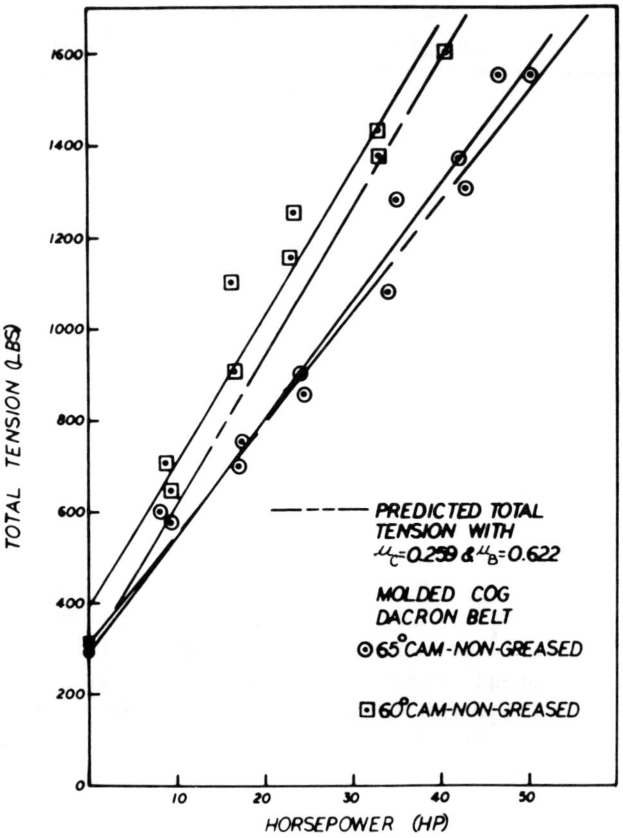

Fig. 5 - Total tension versus horsepower for molded cog-belts at actuator angles of 60 and 65 deg

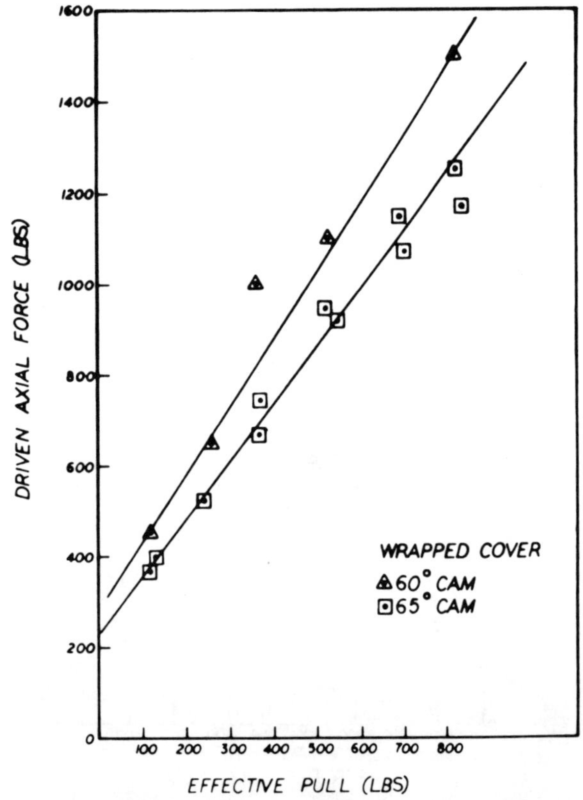

Fig. 4 - Driven axial force versus effective pull for wrapped cover belts at actuator angles of 60 and 65 deg

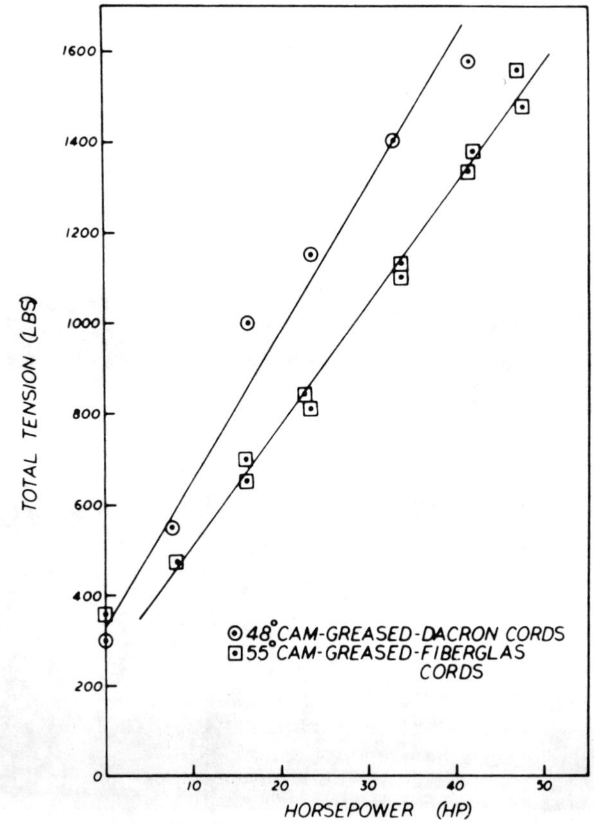

Fig. 6 - Total tension versus horsepower for molded cog-belts at actuator angles of 48 and 65 deg

Fig. 3 illustrates there is an appreciable difference in driver axial force reqüired. According to Eq. 6, a greater T_1 is necessary to produce the higher driver axial forces. These tests indicate that the stiffer wrapped cover belt produces more bending tension; therefore requiring additional input horsepower to operate the drive.

CAM ANGLE - Decreasing the actuator ramp angle will produce a more aggressive torque-sensing drive. The driven axial force equation, Eq. 1, shows that ramp angle β determines the amount of torque-sensing or driven axial force that will be exerted on the belt. By holding the spring force F_{NO}, spring rate K_N, and μ_C constant, various ramp angles will produce the multipliers shown in Table 2.

Table 3 - Actuator Surface Coefficient of Frictions for Different Actuators

Actuator Surface	Mean μ_C	Standard Deviation
"As-cast" without grease, all molded cog-belt	0.259	0.079
"As-cast" with grease, all molded cog-belt	0.216	0.063

Experimental data plotted in Fig. 4 verify that the change from 65 to 60 deg will increase the driven axial force due to torque sensing by 25-30%, which is greater than the 19.5% calculated increase.

Figs. 5 and 6 illustrate that increasing the cam angle will decrease total tension, due to lower total driven axial forces being exerted upon the belt.

CAM SURFACE - The cam actuator ramp surfaces dictate the coefficient of friction between the ramp surfaces. The surface must be smooth, and hard enough to avoid the possibility of the ramps becoming locked together or gouging into each other. Testing has shown that "as-cast" cams to meet these design requirements.

The actuator ramp coefficient of friction was determined by solving the driven axial force equation, Eq. 1, for the coefficient of friction, μ_C, using the measured driven axial force for each individual test. A mean value with its variance was calculated for each ramp surface, with the results recorded in Table 3.

In general, the coefficient of friction tended to increase, as load was applied, and then decrease slightly at high loads. By plotting each calculated μ_C, it was determined that the data fit a normal distribution, with a small amount of skewness.

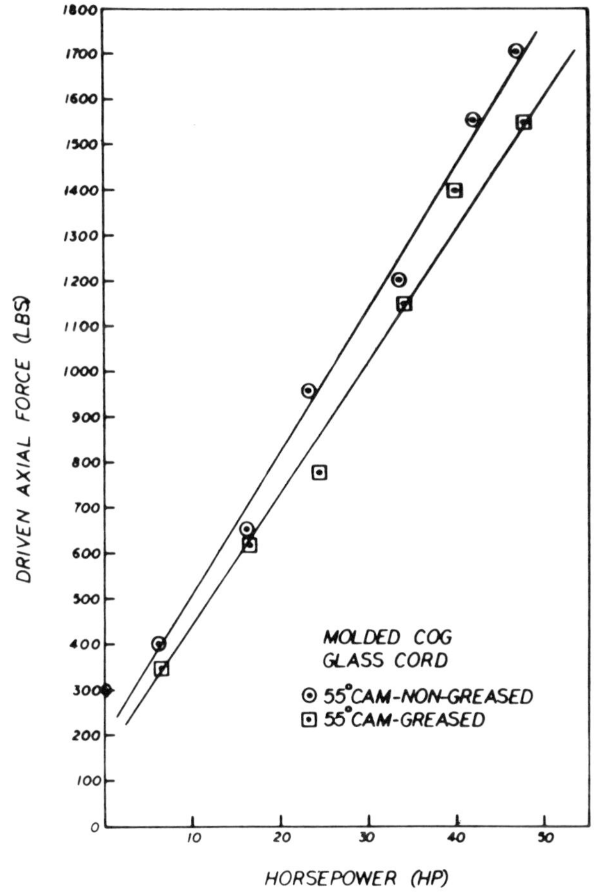

Fig. 7 - Driver axial force versus horsepower for molded cog-belts with fiberglass cords at an actuator angle of 55 deg with "greased" and "non-greased" cams

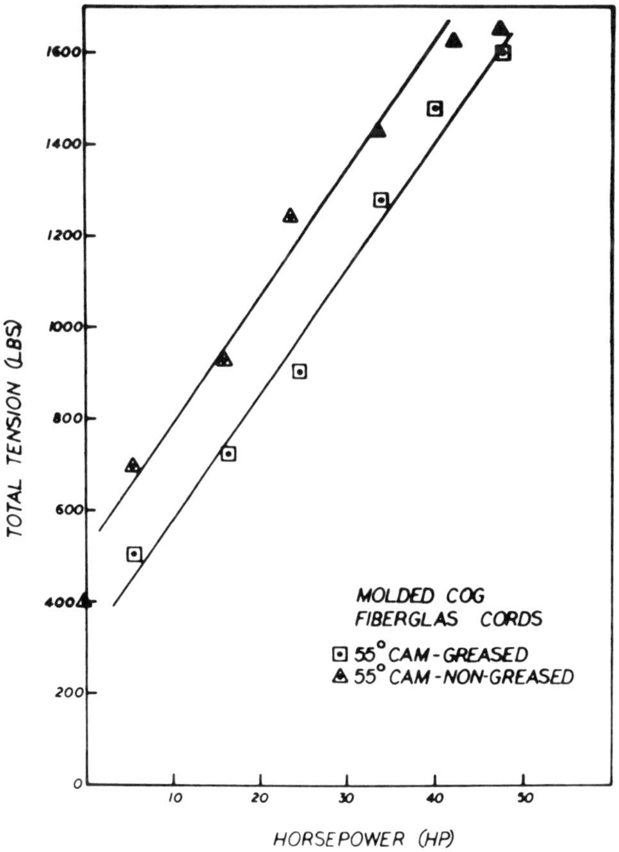

Fig. 8 - Total tension versus horsepower for molded cog-belts with fiberglass cords at an actuator angle of 55 deg with "greased" and "non-greased" cams

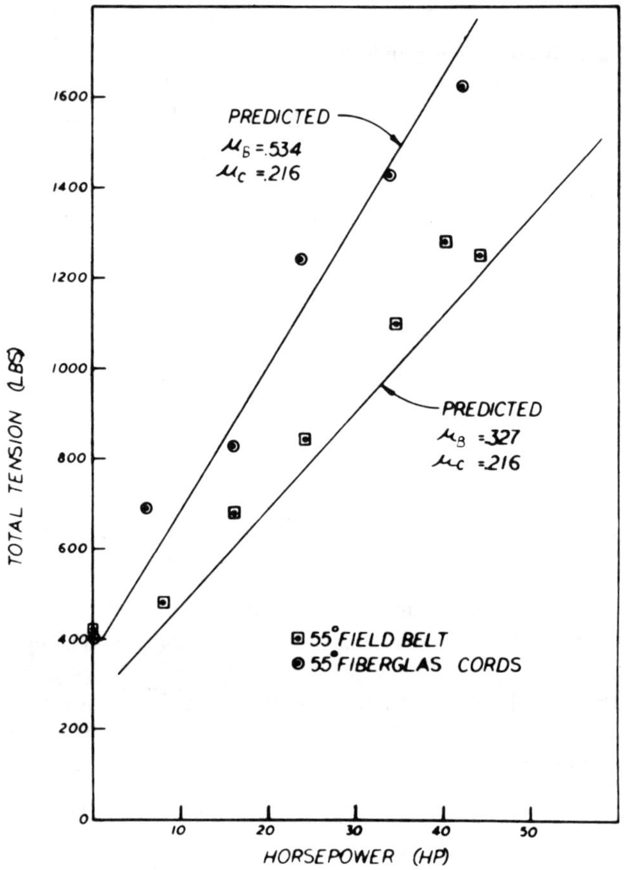

Fig. 9 - Total tension versus horsepower for fiberglass cog-belt with measured tensions compared to predicted tensions

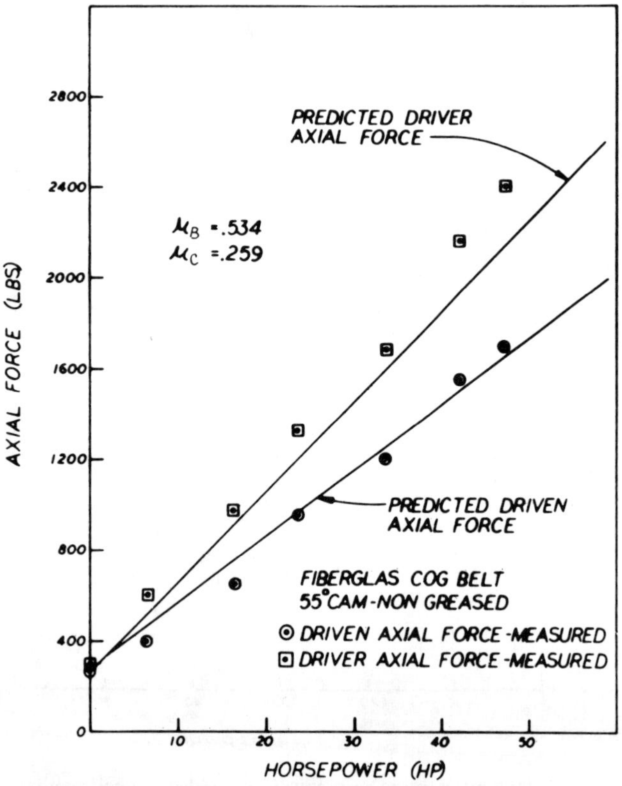

Fig. 10 - Axial force versus horsepower for measured driver and driven axial forces compared to predicted axial forces

Figs. 7 and 8 illustrate that the "greased" cams, with the lower coefficient of friction, reduce the driven axial force applied to the belt, thereby reducing the total tension of the drive.

VERIFICATION OF DAYCO'S EQUATION - Using the experimentally determined coefficients of friction μ_C and μ_B, the Dayco equations were solved for the design parameters (driven and driver axial force, tight and slack side tension). Fig. 5 shows a good correlation between measured total tension and predicted total tension.

The driver axial force equation provides good results for drive conditions when the belt-sheave coefficient of friction is nearly constant. At extremely high loads, when an excessive amount of slippage occurs, the predicted values are lower than measured, due to decreasing μ_B at the driver.

Figs. 9 and 10 verify that the equations predict the drive design parameters, total tension, driven and driver axial forces, with good agreement.

CONCLUSIONS

Within the limits of this investigation, the following conclusions were drawn:

1. Different belt constructions result in different belt to sheave coefficients of friction within a range of 0.327-0.622.
2. Fiberglass cords reduce the amount of belt elongation under heavy output loads.
3. Field operating conditions, due to sand particles and other foreign materials becoming embedded in the side of the belt, can reduce the coefficient of friction nearly 48%, from 0.622 to 0.327.
4. The wrapped cover belt requires more driver axial force to transmit similar loads than the wrapped cover-notched belt.
5. "As-cast" cam actuators provide good ramp surfaces, with a range of coefficients of friction from 0.216 to 0.259, for "with grease" and "without grease," respectively.
6. Reducing the cam actuator ramp angle will increase the torque-sensing aggressiveness of the drive.
7. Dayco's equations predict total tension and driven axial force with a high degree of agreement. The driver axial force equation provides good agreement, except under conditions of excessive driver slip.

NOMENCLATURE

D_A = Diameter of torque sensing ramp

D_N = Pitch diameter of belt at driven sheave

D_{NO} = Maximum belt pitch diameter on driven sheave

D_R = Pitch diameter of belt at driver sheave

D_{RO} = Minimum belt pitch diameter on driver sheave

F_N = Axial force on belt at torque-sensing driven sheave

F_{NO} = Preload compression on driven spring when belt is at its maximum pitch diameter on driven sheave

F_R = Axial force on belt at driver sheave

K_N = Axial spring constant for driven spring

T_1 = Tight side belt strand tension

T_2 = Slack side belt strand tension

α = Driver and driven sheave angle

β = Angle between ramp surface of torque-sensing sheave and plane of belt's path

θ_N = Belt wrap angle on driven sheave

θ_R = Belt wrap angle on driver sheave

μ_B = Coefficient of friction between belt and sheave surfaces

μ_C = Coefficient of friction on ramp of torque-sensing sheave

ϕ = Driven friction angle measured between inlet belt tangent and direction of belt-sheave frictional force

REFERENCE

1. L. R. Oliver and D. D. Henderson, "Torque Sensing Variable Speed V-Belt Drive." Paper 720708 presented at SAE Combined National FCIM and Powerplant Meeting, Milwaukee, September 1972.

Design Equations for a Speed and Torque Controlled Variable Ratio V-Belt Transmission**

L. R. Oliver
Belt Engineering Dept., Dayco Corp.

K. G. Hornung, J. E. Swenson and H. N. Shapiro
Mechanical Engineering Dept., The Ohio State University

**Paper 730003 presented at the International Automotive Engineering Congress and Exposition, Detroit, Michigan, January, 1973.

VARIABLE SPEED V-belt drives requiring an operator to set the operating point mechanically have been used in many engineering applications. Recently, however, sheaves have been designed that can sense torque or speed and then adjust accordingly. When a speed-sensing pulley is used as a driver and a torque-sensing sheave is driven, the result is an automatic transmission that adjusts to changes in input speed and output torque. Currently, this type of variable speed drive is used on recreational vehicles, such as the snowmobile and the all-terrain vehicle, and its use as a transmission for subcompact automobiles is being investigated. In fact, one manufacturer has been producing compact automobiles with variable speed belt transmissions for over 15 yr (1)*. However, the method of controlling speed ratio is somewhat different than the system employed in the transmission analyzed in this paper.

This study involves such a drive. The driver sheave, shown in Fig. 1A, consists of a fixed sheave flange and a moveable flange that can slide axially along the shaft. Its motion is constrained by splines that keep it from rotating relative to the fixed flange, a compression spring that tends to open the pulleys, the belt, and a centrifugal spring that tends to expand radially with increasing speed and pushes against a curved cam housing. The result is a sheave in which increased rotational speed causes outward motion of the V-belt in the groove.

The driven pulley (Fig. 1B) has a moveable flange that can rotate relative to the fixed flange, as well as move axially along the shaft. Load on the driven shaft is converted to axial force on the belt in the groove by three helical torque ramps. A torsion-compression spring also constrains the moveable flange. These components act together and, at equilibrium, all of the forces and torques are balanced and dictate an operating point that depends upon output torque.

As the output torque increases, the force that must be transmitted through the ramps increases. This force has an axial component that either forces the moveable sheave to close, and thereby increases the pitch diameter, or raises the belt tensions. In either event, the torque transmitted by the belt increases.

If the increase in axial force causes the flange to close, the

*Numbers in parentheses designate References at end of paper.

ABSTRACT

A type of V-belt drive which shifts ratios in accordance with input speed and output torque has been successfully employed as the transmission for vehicles ranging from snowmobiles to compact automobiles. Increasing popularity of this kind of transmission has created the need for more reliable criteria to be used in its design. This paper presents a derivation of design equations for a symmetrical V-belt drive in which centrifugal control is governed by an expanding garter spring and torque control results from a helical cam. Experimental results are included which support the validity of these equations.

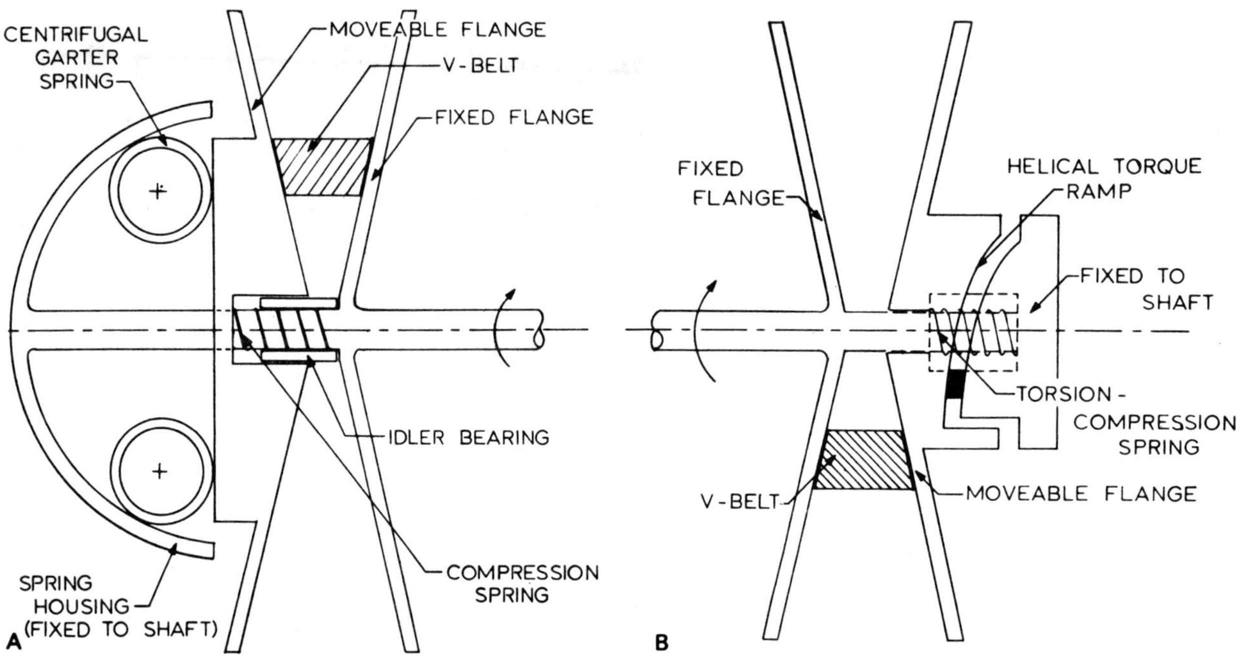

Fig. 1 - Cross-sectional view illustrating drive's operation. A-driver sheave; B-driven sheave

torque in the spring decreases, since as the flange closes it also rotates. The spring torque and the output torque of the drive are balanced by the belt torque.

When the two pulleys are combined in a drive, there is interaction between them. If the pitch diameter of one pulley increases, then the pitch diameter of the other pulley must decrease, since the center distance and the belt length are fixed. The result is an automatic transmission capable of reacting to changes in both prime mover speed and output load.

The present study consists of two parts. First, design equations for the torque-sensing driven and centrifugal, speed-sensing driver sheaves are derived. These equations, along with the equations developed by Worley (2) in the early 1950s for variable speed driver and driven pulleys, are used to generate theoretical predictions of drive performance. Second, an experimental study is described that provided verification of the validity of the analysis.

DERIVATION OF DRIVE EQUATIONS

Presented here is a summary of the analysis of a speed- and torque-sensing V-belt drive performed by Oliver (3). An equation for the axial force applied to the belt by the driven sheave is derived by summing torques and forces. This force is equated to Worley's driven spring force equation (2), thereby allowing solution for the belt tensions. With the belt tensions known, Worley's driver equation (2) is solved for the driver axial belt force. A force analysis of the speed-sensing driver sheave, consisting of radial and axial force balances, is performed and an expression for driver rpm in terms of the two belt pitch diameters and the belt tension difference is obtained. These equations can then be solved on a digital computer and curves generated to predict the drive's operating condition.

DRIVEN PULLEY - In 1955, Worley (2) published a force analysis of a V-belt in a variable pitch driven sheave in which the inlet tension T_2 is less than the exit belt tension T_1. His equation for driven spring force F_n can be used to determine the average direction in which friction will act between the belt and sheave surface around the arc of contact for a given tension difference $T_1 - T_2$ and a given axial force.

$$F_n = (T_1 - T_2)\left[\frac{\cos\frac{\alpha}{2} - \mu \sin\frac{\alpha}{2}\sin\phi}{2\,\mu\cos\phi}\right] \tag{1}$$

where:

α = total groove angle
μ = coefficient of friction between the belt and sheave
ϕ = friction angle

The friction angle is the angle between the inlet belt tangent and the direction of relative motion between the belt and the sheave surfaces (Fig. 2). As pointed out by Hornung (4), the friction angle is not constant around the arc of contact, but it will be treated as such in this analysis. It is important to realize, however, that this simplification is a possible source of error in the model.

Worley's equation for tension ratio can be solved if ϕ is calculated from Eq. 2.

$$\frac{T_1}{T_2} = \exp\left[\frac{\mu\,\theta_n\cos\phi}{\mu\left(\cos\frac{\alpha}{2}\right)(\sin\phi) + \left(\sin\frac{\alpha}{2}\right)}\right] \tag{2}$$

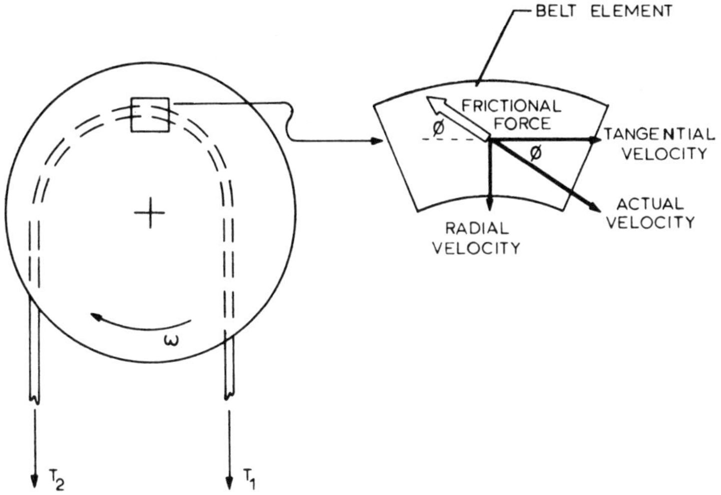

Fig. 2 - Diagram illustrating definition of friction angle

where:

θ_n = total wrap angle of the belt on the driven sheave

Clearly, then, if the axial force F_n can be calculated and the tension difference is known, then the actual belt tensions can be calculated. The actual axial force supplied by a torque-sensing driven sheave operating at a certain belt pitch diameter will now be determined.

The total axial force supplied by the sheave is due to the ramp axial force F_a and the force of the torsion-compression spring F_s

$$F_n = F_a + F_s \tag{3}$$

In order to evaluate F_a, the torque ramp is analyzed as a free body (Fig. 3). The nylon rider supplies a normal ramp force R_a and a frictional component, due to the coefficient of friction on the ramp, μ_a, and the normal force. At equilibrium, these forces are balanced by a tangential force F_t and axial force F_a. A force balance in the axial direction yields

$$-R_a \cos\beta - \mu_a R_a \sin\beta + F_a = 0 \tag{4}$$

where:

β = angle of inclination of the torque ramp from the perpendicular to the shaft centerline

In the tangential direction

$$R_a \sin\beta - \mu_a R_a \cos\beta - F_t = 0 \tag{5}$$

For equilibrium, moments on the moveable flange must also be balanced. The torque, due to the tangential component of the ramp force F_t acting at the ramp radius $D_a/2$, is balanced

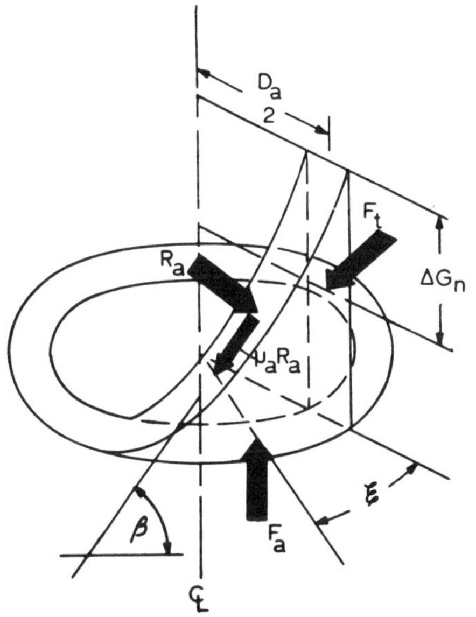

Fig. 3 - Free body diagram of helical torque ramp

by the belt torque, due to the tension difference $T_1 - T_2$ acting at the belt pitch radius $D_n/2$ and the torsion compression spring torque M_n. Symbolically

$$F_t \left(\frac{D_a}{2}\right) = \frac{(T_1 - T_2)}{2} \left(\frac{D_n}{2}\right) + M_n \tag{6}$$

With Eqs. 4-6, R_a and F_t can be eliminated and an expression for F_a determined

$$F_a = \frac{1}{D_a} \left[\frac{(T_1 - T_2)}{2} D_n + 2 M_n\right] \left[\frac{\cos\beta + \mu_a \sin\beta}{\sin\beta - \mu_a \cos\beta}\right] \tag{7}$$

It should be noted that the frictional force $\mu_a R_a$ in Fig. 3 is drawn as if the sheave were in an impending opening condition. Eq. 7 reflects this in the signs of the frictional terms. If impending closing exists, the direction of the frictional forces reverses, changing the signs of the μ_a terms in the equation.

The spring torque M_n in Eq. 7 is best expressed in terms of measurable quantities. To do this, M_n can be evaluated as a preload M_{no} plus a torque equal to the rotational spring constant K_t times the flange rotation ξ. The amount of rotation is related to the axial movement of the flange ΔG_n (see Fig. 3) by the three ramps that cause the flange to move axially when it is rotated.

Fig. 4 relates the axial travel ΔG_n to the difference between the maximum pitch diameter D_{no} and pitch diameter D_n

$$\Delta G_n = (D_{no} - D_n) \tan \alpha/2 \qquad (8)$$

The rotation of the spring is related to ΔG_n by the following equation, derived from Fig. 3.

$$\tan \beta = \frac{\Delta G_n}{(D_a/2)\,\xi} \qquad (9)$$

Using Eqs. 8 and 9, the spring torque can be expressed as

$$M_n = M_{no} + K_t \left[\frac{(D_{no} - D_n) \tan \alpha/2}{(D_a/2) \tan \beta}\right] \qquad (10)$$

The axial spring force F_s of the torsion-compression spring is the preload force F_{no} plus the additional force due to axial compression which equals the compression spring constant K_n times the axial travel ΔG_n. Employing Eq. 8 yields the following relationship for F_s

$$F_s = F_{no} + K_n \left[(D_{no} - D_n) \tan \alpha/2\right] \qquad (11)$$

When Eqs. 7, 10, and 11 are inserted into Eq. 3, the final expression for axial force of a torque-sensing sheave operating at a pitch diameter D_n and a tension difference $T_1 - T_2$ is

$$F_n = \frac{1}{D_a}\left[\frac{(T_1 - T_2)}{2} D_n + 2M_o + \frac{4 K_t (D_{no} - D_n) \tan \alpha/2}{D_a \tan \beta}\right]$$

$$\times \left[\frac{\cos \beta + \mu_a \sin \beta}{\sin \beta - \mu_a \cos \beta}\right] + F_{no} + K_n \left[(D_{no} - D_n) \tan \alpha/2\right] \qquad (12)$$

This relationship can be used to calculate the axial force, F_n, and the resulting value can then be used in Eq. 1 to calculate ϕ, the friction angle. This is easily accomplished by squaring Eq. 1, substituting $(1 - \sin^2 \phi)$ for $\cos^2 \phi$ and then using the quadratic formula to solve for $\sin \phi$. The resulting value for ϕ can then be used to evaluate Eq. 2 for tension ratio, since the wrap angle θ_n is known if D_n is known for a fixed length belt and fixed center drive. Individual strand tensions can then be calculated using

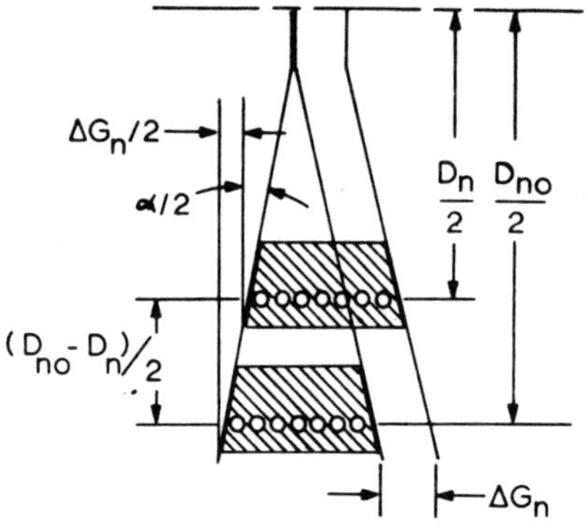

Fig. 4 - Relationship between driven pulley axial travel and belt pitch diameter

$$T_1 = \frac{(T_1 - T_2)}{1 - 1/(T_1/T_2)} \qquad (13)$$

and

$$T_2 = T_1 - (T_1 - T_2) \qquad (14)$$

DRIVER PULLEY - Once the two tensions are known, Worley's driver equation (2) can be used to calculate the driver spring force F_r

$$F_r = \frac{T_1 \theta_r}{2}\left(\frac{1 - \mu \tan \alpha/2}{\mu + \tan \alpha/2}\right) \qquad (15)$$

where θ_r is the wrap angle on the driver sheave.

Since the axial force supplied by the centrifugal, speed-sensing driver sheave under consideration is related to the driver rpm, analysis of the driver will yield an expression for the prime mover speed necessary to achieve a spring force of F_r corresponding to the given output torque requirements. Therefore, with this analysis, the overall drive requirements will be accounted for.

The centrifugal garter spring experiences an axial force F_c equal to the sum of the belt axial force F_r and a force from the driver compression spring. The compression spring force is equal to a preload F_{ro} plus an additional force, due to the axial displacement ΔG_r of the spring and the spring constant K_r. The axial travel of the driver can be related to the pitch diameter of the belt on the driver sheave by considerations similar to those in Fig. 4 to yield

$$\Delta G_r = (D_r - D_{ro}) \tan \alpha/2 \qquad (16)$$

so that the axial force on the garter spring is

$$F_c = F_r + F_{ro} + K_r \left[(D_r - D_{ro}) \tan \alpha/2\right] \tag{17}$$

In order to relate the forces on the centrifugal spring to the driver rpm, it is necessary to examine a free body diagram of a differential element of the spring (Fig. 5). The forces on the element $d\epsilon$ are the normal force (R_c $[(D_c/2)\, d\epsilon]$) and frictional force ($\mu_c R_c$ $[(D_c/2)\, d\epsilon]$) exerted by the curved outside housing, the axial ($F_{c\ell}$ $[(D_c/2)\, d\epsilon]$) and associated frictional ($\mu_c F_{c\ell}$ $[(D_c/2)\, d\epsilon]$) forces from the moveable sheave, the centrifugal force $(W/g)(D_c\, \omega^2/2)[(D_c/2)\, d\epsilon]$ due to the spring's rotation, and the tangential spring force $[K_c \pi (D_c - D_{co})]$. The parameters in these expressions are the normal force per unit length from the outside housing R_c, the spring's middle diameter D_c, the differential arc of spring $d\epsilon$, the coefficient of friction μ_c, the axial force per unit length of spring $F_{c\ell}$, the weight per unit length W, angular velocity of the sheave ω, garter spring constant K_c, and the spring's diameter when at rest D_{co}.

Summation of forces in the outward radial direction requires, for equilibrium, that

$$\frac{W}{g} \omega^2 \frac{D_c^2}{4} d\epsilon - 2K_c \pi (D_c - D_{co}) \sin \frac{d\epsilon}{2} - \mu_c F_{c\ell} \frac{D_c}{2} d\epsilon$$

$$- R_c \cos \gamma \frac{D_c}{2} d\epsilon - \mu_c R_c \sin \gamma \frac{D_c}{2} d\epsilon = 0 \tag{18}$$

For axial equilibrium

$$-R_c \sin \gamma \frac{D_c}{2} d\epsilon + \mu_c R_c \cos \gamma \frac{D_c}{2} d\epsilon + F_{c\ell} \frac{D_c}{2} d\epsilon = 0 \tag{19}$$

where γ is the cam housing angle indicated on Fig. 5.

Eqs. 18 and 19 can be solved for $F_{c\ell}$, the axial force per unit length of garter spring

$$F_{c\ell} = \frac{\dfrac{W}{4g} \omega^2 D_c^2 - K_c \pi (D_c - D_{co})}{\dfrac{D_c}{2} \left(\mu_c + \dfrac{\cos \gamma + \mu_c \sin \gamma}{\sin \gamma - \mu_c \cos \gamma}\right)} \tag{20}$$

The total axial force F_c on the centrifugal spring is found by integrating Eq. 20 around the entire middle circumference.

$$F_c = \int_0^{2\pi} F_{c\ell} \left(\frac{D_c}{2}\right) d\epsilon \tag{21}$$

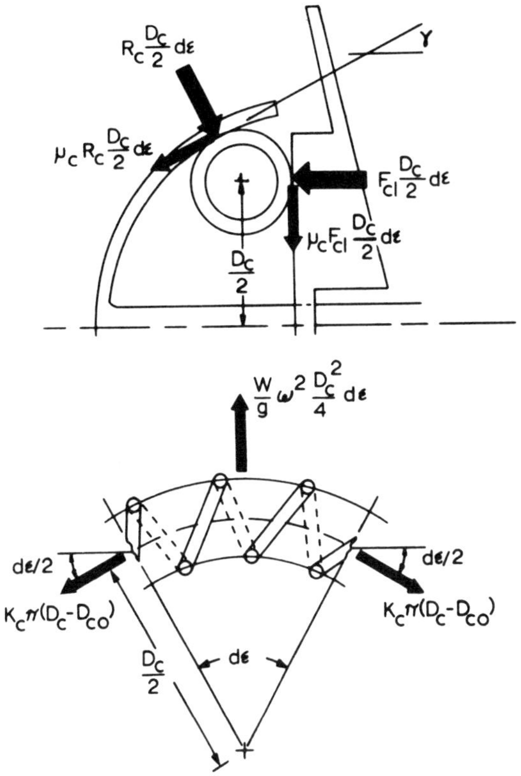

Fig. 5 - Differential element of centrifugal spring

Upon performing the integration and writing angular velocity ω in terms of rpm

$$F_c = 2\pi \left[\frac{\dfrac{W \pi^2 (\text{rpm})^2 D_c^2}{3600\, g} - K_c \pi (D_c - D_{co})}{\mu_c + \dfrac{\cos \gamma + \mu_c \sin \gamma}{\sin \gamma - \mu_c \cos \gamma}}\right] \tag{22}$$

Solving for rpm yields

$$\text{rpm} = \sqrt{\frac{3600\, g}{W \pi^2 D_c^2} \left[\frac{F_c}{2\pi} \left(\mu_c + \frac{\cos \gamma + \mu_c \sin \gamma}{\sin \gamma - \mu_c \cos \gamma}\right) + K_c \pi (D_c - D_{co})\right]} \tag{23}$$

Briefly, it should again be noted that the direction of the frictional forces in Fig. 5 indicates impending closing of the driver sheave. This is consistent with the condition of the driven sheave, and if the situation reverses, the signs of the μ_c terms in Eq. 23 must be changed.

Complete evaluation of rpm can be accomplished if expressions for D_c and γ can be found. With Fig. 6, it can be seen that the diameter D_c and the housing contact angle γ are related to the radius ρ along which the center of the spring is

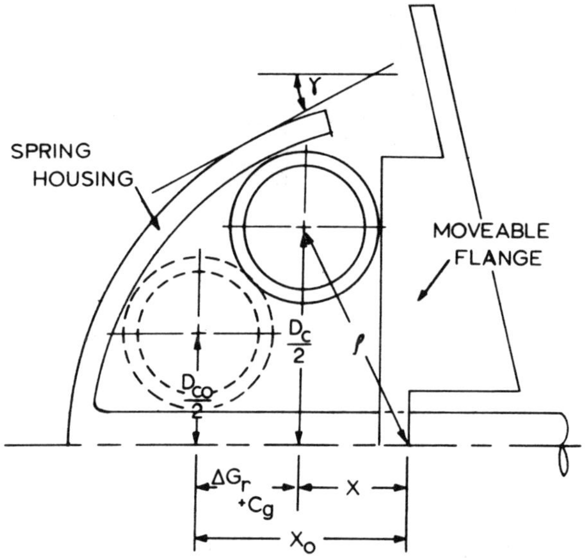

ρ = RADIUS OF SPRING'S MOTION
C_g = CLUTCHING GAP
D_c = MIDDLE DIAMETER OF SPRING
D_{co} = MIDDLE DIAMETER OF SPRING AT ZERO RPM
X = LOCATION OF SPRING'S PLANE
X_o = LOCATION OF SPRING'S PLANE AT ZERO RPM
γ = SPRING-CAM HOUSING CONTACT ANGLE

Fig. 6 - Geometry of garter spring location

constrained to move by

$$\left(\frac{D_c}{2}\right)^2 + X^2 = \rho^2 \tag{24}$$

The axial location of the spring's plane X can be expressed in terms of the location at zero angular velocity X_o, the amount of axial travel of the moveable flange ΔG_r, and the gap through which the flange must move before it clutches the belt C_g.

$$X = X_o - (\Delta G_r + C_g) \tag{25}$$

Using Eq. 16 for ΔG_r and Eqs. 24 and 25, the spring's diameter D_c is found to be

$$D_c = 2\Big[\rho^2 - X_o^2 - C_g^2 + 2X_oC_g + (D_r - D_{ro})(\tan\alpha/2)(2X_o - 2C_g - (D_r - D_{ro})\tan\alpha/2)\Big]^{1/2} \tag{26}$$

Further consideration of Fig. 6 reveals that the housing angle γ is related geometrically to D_c and X by

$$\gamma = \frac{\pi}{2} - \tan^{-1}\left[\frac{D_c}{2X}\right] \tag{27}$$

With Eqs. 16 and 25 this becomes

$$\gamma = \frac{\pi}{2} - \tan^{-1}\left[\frac{D_c/2}{\left[X_o - (D_r - D_{ro})\tan\frac{\alpha}{2} - C_g\right]}\right] \tag{28}$$

Clearly, the rpm Eq. 23 can be solved when Eqs. 17, 26, and 28 are evaluated for a given operating condition. All that must be specified are driver belt pitch diameter D_r and tight side tension T_1, which is known from the driven pulley analysis.

COMPUTER SOLUTION OF EQUATIONS - Solutions to the system of equations derived were obtained in two ways. First, the driven and driver pulleys were analyzed separately. Next, they were put together with the results of the driven pulley being used as the input to the driver pulley equations.

The input variables to the first programmed solution were belt diameters (from measured data), input torque (chosen as constant to generate the data), and total belt tension (measured data). The tension difference was calculated from the input torque divided by the driver sheave belt pitch radius, and the tight side tension was computed from the total tension and the tension difference. The tension difference and belt diameters were sufficient to solve the driven pulley equations for the belt tensions. The driver pulley rpm were determined from the tight side tension and the belt pitch diameters. Thus, curves of total tension and driver speed for constant input torques were obtained.

The second computer solution used belt diameters and input torque to calculate driver speed. The driven pulley equations were solved for the tight side tension, and this result was used in the driver pulley equations to evaluate driver rpm. Thus, predictions of prime mover speed necessary to carry a certain torque load were obtained for given input torques.

These computer results were quite amenable to experimental comparison, and the work done in that area is now described.

EXPERIMENTAL VERIFICATION

A test rig was designed and instrumented to run torque and centrifugal sensing V-belt pulleys and obtain data. The apparatus consisted of a d-c motor-dynamometer that drove the system, and an axial piston pump that absorbed the transmitted energy. The pump was mounted on a table that was free to pivot on bearings. A force transducer was used to restrain the table and also measure the amount of force needed. From a torque balance on the output table, the total belt tension was calculated. Fig. 7 is a schematic diagram of the output table with the forces and torques indicated. The instrumentation enabled the investigators to measure total belt tension, input torque, the two pulley speeds, and the two belt pitch diameters. All other parameters, such as spring constants and physical dimensions, were also measured.

The data were taken with input torque held constant. When torque stabilized at the desired value, all other measurements were taken. In general, data were taken for increasing rpm. However, the load from the pump was not exactly constant,

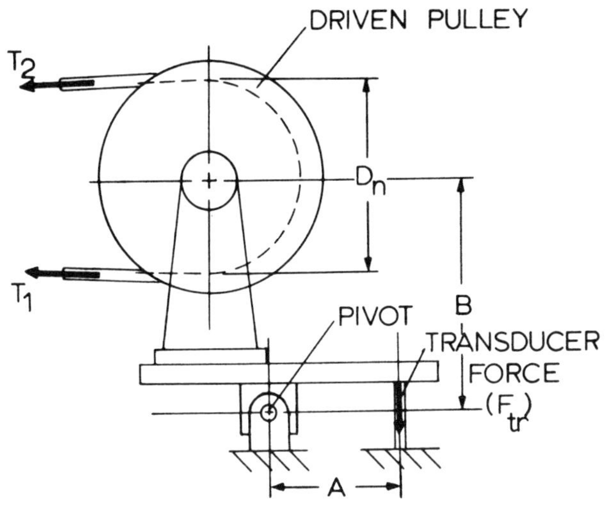

SUM OF MOMENTS ABOUT PIVOT = 0

$$F_{tr} \cdot A + (T_1 - T_2) \cdot \frac{D_n}{2} - (T_1 + T_2) \cdot B = 0$$

Fig. 7 - Diagram of output table

and some trial adjustments were needed to obtain the proper input torque. It was therefore impossible to tell whether the drive was in an impending opening or closing condition.

DISCUSSION OF RESULTS OF EXPERIMENTAL STUDY - The results are presented graphically in two parts. First, theoretical predictions of total tension from the driven pulley analysis and driver rpm from the driver analysis are correlated with measured data independently. Next, the equations for the total drive were solved and the tension difference was plotted as a function of driver rpm for both theoretical and experimental data.

Fig. 8 compares calculated total tension to measured total tension. The 45 deg dashed line would be perfect correlation. The solid line is a least squares fit of the experimental points. This correlation is considered very good and helps validate the driven pulley analysis. The coefficients of friction that were used in the computer solution were $\mu = 0.48$ for the belt-sheave interface and $\mu_a = 0.11$ on the torque ramp.

The driver pulley results are presented in Fig. 9. Here driver speed was calculated for $\mu = 0.48$ and the spring-housing coefficient $\mu_c = 0.04$. Again the solid line representing a least squares fit of the points falls very nearly on the 45 deg line.

Although the results described are quite encouraging, one fact remains somewhat troublesome. The randomness of the loading caused the drive to be continuously shifting from an opening to a closing state and back again the other way. It is to be expected, therefore, that both positive and negative values for μ_a and μ_c should be acting, that is, friction may be changing direction on the ramps and the spring housing. The fact that positive values were obtained would indicate that the data were biased toward opening on the driven sheave and closing on the driver. This may or may not be the case, but the further data help to eliminate this from consideration.

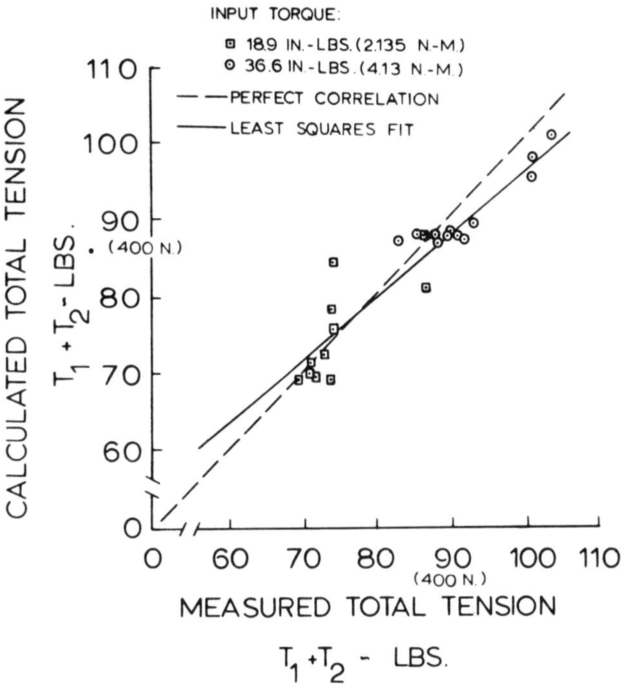

Fig. 8 - Correlation of theoretical and measured total tensions for the driven pulley

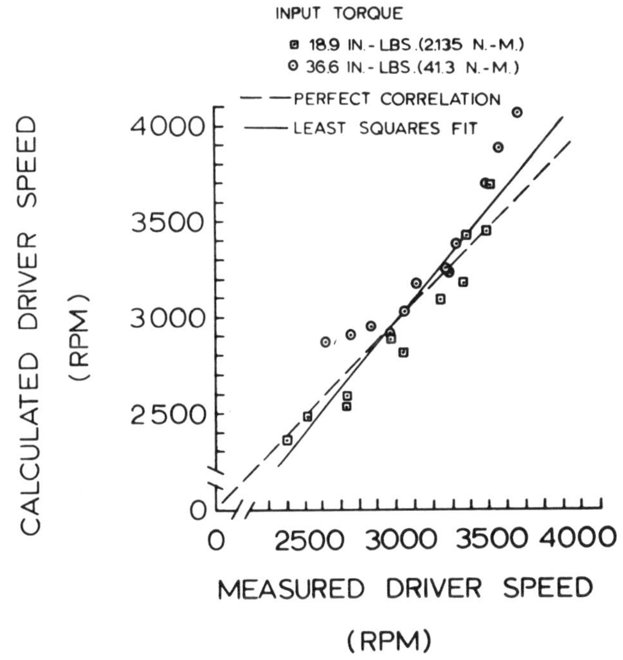

Fig. 9 - Correlation of theoretical and measured driver pulley speeds

When the entire system of drive equations was solved as a unit, coefficients of $\mu_a = \mu_c = 0$ were used. The results are shown in Fig. 10. Here the solid lines represent calculated constant input torque curves, and the points are experimental data. Even the lowest torque curve is close to the data, but as torque increases, the correlation improves. It is suspected that the low torques obtainable in this study may magnify this effect, and that data obtained at higher and higher torques and

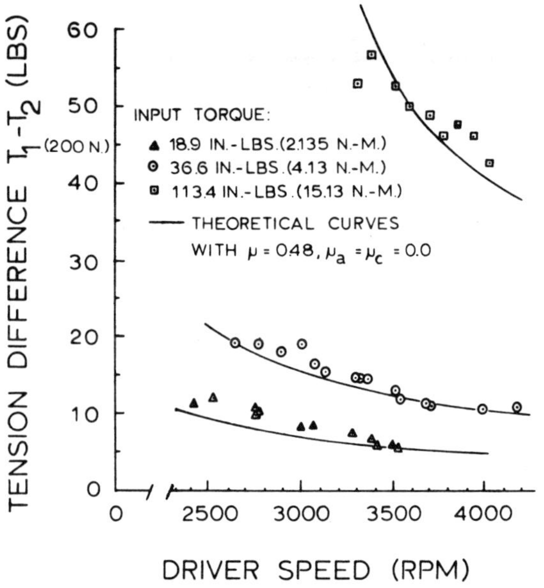

Fig. 10 - Evaluation of total set of drive equations

speeds would correlate best with coefficients of zero. This is further noticed in the fact that Figs. 8 and 9 contain only lower torque data.

The coefficient of friction for the belt-sheave interface of 0.48 was considered reasonable and commensurate with handbook values long used for V-belts in pulleys.

There are a number of possible causes of the bias of the lower torque data. First of all, the lower torque curves were almost a no-load situation when compared with the drive's capacity of 25 hp and 6000 rpm. Due to vibration, the maximum speed obtained for lower torque data was only 4130 rpm, and the maximum horsepower was 2.4. Since the tension difference is small for low-load operation, constant magnitude errors appear to have a more pronounced effect on the low torque results.

One such error is the effect of losses inherent in the drive. Since the tension difference calculated from torques measured at the driver consist of the actual belt tension difference plus the drive losses, the actual measured values should be greater than the predicted values.

In addition, Worley's spring force equations involve assumptions that might result in error. The most glaring assumption is that of constant friction angle, ϕ. There may also be errors introduced by all of the assumptions in the derivation of the equations.

Since each pulley has a moveable flange, error may result from axial forces between the flange and the shaft. These forces were not directly accounted for in the analysis; however, their effect is similar to the spring and ramp friction forces. Including estimates for the forces between the bearing and shaft of the driven pulley and the splines and moveable flange of the driver pulley gave the same results for coefficients of friction about 25% lower on the ramp and cam housing. Since this effect of shaft friction could be accounted for in the ramp and spring-housing friction forces, the analysis was not changed.

There is an additional, more complex force that should be mentioned, although little is known about it. Since there must be clearance between the shaft and moveable flange on each pulley, and since the belt force is exerted on only one side of the flange, it might be cocked relative to the shaft. In such a position, the flange may tend to "climb" the shaft, and this could influence the results in some way.

Although the lower torque data favor positive coefficients, all values were considered reasonable. Thus, the equations yield good predictions about the actual operation of the drive. The equations are, therefore, valid tools to use in designing such transmissions and, indeed, are being used as such. It is felt that the drive described in this paper has much potential, and the analysis presented herein will aid in its development.

NOMENCLATURE

C_g = Clutching gap, distance from moveable flange to belt when driver sheave is not rotating, in (m)

D_a = Center diameter of torque-sensing ramp, in (m)

D_c = Center diameter of centrifugal garter spring in driver sheave, in (m)

D_{co} = Center diameter of centrifugal garter spring when driver sheave is not rotating, in (m)

D_n = Belt pitch diameter at driven sheave, in (m)

D_{no} = Maximum belt pitch diameter on driven sheave, in (m)

D_r = Belt pitch diameter at driver sheave, in (m)

D_{ro} = Minimum belt pitch diameter on driver sheave, in (m)

F_a = Ramp axial force on moveable flange of torque-sensing sheave, lb (N)

$F_{c\ell}$ = Axial force per unit length exerted on centrifugal spring, lb/in (N/m)

F_n = Axial force on belt at driven sheave, lb (N)

F_{no} = Compression preload on driven spring, lb (N)

F_r = Axial force on belt at driver sheave, lb (N)

F_{ro} = Compression preload on driver idle spring, lb (N)

F_t = Tangential force on ramp of torque-sensing sheave, lb (N)

g = Acceleration of gravity, ft/s^2 (m/s^2)

ΔG_n = Axial travel of moveable flange of driven sheave, in (m)

ΔG_r = Axial travel of moveable flange of driver sheave, in (m)

K_c = Spring rate of centrifugal garter spring, lb/in (N/m)

K_n = Spring rate of driven compression spring, lb/in (N/m)

K_r = Spring rate of driver idle spring, lb/in (N/m)

K_t = Torsional spring rate of driven spring, in-lb/rad (N-m/rad)

M_n = Torque on driven spring, in-lb (N-m)

M_{no} = Preload torque on driven spring, in-lb (N-m)

R_a = Normal force on ramp of torque-sensing sheave, lb (N)

R_c = Normal force per unit length of centrifugal spring exerted by curved cam housing, lb (N)
rpm = Rotational speed of driver sheave, rpm, rps
T_1 = Tight side belt tension, lb (N)
T_2 = Slack side belt tension, lb (N)
W = Weight per unit length of centrifugal garter spring, lb/ft (N/m)
X = Distance from center of curvature of cam housing to plane of centrifugal garter spring, in (m)
X_o = Distance from center of curvature of cam housing to plane of centrifugal garter spring when driver sheave is not rotating, in (m)
α = Groove angle of symmetrical sheave, rad
β = Angle between ramp surface of torque-sensing sheave and perpendicular to shaft centerline, rad
γ = Angle between axial direction and tangent at contact point of centrifugal garter spring and driver cam housing, rad
$d\epsilon$ = Differential arc of centrifugal garter spring, rad
ξ = Rotation of moveable flange of torque-sensing sheave, rad
θ = Arc of belt-sheave contact, rad
$\Delta\theta$ = Arc of belt element, rad
μ = Coefficient of friction between belt and sheave surface
μ_a = Coefficient of friction on ramp of torque-sensing sheave
μ_c = Coefficient of friction between centrifugal garter spring and driver
ρ = Radius of curvature of cam housing, in (m)
ϕ = Friction angle, direction opposite to relative motion of belt and sheave surfaces measured from inlet tangent, rad
ω = Angular velocity of centrifugal sensing sheave, rad/s

REFERENCES

1. H. H. J. Ludoph, "Stepless Automatic Transmission for Cars." Published by SAE of Japan Inc., Paper presented at Fédération Internationale des Sociétés d'Ingénieurs des Techniques de l'Automobile Tenth International Automobile Technical Congress, Japan, 1964.
2. W. S. Worley, "Designing Adjustable-Speed V-Belt Drives for Farm Implements." SAE Transactions, Vol. 63 (1955), pp. 321-333.
3. L. R. Oliver, "Steady-State Force Analysis of a Variable Speed Drive with Torque Sensing Driven and Centrifugal Sensing Driver Sheaves." Dayco Report, June 1969.
4. K. G. Hornung, "Factors Influencing the Fatigue Characteristics of Rubber-Textile Machine Elements." Ph.D. Dissertation, The Ohio State University, 1959.
5. J. E. Swenson, "Performance of Torque and Speed Controlled Variable Speed V-Belt Drives." M.Sc. Thesis, Mech. Engrg. Dept., The Ohio State University, 1972.
6. L. R. Oliver, K. G. Hornung, and H. N. Shapiro, "An Automatic V-Belt Transmission With an Asymmetrical Belt." Paper 72-WA/DE-10 presented at ASME Winter Annual Meeting, New York, November 1972.
7. L. R. Oliver and D. D. Henderson, "Torque Sensing Variable Speed V-Belt Drive." Paper 720708 presented at SAE Farm, Construction and Industrial Machinery Meeting, Milwaukee, September 1972.

Continuously Variable Transmissions for Cars with High Ratio Coverage*

P. Baudoin
Regie Nationale des
Usines Renault
(France)

*Paper 790041 presented at the Congress and Exposition, Detroit, Michigan, February, 1979.

ABSTRACT

Starting from a classification of car transmissions and from the manufacturers requirements, especially for fuel savings, the potential advantages of continuously variable transmissions permitting to associate low engine RPM with high vehicle speed, are shown.

The control of the engine, transmission and clutch to reach a given optimum, is explained.

Here are given results of simulation or testing of vehicles equipped with a Renault hydrostatic prototype transmission, a 2-mode hydromechanical transmission and a Van Doorne prototype Transmatic.

Highway fuel economy trends are quantified in case of substitution of the conventional S.I. engine for Diesel and/or turbocharged engines.

INTRODUCTION

FOREWORD ON TRANSMISSIONS - We shall only discuss transmissions for cars powered by a thermal prime mover:

. spark ignited or Diesel conventional engine
. atmospheric or turbocharged
. gas turbine or Stirling engine.

These engines have in common, at least two characteristics:

- The maximum power they can deliver increases with RPM
- Their efficiency, noise and polluting emissions vary considerably with power and RPM (or torque and RPM).

These two characteristics make it necessary to use a multiratio or continuous transmission between the engine and the wheels in order to:

A) Satisfy the drivers' demand for various positive or negative accelerations at various vehicle velocities
B) Optimize the engine operating conditions in terms of minimum fuel consumption, minimum noise, or minimum polluting emissions etc...
Only an engine capable of delivering its maximum power at any RPM between zero and maximum power RPM, and whose efficiency, noise and emissions would only depend on the delivered power, could be rigidly connected to the wheels.

Statement A can be verified by anybody driving a car with a manual transmission:
It is impossible to reach the vehicle maximum speed in first gear, or to go up a steep grade in top gear...

Statement B is partly supported by the following data which indicate the fuel economy of a Renault 14 TL at a steady speed of 55 MPH with the same stock engine but with different transmission ratios corresponding to different engine RPM:

gear	3rd	4th	/
RPM	4980	3455	1750
MPG	27.0 -23%	35.0 0%	44.8 +28%
noise	-	↘	↘↘

We define the transmission ratio by:

$$\text{transmission ratio} = \frac{\text{transmission output speed}}{\text{transmission input speed}}$$

The higher the transmission ratio, the higher the engine torque but the lower its RPM and noise.

In the following chapters, we shall use the expression "overdrive ratio", defined by

$$\text{overdrive ratio} = \frac{\text{maximum transmission ratio}}{\text{transmission ratio permitting to reach the maximum speed of the vehicle}}$$

CLASSIFICATION - Amongst many possible classifications, we have chosen the following one, which we believe reasonably exhaustive: See Table 1.

TYPICAL TRANSMISSION REQUIREMENTS OF CAR MANUFACTURERS - They include:

- Mass producibility at low cost
- High efficiency or more generally, ability to give a good fuel economy
- Comfort: no noise and limited jerk when shifting occurs
- Limited weight and volume
- Limited maintenance
- Good reliability and life

The reference values are implicitly given by 4 speed manual transmissions in Europe and Japan, and conventional 3 speed automatic transmissions in the U.S.A.

Hybrid transmissions we have investigated show no promise to fulfill these specifications in the near future; they only present small advantages for strictly urban vehicles (prime mover downsizing, reduced emissions and fuel consumption but with a high cost penalty for the consumer).

We previously mentioned that high overdrive ratios could give a better fuel economy and less noise. But they mean increased ratio coverage and for conventional transmissions:

- More ratios
- More shiftings and either automatism, or education of drivers and their acceptance of more constraint.

This means less comfort.

In view of the above requirements, this also means a tough job for the designers of conventional tranmissions, and this is one of the reasons why we became involved in CVT programs.

In the following sections, we shall outline the control principles (which are described with more details in patents filed by Renault) and give some results obtained by

Table 1

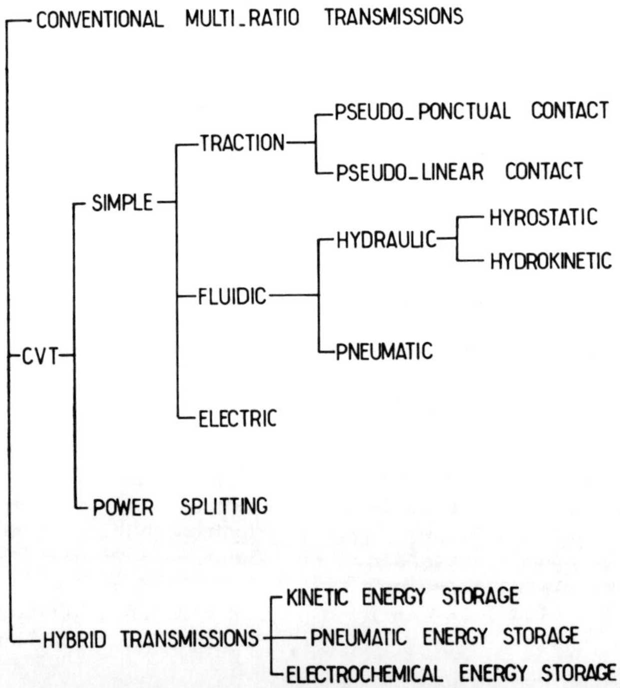

computer simulation and/or by testing on benches or road with:
- A purely hydrostatic transmission
- A hydromechanical transmission
- and a Van Doorne prototype Transmatic transmission.

Most of the research was made under contracts of the "Institut de Recherches des Transports.

ENGINE - CLUTCH - CVT SYSTEM CONTROL

PROBLEM - The problem is to control a system (see Fig. 1) consisting of:
- An engine
- A clutch
- A CVT.

We shall describe the most general case in which a clutch is located between the engine and the transmission. The control of a simplified system with no clutch can easily be deduced by elimination from the following.
- The parameters to control are:
  . On the heat engine, the air-fuel mixture flow rate, or any connected parameter such as the position of the carburettor throttle on a spark ignited conventional engine.
  . On the clutch, the torque transmitted and completion of clutching or declutching.
  . On the transmission, the transmission ratio.
- The general objectives are in order of priority:
  . Fulfill the demand of the driver, expressed either through the displacement of the accelerator pedal or through the effort exerted on the brake pedal.
  . While optimizing the engine's working conditions for a given criterion:
    - Either chosen by the manufacturer
    - Or variable, and chosen by the driver or by the D.O.T. through some means incorporated to the road.

For any given criterion (minimal fuel consumption, minimal noise, minimal emissions or a compromise between them), we can define an ideal operating line in the power-RPM or torque-RPM axis of the engine. See Fig. 2.

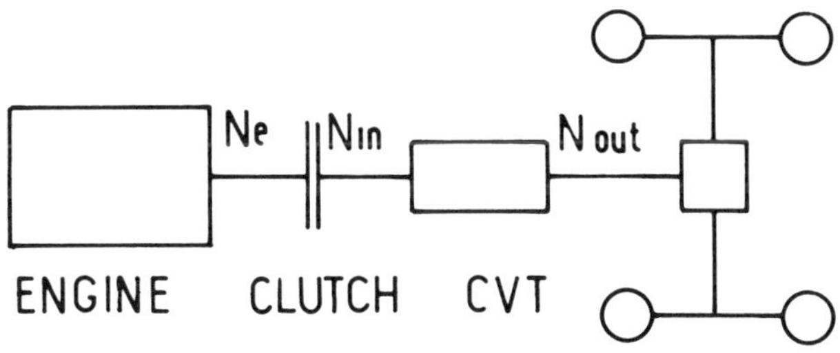

Fig. 1 - Vehicle propulsion system sketch

CONTROL LOGIC

A) Under normal operating conditions, out of the clutching phase, the position of the accelerator pedal commands the power the engine has to deliver and consequently its ideal speed. The control system pilots the transmission ratio to decrease the difference between the actual engine RPM and the ideal RPM within the limits of the transmission possibilities. Meanwhile, the air-fuel mixture flow, or a connected parameter (position of the carburettor throttle for instance), is controlled to make the engine deliver the required power. When the driver pulls the brakes, the transmission ratio is controlled to decrease the difference between the actual engine RPM and another optimum RPM, function of the effort exerted on the brake pedal and/or the vehicle speed.

B) During the clutching phase, the control system makes the accelerator pedal command the angular acceleration of the engine as long as the vehicle is motionless and then the actual acceleration of the vehicle wheels whatever the outside conditions providing no wheel slip. During this phase, the transmission remains on its low ratio.

The following notations are used:

a	=	Displacement of the accelerator pedal
b	=	Effort exerted on the brake pedal
Ne	=	Engine RPM
Nin	=	Transmission input RPM
Nout	=	Transmission output RPM
R	=	$\frac{Nout}{Nin}$ = Tranmission ratio
θ	=	Carburettor throttle rotation (or injection rate)
Tc	=	Torque transmitted by the clutch
Te	=	Engine torque
Tr	=	Resisting torque on the transmission input shaft

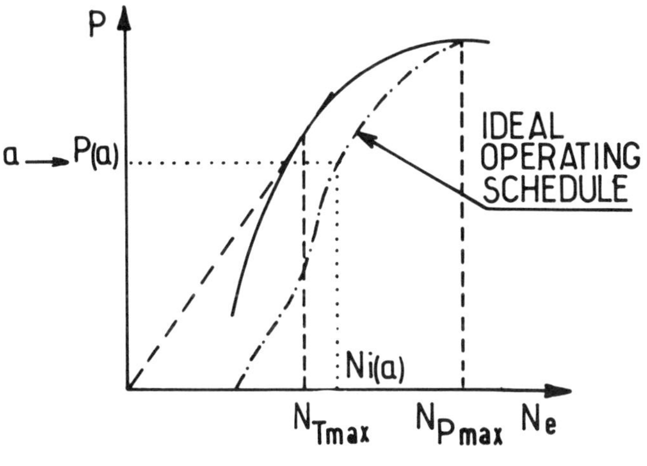

Fig. 2 - Thermal engine ideal operating schedule

R_{min} = Minimum transmission ratio

R_{max} = Maximum transmission ratio

N_{idle} = Idle engine RPM

N_{Tmax} = Maximum torque engine RPM

N_{Pmax} = Maximum power engine RPM

- At time t = 0, the driver moves the accelerator pedal, the vehicle being motionless.

The torque Tc transmitted by the clutch is controlled such as to get:

$$\frac{dNe}{dt} = f\ (a)$$

The function f, chosen by the manufacturer, has to be positive for any positive value of the parameter a.

The carburettor throttle rotation θ is controlled to equal θ1(a, Ne). This function θ1 has to be defined to give an engine torque increasing with the engine speed for any value of a.

- At time t = to, Tc reaches the value Tr of the resisting torque on the transmission input shaft. The vehicle begins to move. Tc is then controlled to maintain the engine RPM constant, or such as to get:

$$\frac{dNe}{dt} = 0, \quad N = No$$

which automatically gives:

$$\frac{dNin}{dt} = k.f\ (a)$$

k being a function of the inertias on both sides of the clutch, on the low transmission ratio.

This means that the actual acceleration of the vehicle, providing no wheel slip, is a function of the driver's demand.

- At time t = t1, N1 reaches the value Ne; this means the end of the clutching phase, the total duration of which is:

$$t1 = \frac{No\ (1 + \frac{1}{k}) - Nr}{f\ (a)}$$

To prevent any damage of the clutch, the function f has to be selected carefully.

The function θ1 can be replaced by the constant max to minimize the heat generation in the clutch, with no penalty on performance.

TECHNOLOGY

The sensors are of different well-known types:

- Speed sensors : Ne
  Nout
  Nin or Ne - Nin
- Displacement sensor : a
- Stress sensor : b
- Temperature sensor : if the idle engine RPM varies with its temperature

If there is a clutch the engine speed sensor signal has to be derived to get:

$$\frac{dNe}{dt}$$

The actuators:

- One on the engine to control the air-fuel mixture flow; for instance, an electrical motor to drive the throttle.
- One on the clutch; for instance, a hydraulic or pneumatic jack with an electrovalve.
- One on the transmission, of the same type.

The logic unit has basically:

- To make comparison tests.
- To calculate (generally by interpolation of memorized data) the following functions of one or two variables:

θ1(a,Ne)
θ2(a,Ne)
Ni(a)
Nj(b,Nout)
f(a)
N_{idle}(TEMP)

This can be done by a microprocessor.

Figures 3 to 7 show a sketch of the propulsion system and the logic diagrams of the whole system control.

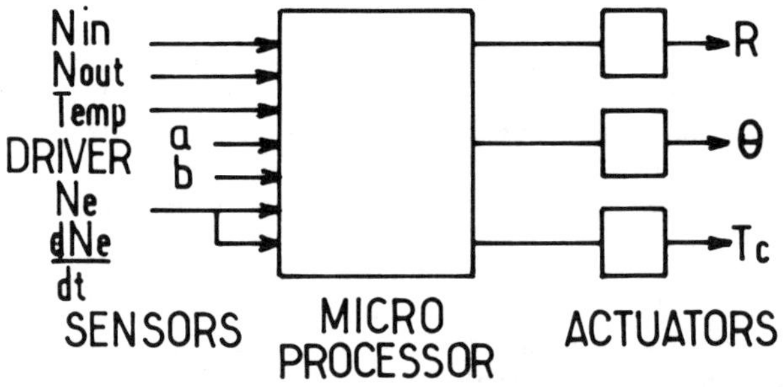

Fig. 3 - Control system schematic diagram

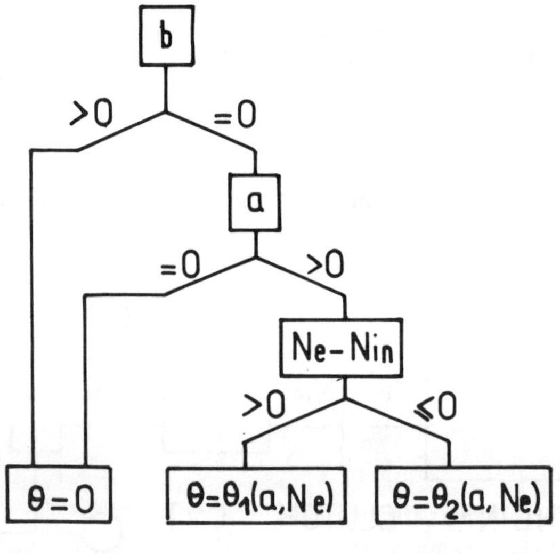

Fig. 4 - Throttle position control logic diagram

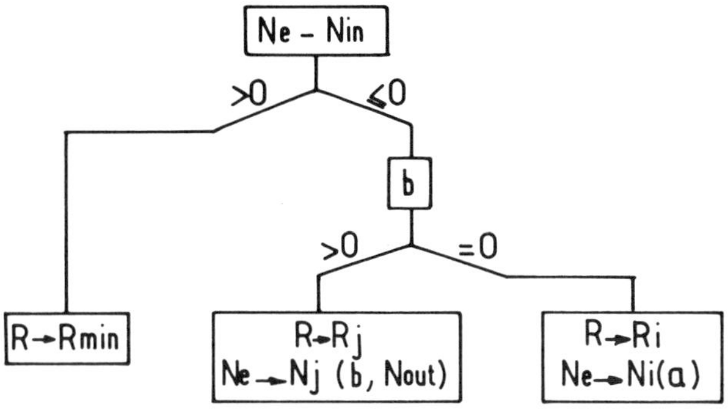

Fig. 5 - Transmission ratio control logic diagram

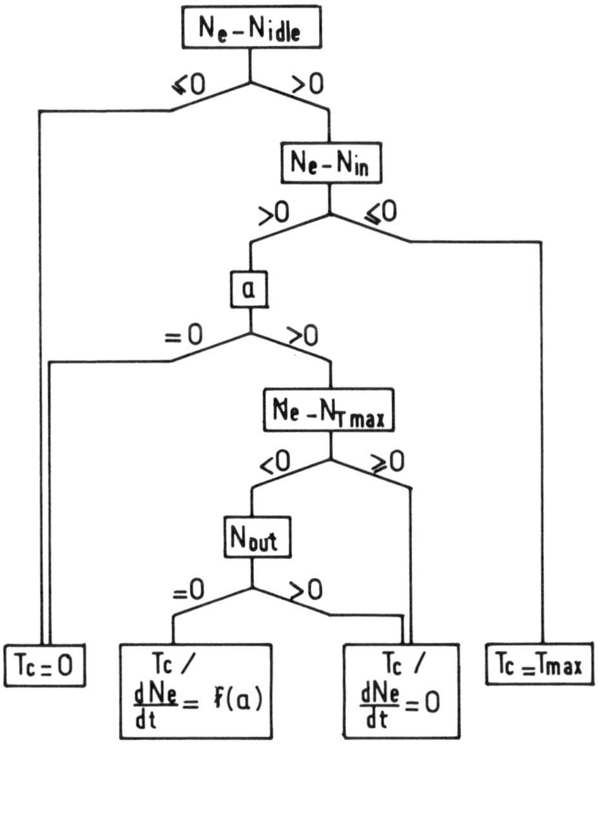

Fig. 6 - Clutch control logic diagram

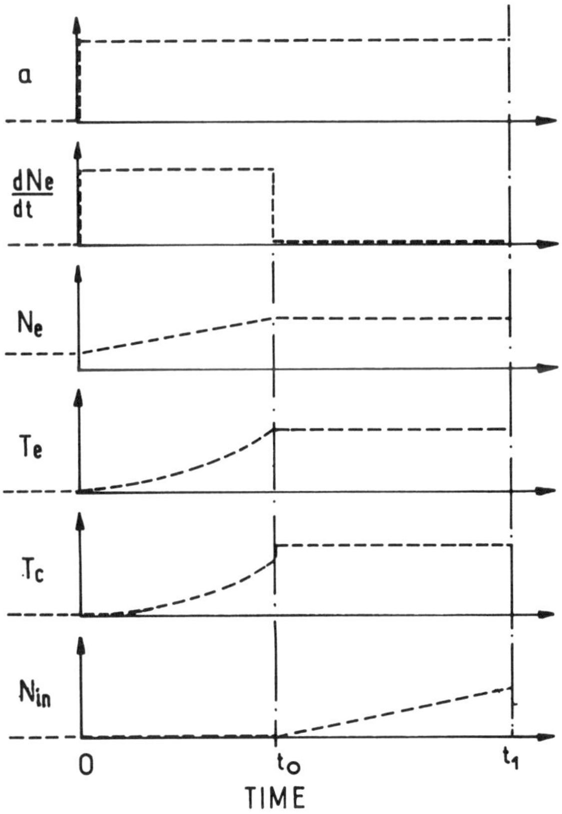

Fig. 7 - Clutching phase. Typical parameters evolution as controlled

Mass production at a reasonable cost should be possible in a near future.

COMPUTER SIMULATION AND ACTUAL VEHICLE TESTING RESULTS

We are currently working on three different projects. They will reach completion towards the middle of this year:
- Bench testing of a hydrostatic transmission, developed by Renault.
- Optimization calculation and computer simulation of a hydromechanical transmission.
- Test driving of an experimental car equipped with a metallic belt variable pitch drive developped by the Dutch company Transmissie B.V.

HYDROSTATIC TRANSMISSION - As this transmission will be the subject of another technical paper, we shall describe only its main characteristics:
- It is composed of:
  . One pump
  . Two motors.

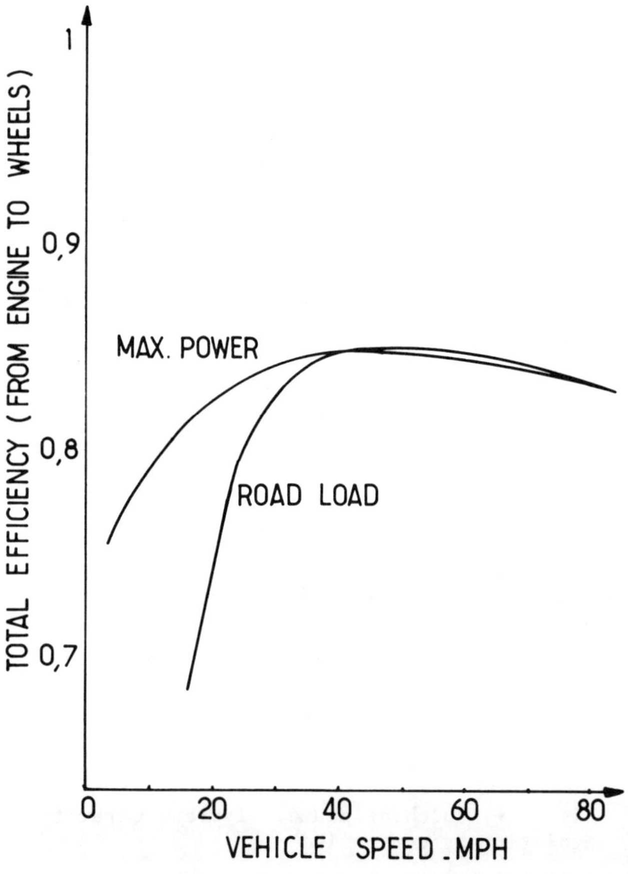

Fig. 8 - Renault hydrostatic prototype transmission efficiency. Steady state mock-up test data

These three modules are identical
axial piston type
and variable displacement.

With a maximal displacement of 22 cm3/rev
(1,34 cu.in./rev)

The motors are mounted opposite, for a better balance and to reduce some of the mechanical losses.
A housing incorporating the three modules and the ducting has been designed for mounting on a Renault 12 chassis without structural modification.
- Design power = 35 kW (45 HP)
-

input power kW	overall ratio
67.4	2
44.9	4.5
33.7	6

- Maximum pressure = 350 bars (5000 PSI)
- Maximum speed = 5200 RPM
- Efficiency = Fig. 8 shows the total efficiency of the transmission, from the engine to the wheels, under different working conditions. The average efficiency during a typical driving cycle is high enough to improve the fuel economy of the vehicle, when compared with a manual transmission.

2-MODE HYDROMECHANICAL TRANSMISSION - A paper study has been made by Renault and "Le moteur moderne" to define a hydromechanical transmission for a R 20 TS. The main features of this car (See the 1st photography), currently produced, with either 4 or 5 speed manual transmissions, or a 3 speed automatic transmission are:

1995 cm3 (122 cu.in.) displacement S.I. engine delivering 81 kW or 110 HP at 5500 RPM.
Curb weight = 1250 kg (2760 lbs).
The best compromise in terms of fuel economy-performance-cost, appeared to be a 2-mode transmission with a shifting at zero hydrostatic power.

The one-mode transmissions that we considered had too low a volume reduction and efficiency improvement compared with a purely hydrostatic transmission. The three mode (or more) transmissions were too complicated.

The other main characteristics of the transmission are:

- Nominal power = 81 kW (110 HP)
-

input power kW/RPM	overall ratio
81/5500	3.31
52/3000	4.56
22/1500	7.29

Fig. 9 shows its theorical hydrostatic power to input power ratio versus transmission ratio.
This transmission is now under detailed designing and could incorporate the hydrostatic modules described above.
Fig. 10 shows the computed total transmission efficiency versus vehicle speed:

Photograph 1

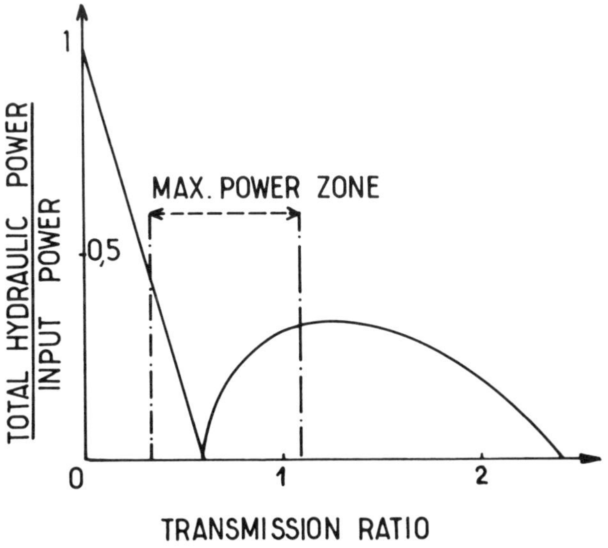

Fig. 9 - Renault - Le moteur moderne hydromechanical transmission for R 20 TS. Total hydraulic power to input power ratio vs. transmission ratio

. At full throttle
. At constant speed.

Fig. 11 shows the fuel economy of the Renault 20 TS at constant speed in three versions:

. 4 speed manual transmission
. 3 speed automatic transmission
. Hydromechanical transmission.

The computed performance and fuel economy of the three versions are listed below:

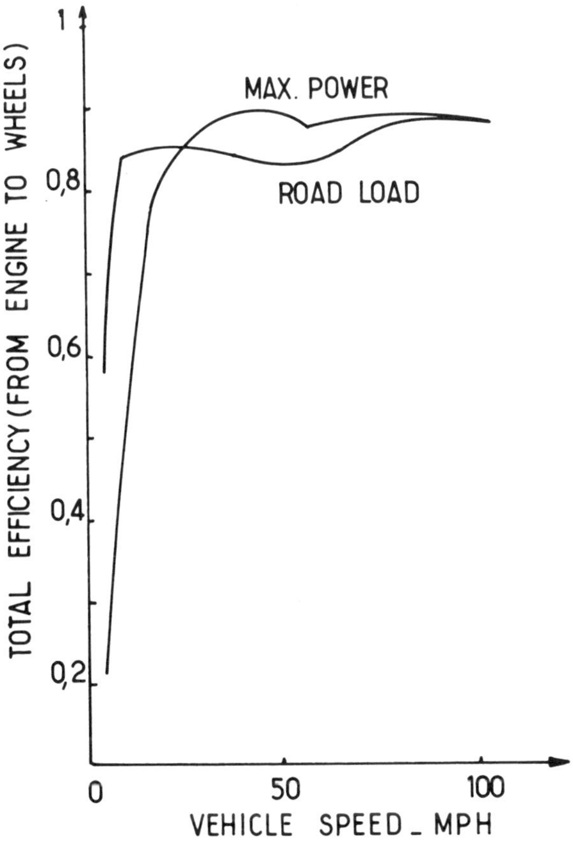

Fig. 10 - Renault - Le moteur moderne hydromechanical transmission for R 20 TS. Efficiency curves. Computer simulation results

	manual	automatic	hydro-mechanical
Maximum speed, MPH	105	102	103
0 to 60 MPH, s	12.2	14.4	11.3
standing 400 m	18.2	19.7	17.8
standing 1000 m	34.2	36.5	33.8
MPG ECE cycle	19.0	20.6	22.5
MPG EPA cycle	22.7	..	27.1

The hydromechanical transmission shows a potential for improvement of the fuel economy and of the acceleration of the stock vehicles.

VAN DOORNE'S TRANSMATIC TRANSMISSION - An experimental vehicle has been assembled:

. On a Renault 14 basis

. With a prototype Transmatic transmission, supplied by the Dutch company Transmissie B.V.

- The Renault 14 is an average European car (See the 2nd photography). Its main features are:

. Doors: 5
. Seating capacity: 5
. Front Wheel drive
. Curb weight = 1910 LBS
. Stock maximum speed = 89 MPH.

- The engine, with no modification, is a 4 cylinders in-line, mounted transversely:
  . 1218 cm3 (74 cu. in. displacement)
  . delivering 42 kW (57 HP) at 6000 RPM
  . 92 N.m at 3000 RPM

- The transmission is a variable pitch drive, with a metallic belt. The original gear-box, mounted beneath the engine, has been kept and is used as a speed reducer, in 3 RD gear, and to get a reverse. The Transmatic is mounted vertically between them. The transmission ratios give the following vehicle MPH/1000 RPM:

stock	4.3 - 7.2 - 11.1 - 15.9
Transmatic	5.9 continuously → 27.7

The ratio coverage of the Transmatic prototype is only 4.7; increased coverage would have required more special parts, and we chose the described solution featuring a nearly optimum overdrive ratio, and a low ratio sufficient for the tests if not for production.

- The control system was first designed by Transmissie B.V., and was purely mechanical and hydraulic. We are now sophisticating the control system according to the above description.

- Performance and fuel economy:

The vehicle has been tested comparatively
  . with its manual transmission
  . with the Transmatic transmission and at that time the Transmissie B.V. control system.

The main results are listed below

	4 speed manual	Transmatic
maximum speed (MPH)	89.4	87.4
standing 100m (s)	8.4	8.7
standing 400m (s)	20.4	20.6
40 to 80 KPH (s)	9.5 in 3rd gear 15.1 in 4th gear	7.6
60 to 100 KPH (s)	10.6 in 3rd gear 15.9 in 4th gear	11.4
80 to 120 KPH (s)	20.1 in 4th gear	19.1
fuel economy (MPG) ECE cycle	21.3	25.7 (+20.7%)
fuel economy (MPG) urban EPA cycle	25.1	32.2 (+28.3%)

Fig. 12 shows the fuel economy of both vehicles at constant speed.

Photograph 2 - Renault 14 TL

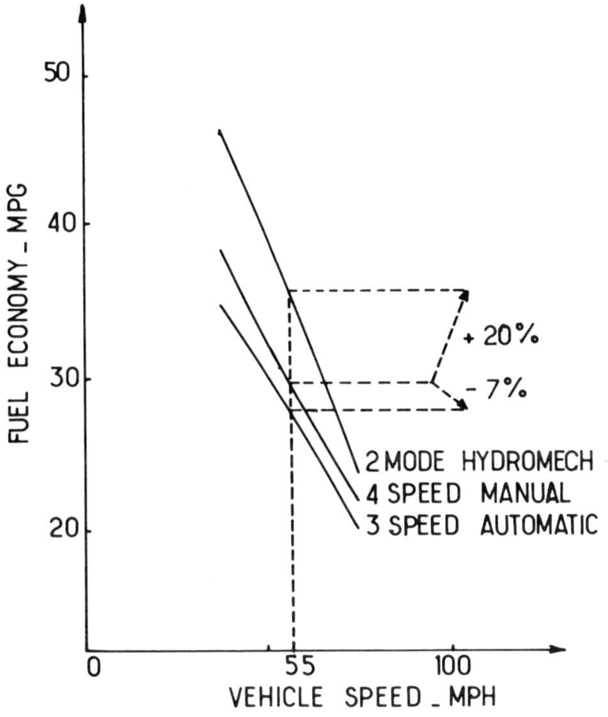

Fig. 11 - Road load fuel economy for a R 20 TS equipped with three different types of transmission. Computer simulation results

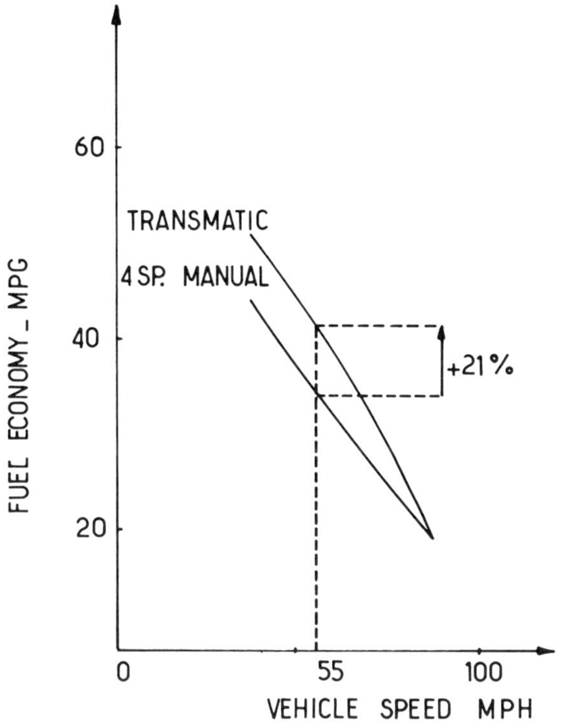

Fig. 12 - Road load fuel economy for a R 14 TL equipped with a 4-speed manual stock transmission or a Transmatic prototype transmission actual vehicle test results

The fuel savings permitted by the Transmatic with a high ratio coverage are significant. Measurements are going on to determine the emissions with the Transmissie B.V. control system and the influence of our new control system, especially in transient.

SUMMARY AND TRENDS

ADVANTAGES OF CVT'S - allowing a high ratio coverage and consequently high overdrive ratios:

We have just seen the inherent advantages of such transmissions when coupled with spark-ignited engines, and controlled by systems such as those we described.

There are three advantages:

- Driver's comfort thanks to the automatism and for the passengers as well, as the jerking movements which are usual with normal transmissions, are suppressed
- Improved fuel economy compared to conventional manufactured transmissions
- In average, less noise as the engine runs with increased load levels but lesser RPM, which is of prime importance in this respect.

COMPARATIVE TRENDS WITH DIESEL AND/OR TURBOCHARGED ENGINE - What about these three advantages if the prime mover is no longer a spark ignited atmospheric engine but a Diesel and/or a turbocharged engine?

We have the same better comfort and lesser noise. As for the fuel economy the graphs 13 to 16 show some trends: Fig. 13 and 14 indicate the fuel consumptions of two cars, R 14 TL and R 20 TS, on two almost identical highway cycles, with average speeds respectively 52 and 55 MPH and the same minimum and maximum speeds, 37 and 81 MPH.

The R 20 TS is alternately equipped with four engines delivering the same maximum power (81 kW):

- an atmospheric S.I. engine
- an atmospheric Diesel engine
- a turbocharged S.I. engine
- a turbocharged Diesel engine.

The R 14 TL is alternately equipped with only three of them, the dimensions of a turbocharged S.I. engine capable of 42 kW being presently too small.

We made some important assumptions:

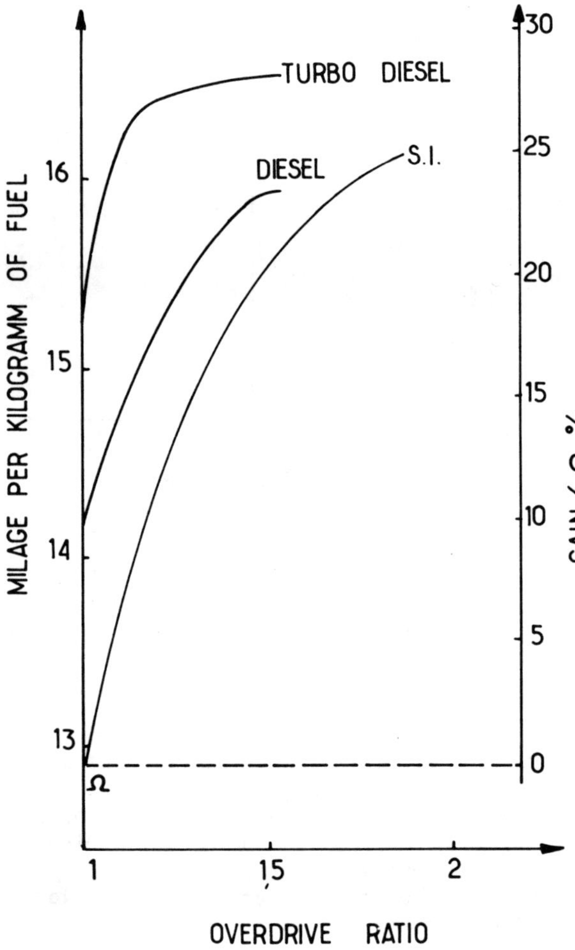

Fig. 13 - Highway fuel economy for a R 14 TL equipped with a spark-ignited engine, a Diesel engine and a turbocharged Diesel engine of equal maximum power. Computer simulation results

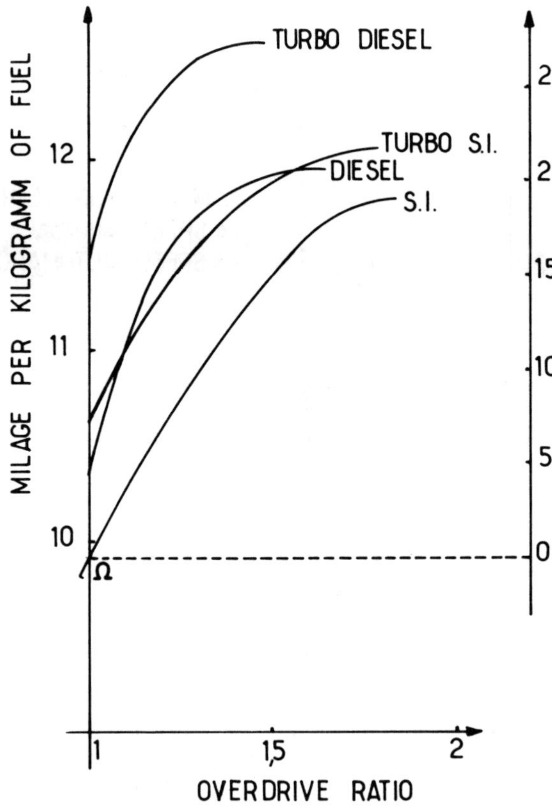

Fig. 14 - Highway fuel economy for a R 20 TS equipped with spark-ignited engines, atmospheric or turbocharged, and Diesel engines, atmospheric or turbocharged, all engines of equal maximum power. Computer simulation results

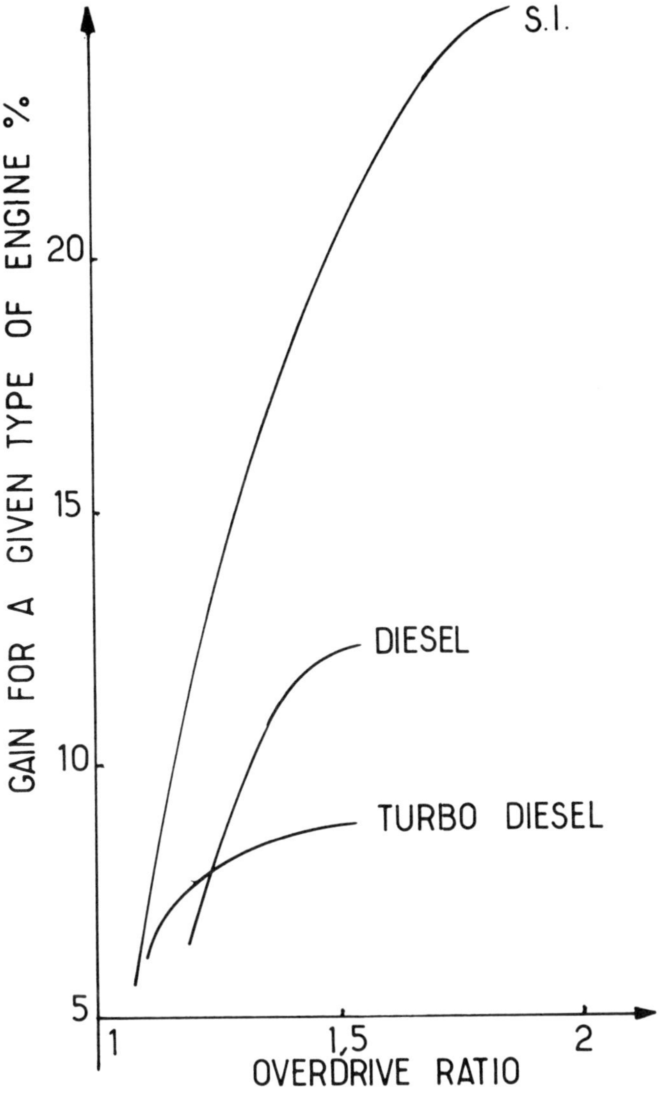

Fig. 15 - R14 TL highway fuel economy improvement vs. overdrive ratio for each of the 3 above mentioned types of engine. Computer simulation results

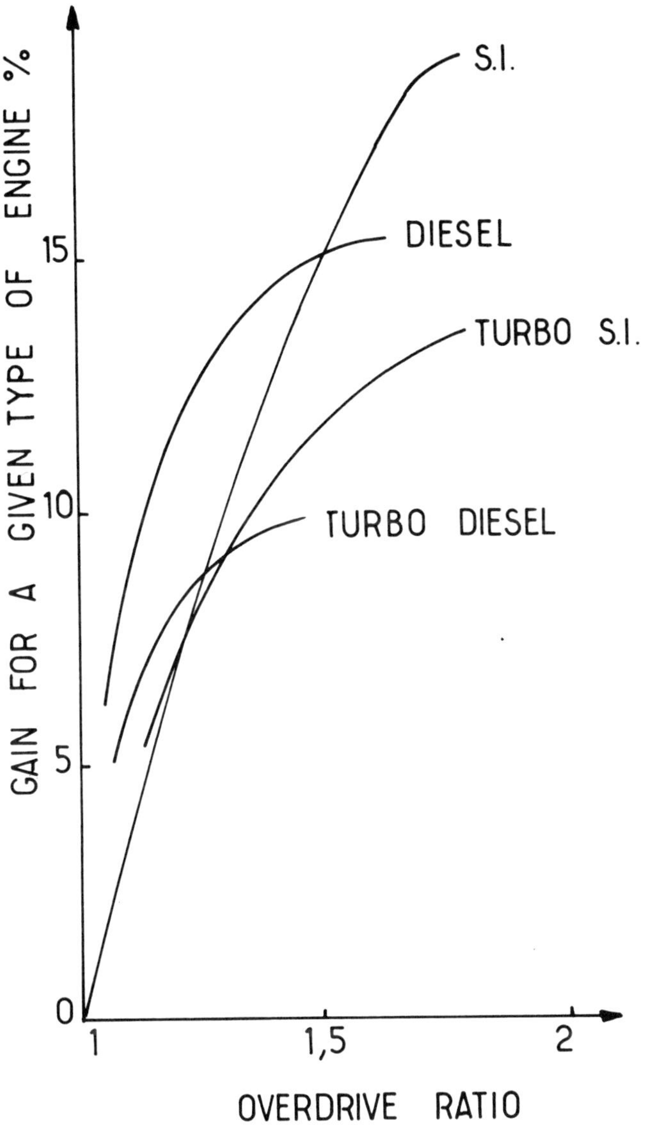

Fig. 16 - R20 TS highway fuel economy improvement vs. overdrive ratio for each of the 4 above mentioned types of engine. Computer simulation results

- The mere substitution of an engine with one other is without effect on the weight of the vehicle
- The transmissions are continuous or with a very high number of ratios and with a constant 93% efficiency independently of the maximum overdrive ratio
- The engines, including features of the near future technology, when running at low RPM, are never used in zones with critical vibratory levels.

It should also be pointed out that some types of engine included only a few specimens and those results could have been slightly modified by a more exhaustive investigation.

Fig. 15 and 16 show that for a conventional overdrive ratio equal to 1, the fuel economy depends largely on the engine type:
The best and worst results are respectively obtained by turbocharged Diesel engines and atmospheric S.I. engines, with a difference in the order of 15% on both cars. The difference is considerably reduced, down to 3% when maximum overdrive ratio increases.

Fig. 15 and 16 show fuel economy improvements versus overall ratio, on both cars, for each type of engine.

The optimum overdrive ratios and corresponding fuel economy improvements are roughly:

1.6 and 20%	for atmospheric S.I. engines
1.4 and 12%	for turbocharged S.I. engines and for atmospheric Diesel engines
1.2 and 8%	for turbocharged Diesel engines.

These figures seem to be relatively independent from the power to weight ratio of the vehicle, at least in the considered range.

Optimization of a Continuously Variable Transmission with Emission Constraints*

R. R. Radtke, L. E. Unnewehr and R. J. Freedman
Scientific Research Lab
Ford Motor Co.

*Paper 810107 presented at the International Congress and Exposition, Detroit, Michigan, February, 1981.

CONTINUOUSLY VARIABLE TRANSMISSIONS (CVT's) have been used for many years in the machine tool industry and as variable-speed transmissions in many types of small vehicles such as snowmobiles and self-propelled lawnmowers. There has developed in recent years considerable interest in application of CVT's to automobiles, and technical developments plus the reduced weight of current vehicles have brought this application closer to reality. Potential CVT's under consideration include the Van Doorne, Excelermatic, Vadetec and Perbury units (Ref. 1-9).

The benefits claimed for CVT's in comparison to conventional transmissions include:

1. Improved fuel economy
2. Improved acceleration performance due to the wider ratio range of a CVT
3. More flexible engine and vehicle control strategy
4. Possible weight and/or size reduction

Most of the present interest in CVT's centers around the first of these claims. CVT's could in theory improve vehicle fuel economy through three basic mechanisms:

1. Achieving ratio coverage to permit engine operation at low BSFC under all driving conditions
2. Achieving higher average transmission efficiency than that in the conventional torque converter automatic transmission
3. Reducing torque at engine idle over some current designs

ABSTRACT

An optimization technique has been developed which can be used to compute the optimal control strategy for an engine/CVT configuration subject to emission constraints. (CVT stands for Continuously Variable Transmission). The engine/CVT system is represented as a function of the four independent variables: spark advance, air/fuel ratio, exhaust gas recirculation percentage, and CVT gear ratio. The Generalized Lagrange Multiplier Technique is used to optimize these variables at a set of drive shaft torque/speed points which represent operation of the vehicle over a drive cycle. Results are presented for idealized CVT's.

The evaluation of CVT's as a means for improving vehicle fuel economy has been hampered by a number of problems in the past: The great diversity of CVT configurations has not fostered development of generalized analytical models for use in vehicle simulations; there has been almost no verifiable, measured data describing CVT efficiency as a function of speed and torque; and there has been almost no cognizance of emissions constraints in making fuel economy predictions. This paper describes the development of analytical tools which can be used to evaluate the CVT's potential for improving vehicular fuel economy with emissions constraints. An optimization technique has been applied to the vehicle-engine-CVT system which projects the best possible fuel economy that can be achieved within prescribed emission standards. A similar program projects the best possible fuel economy under emission constraints for a vehicle with a more conventional transmission (automatic, manual, lock-up, etc.). Calculations using these optimization methods will be presented which compare CVT's with conventional transmissions for several vehicle test weights, emissions levels, and CVT ratio ranges. Comparisons will be made on the basis of an idealized CVT model.

CONSTRAINED OPTIMIZATION TECHNIQUE

SELECTION OF TECHNIQUE - In order to evaluate fuel economy improvements afforded by a CVT, a constrained optimization technique was developed. An optimal engine/CVT control strategy was determined by minimizing fuel consumption constrained by different feedgas emissions levels. The optimization procedure picks the optimal air-fuel ratio (A/F), spark advance timing (SA), exhaust gas recirculation percentage (EGR), and CVT ratio associated with each driveshaft speed/load point.

Three constrained optimization techniques were evaluated. The dynamic programming technique used for optimizing engine calibrations (10) was eliminated because of the excessive amount of computer memory needed to represent the large number of competitive strategies.

Continuous optimization techniques were eliminated because these techniques converge to local rather than global solutions. Based on previous results (11) it can be shown that there is more than one local minimum for the CVT optimiztion program.

The third optimization technique considered is the Generalized Lagrange Multiplier Technique (GLMT) for cell problems. This technique has neither of the limitations of the other two in that it requires low computer memory and converges to a global minimum.

OUTLINE OF GLMT - Figure 1 depicts the overall outline of the GLMT solution procedure. Terms and nomenclature are defined in Appendix I. The driving cycle for a particular vehicle was described approximately by a finite sequence of driveshaft speed/load points, one speed/load point for each second of time. This sequence of speed/load points (η_i, τ_i) i = 1, . . ., T(e.g., T=1372 for the EPA Metro cycle) is fixed independent of the engine/CVT operating strategy. Thus the optimization technique can be reduced to the determination of the set of 4-tuples $(A/F_i, SA_i, EGR_i, r_i)$ i=1, . . ., T which minimize fuel consumption subject to the NOx, CO, and HC emission constraints.

A powertrain simulation technique similar to the one described in (12) was used to generate the sequence of driveshaft speed/load points. To reduce the amount of computation, a representative set of driveshaft speed/load points was chosen. Each of the (η_i, τ_i) i=1, . . ., T was mapped onto the new set (η_i', τ_i', t_i) i=1, . . ., N. The mapping algorithm computes an amount of time t' associated with each driveshaft speed/load point so that total energy is conserved. The problem is then reduced to finding the set of 4-tuples $(A/F_i', SA_i', EGR_i', r_i')$ i=1, . . ., N. Taking N to be 25 gives good prediction and in the case of the Metro cycle, results in a factor of more than 50 reduction in program execution time.

The constrained minimzation becomes:

$$\underset{\begin{bmatrix} r_i \varepsilon R \\ A/F_i \varepsilon A \\ SA_i \varepsilon S \\ EGR_i \varepsilon E \end{bmatrix}}{\text{minimize}} \sum_{i=1,N} f_{flow}\left(\eta_i', \tau_i', r_i, A/F_i, SA_i, EGR_i\right) t_i' \qquad [A]$$

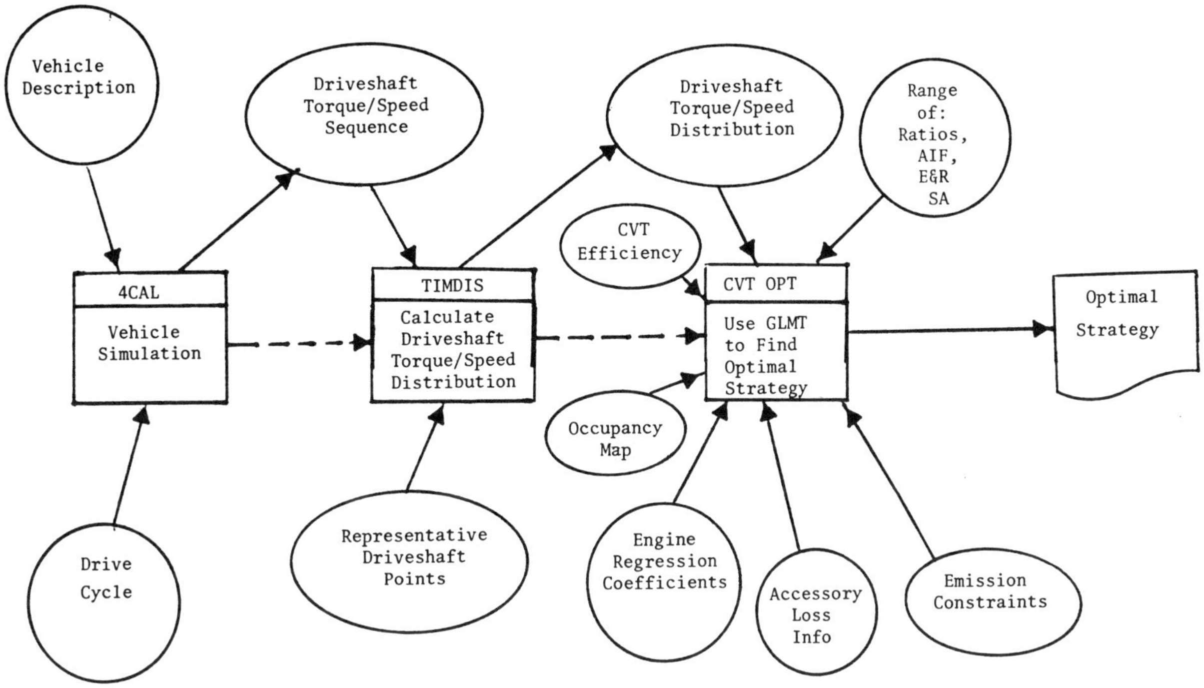

Fig. 1 - GLMT solution procedure

Subject to:

$$\sum_{i=1}^{N} NOX_{flow}\,(\eta_i', \tau_i', r_i, A/F_i, SA_i, EGR_i)\, t_i' \le NOX_{max}$$

$$\sum_{i=1}^{N} HC_{flow}\,(\eta_i', \tau_i', r_i, A/F_i, SA_i, EGR_i)\, t_i' \le HC_{max}$$

$$\sum_{i=1}^{N} CO_{flow}\,(\eta_i', \tau_i', r_i, A/F_i, SA_i, EGR_i)\, t_i' \le CO_{max}$$

Using the Langrange Optimality Thereom (13) problem (A) can be rewritten:

Find $\hat{\lambda}_1, \hat{\lambda}_2, \hat{\lambda}_3, \hat{r}_i, \widehat{A/F}_i, \widehat{SA}_i, \widehat{EGR}_i \quad i=1, \ldots, N$

which minimize

$$\sum_{i=1}^{N} f_{flow}\,(\tau_i', \eta_i', r_i, A/F_i, EGR_i)\, t_i' \qquad [B]$$

$$+ \lambda_1 \sum_{i=1}^{N} NOX_{flow}(\tau_i', \eta_i', r_i, A/F_i, SA_i, EGR_i)\, t_i'$$

$$+ \lambda_2 \sum_{i=1}^{N} HC_{flow}\,(\tau_i', \eta_i', r_i, A/F_i, SA_i, EGR_i)\, t_i'$$

$$+ \lambda_3 \sum_{i=1}^{N} CO_{flow}\,(\tau_i', \eta_i', r_i, A/F_i, SA_i, EGR_i)\, t_i'$$

where $r_i \in R$, $A/F_i \in A$, $SA_i \in S$, and $EGR_i \in E$

and satisfy the conditions:

$$\sum_{i=1}^{N} NOX_{flow}\,(\tau_i', \eta_i', \hat{r}_i, \widehat{A/F}_i, \widehat{SA}_i, \widehat{EGR}_i)\, t_i' = NOX_{max} \text{ if } \lambda_1 > 0$$

$$\sum_{i=1}^{N} NOX_{flow}\,(\tau_i', \eta_i', \hat{r}_i, \widehat{A/F}_i, \widehat{SA}_i, \widehat{EGR}_i)\, t_i' \le NOX_{max} \text{ if } \lambda_1 = 0$$

$$\sum_{i=1}^{N} HC_{flow}\,(\tau_i', \eta_i', \hat{r}_i, \widehat{A/F}_i, \widehat{SA}_i, \widehat{EGR}_i)\, t_i' = HC_{max} \text{ if } \lambda_2 > 0$$

$$\sum_{i=1}^{N} HC_{flow}\,(\tau_i', \eta_i', \hat{r}_i, \widehat{A/F}_i, \widehat{SA}_i, \widehat{EGR}_i)\, t_i' \le HC_{max} \text{ if } \lambda_2 = 0$$

$$\sum_{i=1}^{N} CO_{flow}\,(\tau_i', \eta_i', \hat{r}_i, \widehat{A/F}_i, \widehat{SA}_i, \widehat{EGR}_i)\, t_i' = CO_{max} \text{ if } \lambda_3 > 0$$

$$\sum_{i=1}^{N} CO_{flow}\,(\tau_i', \eta_i', \hat{r}_i, \widehat{A/F}_i, \widehat{SA}_i, \widehat{EGR}_i)\, t_i' \le CO_{max} \text{ if } \lambda_3 = 0$$

Problem (B) can be solved with the following algorithm:

1) Pick $\lambda_1^o, \lambda_2^o, \lambda_3^o$.

2) Find r_i, A/F_i, SA_i, EGR_i, which minimizes

   $f_{flow}() + \lambda_1\, NOX_{flow}() + \lambda_2\, HC_{flow}() + \lambda_3\, CO_{flow}()$

   for each i.

3) Check for convergence.

4) Pick $\lambda_1^k, \lambda_2^k, \lambda_3^k$; go to 2.

REPRESENTATION OF ENGINE - The functions $f_{flow}()$, $NOX_{flow}()$, $HC_{flow}()$, and $CO_{flow}()$ are computed using five dimensional regression equations. These regression equations were developed from extensive steady state data (14). Occupancy maps were also constructed to define areas where the regression equations are valid. These regression equations allow the engine to be represented compactly in the computer. The regression equations are also necessary because the engine operating conditions cannot be defined a priori.

COMPUTATION OF LAMBDAS - Several algorithms are available for computing the sequence of lambdas. The algorithm used in this study was developed by Hugh D. Everett, III (15).

COMPUTER REQUIREMENTS - The CVT optimization program was run on a CYBER 176, requiring less than 30K words of memory. A typical problem converges in 13 iterations of lambda and takes about 3 minutes of CPU time.

CVT MODELS - Initial runs were made assuming that the CVT efficiency was constant over its entire operating range. As data become available, the simplified CVT model will be replaced with real efficiency data.

The optimization procedure described here assumes that the CVT dynamic response time is instantaneous. No hard data is currently available on actual response times of CVT's although manufacturers claim nearly instantaneous response.

FURTHER MODEL DEVELOPMENT POSSIBILITIES - Future studies with the current model will use experimental efficiency data taken on prototype CVT's to determine estimated benefits of CVT use over various automatic and manual discrete-gear transmissions.

A possible change to the current model would involve improved tailpipe emission simulation. Another model developed at Ford Research describes efficiency of oxidation catalysts as a time independent function of engine operating conditions. Therefore, this simplified catalyst model could be added to the CVT optimization program, and tailpipe emissions could be used as the problem constraints.

If CVT ratio response time is found to be a critical factor, it will be necessary to use a different numerical procedure to solve the CVT optimization problem. One possible candidate is the sequential decision technique (17) which has been used on similar problems. The penalty paid for using this more general technique is increased CPU time.

FUEL ECONOMY PROJECTIONS FOR IDEALIZED CVT

The optimization program described in the previous section will operate with a relatively simple model of the CVT. The required model is basically a steady-state, two-port network model. The output port parameters are driveshaft speed and torque and the input port parameters are engine speed and torque. (The term "driveshaft" is used to represent the output CVT port for either front or rear-wheel drives, although it is strictly meaningful only for the latter). The CVT model, therefore, requires a knowledge of CVT efficiency over the entire range of engine and driveshaft speed and torque. At the time this optimization program was developed, such data were not available in the literature for any known CVT configurations. Therefore, an idealized model was used in the studies reported in this paper. The idealized model consists of:

1. An assumed CVT efficiency, constant for all input and output speeds and torques (This efficiency can also be termed CVT energy efficiency)
2. CVT ratio range
3. Input torque at engine idle
4. Maximum CVT torque
5. Maximum CVT speed.

Since this is a steady-state model, CVT dynamic response is ignored. This has raised some questions concerning the ability of a practical CVT to meet the optimum strategy derived by the optimization program. At present, these questions remain unanswered, although most CVT's would appear to have very fast response time compared to the high-inertia systems connected to each port. For example one traction type CVT is said to be able to change through its entire ratio range within eight revolutions of the input shaft (3).

In the CVT optimization studies reported below, inertia or test weights, road load coefficients, engine fuel and emissions maps, and axle ratios used are based on production vehicles. Fuel economy and emissions calculations are made for the vehicle driving over the Federal Urban (Metro) or Highway Drive Cycles. Emissions constraints are in terms of feedgas emissions levels. Studies have been performed with both variable and fixed A/F ratios. In studies reported here, the A/F ratio is fixed at stoichiometry.

The principal interest in the data to be presented is probably the comparisons between fuel economy benefits possible with a CVT controlled ideally and the fuel economy with other types of transmissions. The comparisons are made using the same engine, vehicle test weight, and emissions constraints in all cases. The "base" case for the comparisons is always the conventional transmission at the 1980 stage of development. The comparisons do not acknowledge potential improvements to the 1980 base. As in the case of other new or alternate automotive components, the CVT must be compared with a moving baseline that results from further improvements in the hardware.

Table 1 lists some comparisons between a CVT of varying energy efficiency and a conventional three-speed transaxle automatic transmission in a subcompact passenger car. The CVT efficiences of 1.0 and 0.91 are not meant to reflect a producible product but are shown only to place an upper bound on fuel economy improvement with emissions constraints resulting from the use of a CVT. For the remaining CVT studies to be shown, a more realistic energy efficiency of 0.82 is assumed.

In one of the very few published studies of CVT fuel economy improvement (16), an unconstrained fuel economy improvement of 28% over a four-speed manual transmission was projected for a European subcompact car of similar test weight to that in Table 1. This is somewhat larger than the projection with 0.82 efficiency shown in Table 1.

Table 2 illustrates the optimum fuel economy using a CVT for three vehicles of different test weights. Table 3 shows optimum CVT fuel economy as a function of emissions constraints. For this case the percent improvement in fuel economy is relatively independent of emission constraints. Table 4 compares optimum fuel economy of the CVT with several conventional transmissions: three-speed automatic, a four-speed manual, and a four-speed automatic with lock-up in overdrive.

Table 5 summarizes the average BSFC over the drive cycle for the five transmissions used in Table 4. This demonstrates one of the advantages for CVT's referred to earlier in this paper; namely, the ability of the CVT to allow engine operation at low BSFC under all driving conditions.

CONTROL

The optimum strategy which results from application of this GLMT program requires rapid and sometimes continuous ratio changes. It is recognized that when prototype CVT

Table 1 - Effect of CVT Efficiency on Fuel Economy

		Metro		Highway	
Transmission	Efficiency	MPG	% Over Base	MPG	% Over Base
Base	-	26.3	-	33	-
CVT	1.0	35.7	36	45.3	37
CVT	0.91	33.8	28	42	27
CVT	0.82	31.2	18	37.7	14
CVT	0.75	28.9	10	34	3

Conditions
Base = 3-Speed Automatic
CVT Ratio Range = 0.5 - 2.3
Test Wt. = 1134 kg
HC = 2.0 gm/mi
NO_x = 1.75 gm/mi

Table 2 - Effect of Vehicle Weight on Fuel Economy

Weight kg	Metro Base MPG	Metro CVT MPG	Metro CVT % Over Base	Highway Base MPG	Highway CVT MPG	Highway % Over Base
1140	26.3	31.1	18	33	37.7	14
1360	22.6	28.4	6	27	32.4	20
1815	14.3	19.5	36	22	25	14

Conditions
Base = 3-Speed Automatic
CVT Ratio Range = 0.5 - 5.0
CVT Efficiency = 0.82
HC = 2.0 gm/mi
NO_x = 1.75 gm/mi

Table 3 - Effect of Emission Constraints on Fuel Economy

HC gm/mi	NO_x gm/mi	Base MPG	CVT MPG	CVT % Over Base
Unconstrained		22.6	28.4	26
2.0	1.75	22.6	28.4	26
1.5	1.5	21.8	26.7	22
1.2	1.2	20.0	25.4	27

Conditions
Base = 3-Speed Automatic
Metro Cycle
Weight = 1360 kg
CVT Ratio Range = .5 - 5.0
CVT Efficiency = .82

Table 4 - Effect of Transmission on Fuel Economy

Weight kg	Transmission Type	Metro MPG	Metro % Over Base	Highway MPG	Highway % Over Base
1815	Base	14.2	-	22	-
1815	4-Speed Lock-Up	15.6	9	25.6	16
1815	CVT	19.5	36	25	13
1360	Base	22.6	-	27	-
1360	4-Speed Manual	24.4	8	30.6	13
1360	CVT	28.4	26	32.4	20

Conditions
Base = 3-Speed Automatic
CVT Ratio Range = .5 - 3.2
CVT Efficiency = .82
HC = 2.0 gm/mi
NO_x = 1.75 gm/mi

Table 5 - Effect of Transmission on Engine BSFC

Weight kg	Transmission Type	BSFC Metro	BSFC Highway
1815	Base	.674	.533
1815	4-Speed Lock-Up	.665	.513
1815	CVT	.496	.456
1360	Base	.536	.507
1360	4-Speed Manual	.545	.520
1360	CVT	.472	.455

Conditions
Base = 3-Speed Automatic
CVT Ratio Range = .5 - 3.2
CVT Efficiency = .82
HC = 2.0 gm/mi
NO_x = 1.75 gm/mi

hardware becomes available these assumed ideal strategies may be found impractical. A companion computer program has been developed which computes optimum engine calibrations (SA, A/F, EGR) for a non-ideal CVT shifting strategy and minimizes fuel consumption. This can then be compared to the ideal strategy case to determine how much of the fuel economy gain can be realized. Fig. 2 shows a typical ideal shift schedule, while Fig. 3 indicates possible smoothing which would be introduced if a particular CVT's operating dynamics were known, or if a simplified controller were contemplated.

Further work will concentrate on development of practical shift strategies as experimental data is taken on CVT's.

In particular, it may be desirable to have the shift strategy dependent on engine speed and torque rather than driveshaft values as indicated in Figs 2 and 3. This would require the development of a suboptimal control which inferred desired driveshaft speed and torque from driver demand at the engine.

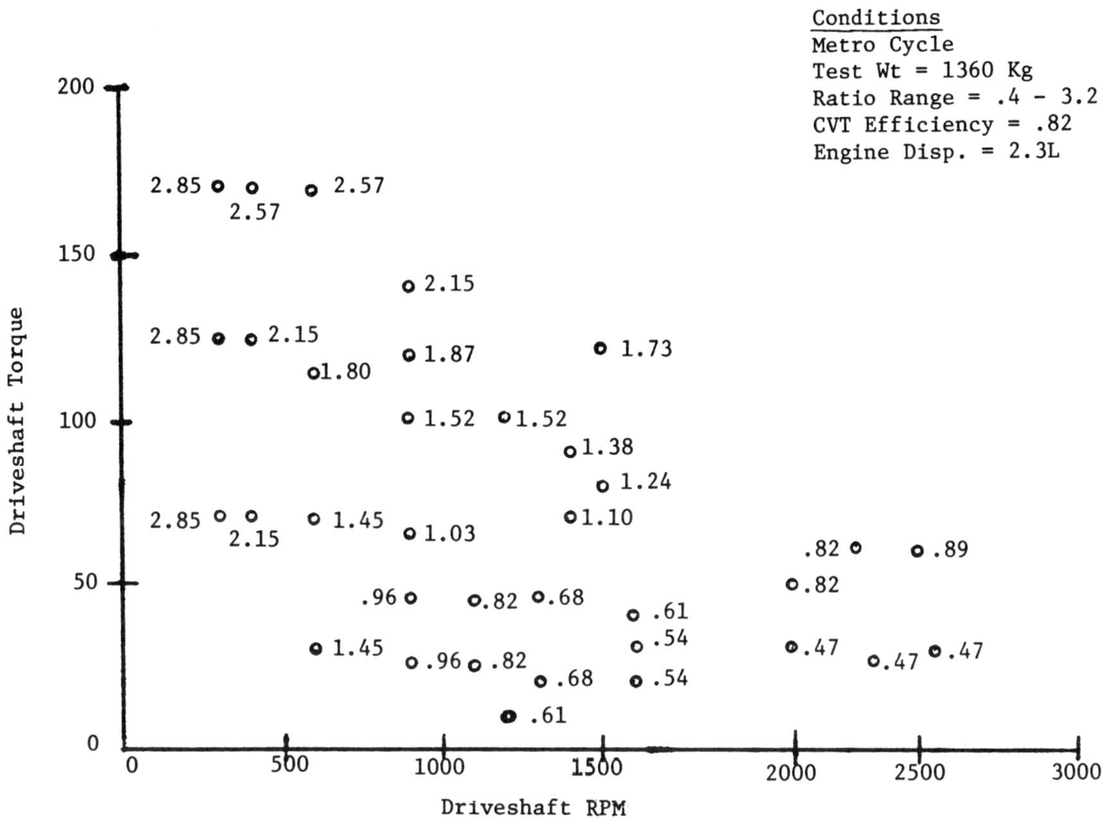

Fig. 2 - Shift schedule computed by GLMT

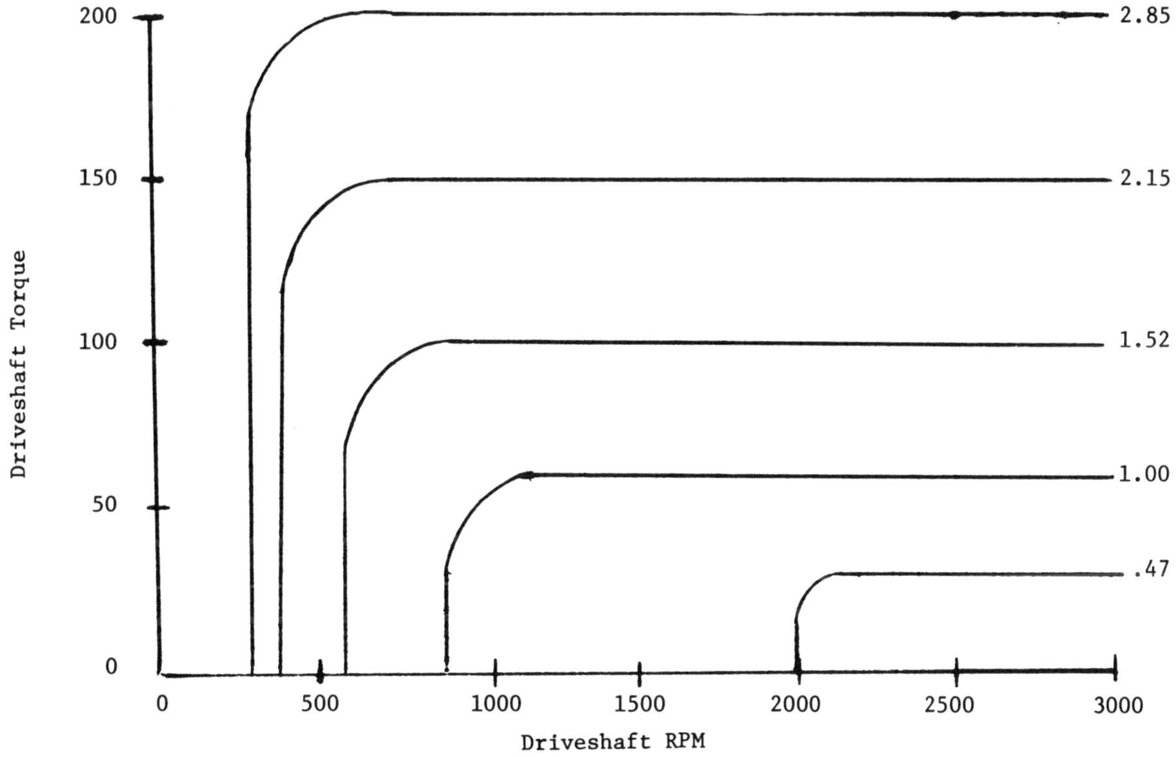

Fig. 3 - Smoothed shift schedule

SUMMARY

A mathematical optimization technique has been developed to calculate the maximum fuel economy improvement with emissions constraints that can be achieved by means of a continuously variable transmission (CVT). The computer simulation program calculates the optimum engine and CVT strategies for maximizing fuel economy with constrained emissions for a vehicle driven over a specified drive cycle. The Generalized Lagrange Multiplier Technique has been adapted for this optimization program, and has resulted in relatively low CPU times and a small amount of computer memory.

The engine map is represented as a set of polynomial coefficients and an occupancy map. Thus the engine operating conditions can be restricted to those that are driveable.

The two port network model allows any CVT to be represented once its input requirements are determined as a function of output speed, output torque, and ratio.

Initial results have been obtained for idealized CVT's with constant efficiencies and instantaneous ratio changes.

Current experimental effort is focussed on obtaining efficiency maps and ratio change response times of prototype CVTs. This data will be used by the optimization technique to predict fuel economy benefits which could be expected under emission constraints with prototype CVTs, and an ideal control strategy.

Implementable control strategies will be deduced from the ideal control strategy. These suboptimal control strategies will be analyzed, and the resulting fuel economy will be compared to the optimum fuel economy.

REFERENCES

1. R. H. MacMillan, "The Control of Stepless Variable Speed Transmissions in Automobiles and Their Possible Application to Regenerative Systems", Fisita, 1964.

2. David Scott, "Belt-Drive CVT for 1982 Model Year", Automotive Engineering, February, 1980.

3. J. H. Kraus, L. E. Kraus, and M. E. Gres, "Continuously-Variable Transmissions", Automotive Engineering, December, 1975.

4. Y. Kemper, "A High Power Density Traction Drive", SAE Paper #790849, March, 1979.

5. P. W. R. Stubbs, "The Development of a Perbury Traction Transmission for Motor Car Applications", Proceedings of the ASME Third International Power Transmission and Gearing Conference, August, 1980.

6. "Electric and Hybrid Vehicle Program", Vol. 2, Third EHV Program Contractors Meeting Transactions, Section 34, June, 1979.

7. E. Behrin, et al., "Energy Storage Systems for Automobile Propulsion", Lawrence Livermore Laboratory Report VCR L-52553, Vol. 2, December 15, 1978.

8. A. A. Frank and N. H. Beachley, "Evaluation of the Flywheel Drive Concept for Passenger Vehicles", SAE Paper #790049, February, 1979.

9. N. H. Beachley, and A. A. Frank, "Continuously Variable Transmissions: Theory and Practice", Lawrence Livermore Laboratory Report VCRL-15037, August, 1979.

10. J. E. Auiler, J. D. Zbrozek and P. N. Blumberg, "Optimization of Automotive Engine Calibration for Better Fuel Economy-Methods and Applications", SAE Paper #770076, 1977.

11. Zdenek Mencik, "Regression-Based Optimizer of Engine Control Strategy with Emission Constraints", Tech. Report No. SR-80-26.

12. P. N. Blumberg, "Powertrain Simulation: A Tool for the Design and Evaluation of Engine Control Strategies in Vehicles", SAE Paper #760158, 1976.

13. G. P. McCormick and A. V. Fiacco, "Nonlinear Programming: Sequential Unconstrained Minimization Techniques", John Wiley and Sons, p. 139.

14. Z. Mencik and P. N. Blumberg, "Representation of Engine Data by Multi-Variate Least-Squares Regression", SAE Paper #780288, 1978.

15. G. P. McCormick and A. V, Fiacco, "Nonlinear Programming: Sequential Unconstrained Minimization Techniques", John Wiley and Sons, p. 139.

16. P. Baudoin, "Continuously-Variable Transmissions for Cars with High Ratio Coverage", SAE Paper #790041, February, 1979.

17. F. G. Willis, W. F. Kaufman and G. A. Kern, "Mechanical Hybrid Vehicle Simulation." SAE Paper No. 790015, February, 1979.

APPENDIX I

Glossary

η_i - driveshaft speed at ith second

τ_i - driveshaft torque at ith second

T - number of seconds in drive cycle

A/F - Air to fuel ratio

SA - Spark advance

EGR - Percent exhaust gas recirculation

η_i' - ith representative driveshaft speed

τ_i' - ith representative driveshaft torque

N - Number of representative speed/torque points

A - Discrete set of A/F ratios

S - Discrete set of spark advances

E - Discrete set of EGR settings

R - Discrete set of CVT ratios

NOX_{max} - Total NOX allowed over drive cycle

HC_{max} - Total HC allowed over drive cycle

CO_{max} - Total CO allowed over drive cycle

f_{flow} (η, τ, r, A/F, SA, EGR) - fuel flow in lbs./sec.

NOX_{flow}(η, τ, r, A/F, SA, EGR) - NOx flow in gm/sec

HC_{flow} (η, τ, r, A/F, SA, EGR) - HC flow in gm/sec

CO_{flow} (η, τ, r, A/F, SA, EGR) - CO flow in gm/sec

λ_1^k - Lambda multiplier for NOX at k'th iteration

λ_2^k - Lambda multiplier for HC at k'th iteration

λ_3^k - Lambda multiplier for CO at k'th iteration

K - 1024

Optimal Control for Maximum Mileage of a Flywheel Energy-Storage Vehicle**

Fereidoon S. Jamzadeh and Andrew A. Frank
Dept. of Electrical and Computer Engineering
Univ. of Wisconsin - Madison
Madison, WI

**Paper 820747 presented at the Passenger Car Meeting, Troy, Michigan, June, 1982.

ABSTRACT

The flywheel energy-storage unit is examined as a tool for engine load management. The control decision to store or retrieve energy is formulated and discussed. Vehicle dynamics are simulated on a digital computer in combination with dynamic programming techniques to obtain optimal operation policy. The simplified algorithm is explained, as well as the cost-function criteria and optimization constraints. The sensitivity of the optimal path and the vehicle gas-mileage improvements are elaborated. The study of losses indicates that the transmission is the largest energy sink in the power train. The result of this study provides an indication of the appropriate real-time control policy.

FLYWHEEL ENERGY-STORAGE VEHICLES have shown great potential city fuel-economy improvement (1,2).* The three reasons behind this fact are (1) very efficient operation of the engine, (2) regenerative recovery of braking energy, (3) elimination of engine idle. Of these three, the last two sources of fuel savings are obvious.

The problem of optimal scheduling for flywheel energy-storage vehicles over the Federal Urban Driving Cycle (FUDC) has been formulated as a dynamic programming process and simulated on a digital computer. The cost function to be minimized consists of system losses. The initial boundary condition (flywheel speed) is assumed free, while the terminal point is fixed.

SYSTEM DESCRIPTION

Figure 1 is the system diagram of a typical flywheel energy-storage vehicle, and Fig. 2 shows the engine map of the 2.5-liter, 4-cylinder Chevrolet Citation (3) used for simulation.

A continuously variable transmission (CVT) is required in the power train for stepless transfer of energy between the flywheel and the car wheels. The CVT allows the engine throttle to be fixed for best brake-specific fuel consumption (BSFC). The double-lined curve in Fig. 2 shows the engine-operation region. Unlike conventional power trains, in this vehicle the torque is commanded through the CVT and not the engine. The average positive (driving) power on the drive shaft of a 3000-lb vehicle driven over the FUDC is less than 9 hp. The engine average BSFC corresponding to this power level is 0.71 lb/(hp•hr), assuming an average speed of 1500 rpm. Generation of this average power through a CVT power train will cause the engine to operate at a very low speed (900 rpm) and better BSFC (0.52 lb/(hp•hr)), but the engine runs very roughly at these loads and speeds. The engine-operation efficiency can be further improved by

*Numbers in parentheses designate References at end of paper

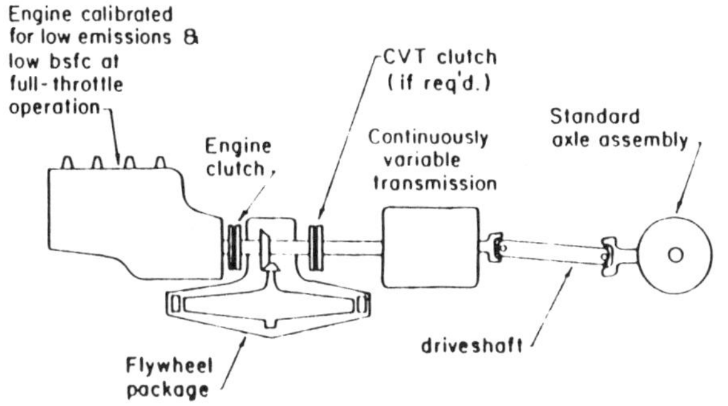

Fig. 1 - Schematic diagram of a flywheel energy-storage automobile

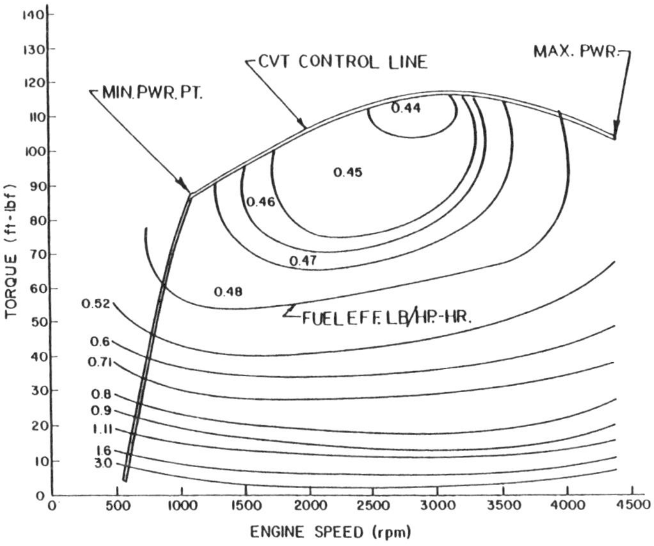

Fig. 2 - Engine BSFC map and wide-open-throttle operation line for a Chevrolet Citation with a 2.5-liter, 4-cylinder engine

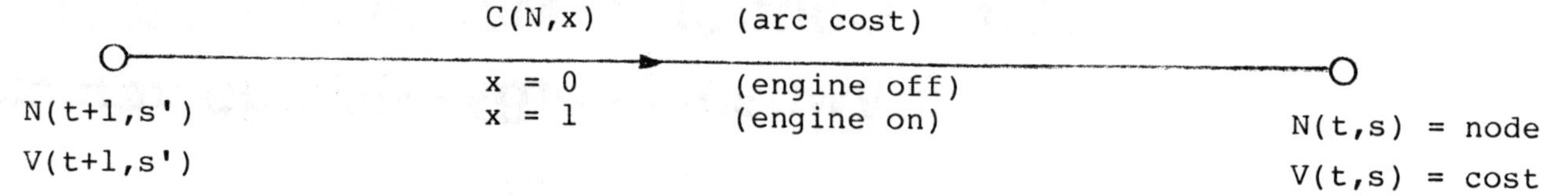

$$V(t,s) = Min(C(N(t,s),x) + V(t+1,s'))$$

Fig. 3 - Nodes, arc and recursive equation of dynamic programming

including a flywheel in the drive train. Now the engine can operate intermittently. Limiting the flywheel minimum speed to 1000 rpm restricts the worst-case BSFC to 0.48 lb/(hp•hr).

As mentioned, the driver controls the vehicle torque by means of the CVT; thus he has no control over engine operation. This raises two questions: (1) How often can the engine come on without overspinning the flywheel when braking? (2) How frequently can the engine be turned off without underspinning the flywheel at the times of high power demand? The similarities between these questions and those raised by a typical inventory-control problem lead us to incorporate dynamic programming techniques (4,5). However, notice that the control variable in our problem has only two states. That is, either the engine is on or it is off. This will eventually result in the drastic simplification of the vehicle-control-system implementation.

Early control policies applied to an experimental flywheel energy-storage vehicle were very simple (1). Whenever the flywheel speed dropped below a certain level, dependent upon vehicle speed, the engine would be turned on until the flywheel was fully charged. Through the use of dynamic programming, it is shown that the optimum control policy is far different from this simplified approach.

In the course of optimization, the cost function to be minimized consists of the sum of energy losses. These losses are (1) engine not operating at best BSFC, (2) engine turn-on/turn-off loss, (3) flywheel-drag loss, (4) transmission loss. Items 1 and 2 are included depending on whether the engine is operating or turned off, respectively. Minimum BSFC for the engine considered is 0.44 lb/(hp•hr). Thus, anytime the engine operates with a BSFC higher than 0.44 lb/(hp•hr), it is wasting fuel. Engine turn-off/turn-on loss consists only of its kinetic energy before shutoff because other losses are considered already in the BSFC number. The flywheel loss is considered proportional to its speed raised to a power of 2.4. The transmission loss is a function of three variables: input speed, output speed, and input or output torque.

SIMULATION

Backward dynamic programming is applied. This means that the optimum path from time t+1 to the end of the driving cycle is required in order to obtain the best decision to arrive at time t. For each of these time durations, there is only one decision, i.e., should we let the engine be off or on? The one with the minimum total cost (loss) from the end of the cycle up to this time t is chosen. Then this cost is compared to those with different initial states at time t+1 (i.e., different flywheel speeds) whose final states at time t are the same. Finally, the absolute minimum-cost path from the end of cycle up to time t is determined.

Figure 3 shows the transition process from one time to another. Each node N(t,s) is identified with two numbers: time t and flywheel speed s. The cost of progressing from one node to the other, which consists of the four items discussed above, is represented by C(N,x), where x is the bi-level decision variable indicating the engine state (when the engine is working, x = 1; otherwise, x = 0). If the cost of each node is represented by V(t,s), the dynamic-programming recursive optimization process can be summarized by the following equation:

$$V(t,s) = Min\{C[N(t,s),x] + V(t+1,s')\}$$

The difference between s and s' accounts for the energy produced by the engine and the energy consumed to drive the car.

The state of the vehicle is simulated second-by-second over the driving cycle. But only 1353 sec out of 1369 sec of the FUDC are important to us. The last braking and idle segments will have no effect on fuel economy if the engine is left off. In fact, the energy left in the flywheel just before the last time the brake is applied is considered wasted. Since there is no realistic way to recover this energy, it is desired to minimize the energy level of the entire system at this stage. Thus it is proposed that the flywheel should be at its minimum (1000 rpm) before the last brake application. This assumption fixes the end boundary point.

There are certain restrictions that should be kept in mind while progressing backwards through the cycle. First, there are limits on the system variables, specifically the engine and flywheel speeds:

$$1000 \text{ rpm} \leq \text{flywheel speed} \leq 3500 \text{ rpm}$$

$$0 \text{ rpm} \leq \text{engine speed} \leq 3500 \text{ rpm}$$

Second, infeasible answers should be ignored. To leave some flexibility for the consideration of almost feasible solutions, the limits are set wider than the absolute limits on the variables. That is, only answers with either of the following conditions are ignored:

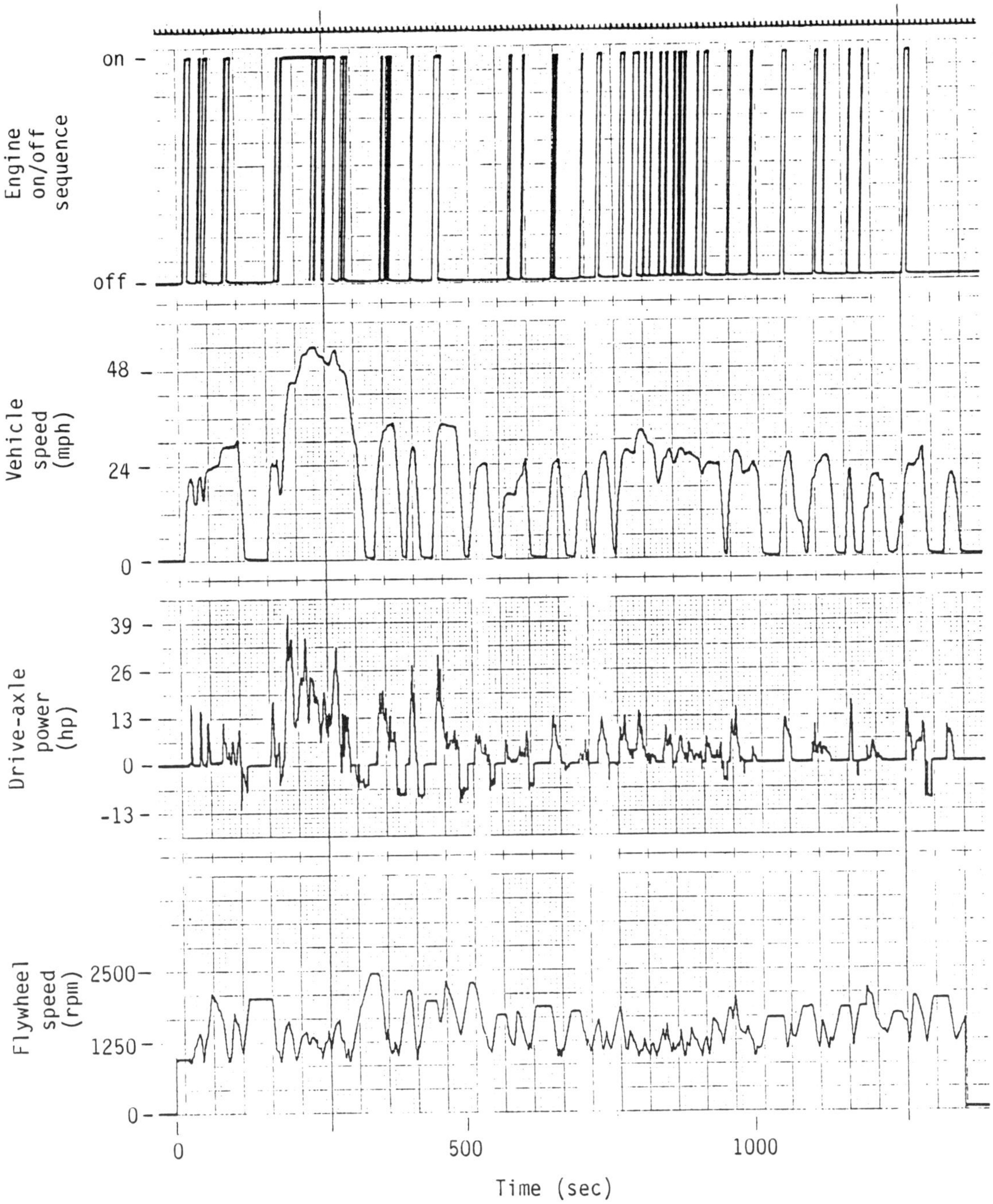

Fig. 4 - Simulation results with no transmission loss and low initial flywheel charge (Run #1)

flywheel speed < 900 rpm

flywheel speed > 4200 rpm

As the optimization progresses, the number of possible paths with different flywheel speeds increases drastically. Since the range of reasonable solutions is limited, many of these paths result in similar conclusions. Therefore, the maximum number of paths to be kept for each time stage is limited to 50. At any time stage, we have a number of possible flywheel speeds (nodes); the nodes are separated from each other by at least 20 rpm.

Once the dynamic-programming process has reached the start of the driving cycle, we are faced with the decision of picking one path among many possible (normally 20 or so, but always less than 50). Clearly, all these paths have different starting flywheel speeds. These represent the initial charges needed in the flywheel before starting the driving cycle. Naturally, the higher the initial charge, the more fuel must be consumed for cycle preparation. The final optimum path is the one that has the minimum total fuel consumption, including that required for this start-up charge.

RESULTS AND DISCUSSION

Figures 4, 5 and 6 contain the results of three different optimum paths. The top strip of

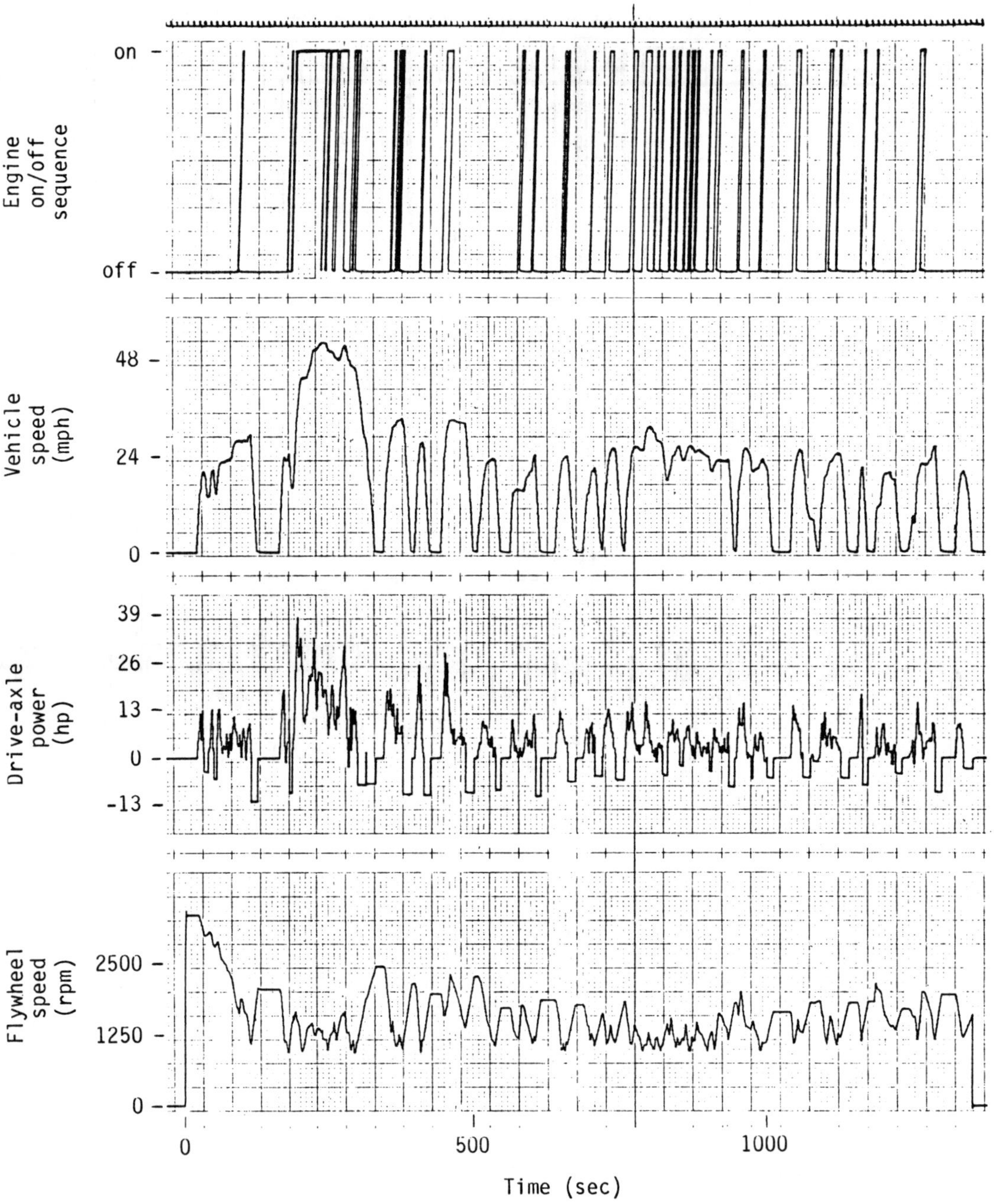

Fig. 5 - Simulation results with no transmission loss and high initial flywheel charge (Run #2)

these charts indicates when the engine is turned on or off. The second and third strips correspond to the vehicle speed and the instantaneous power on the drive axle. The last strip represents the flywheel speed. Runs #1 and #2 (Figs. 4 and 5), with different initial speeds, do not include the transmission loss. Runs #1 and #3 (Figs. 4 and 6) have low (944 rpm) initial flywheel speeds, while the initial speed of Run #2 (Fig. 5) is 3394 rpm. Run #3 (Fig. 6) includes the losses of the transmission. Efficiency tables for a Volvo variable-displacement pump and motor (V20A-180 scaled down by a factor of 4) are used (6).

Comparing the flywheel speed in Runs #1 and #2 (Figs. 4 and 5) reveals the dominance of the optimum path. Although these two paths start with different conditions, they converge after a short period to an identical path. The same has been experienced with other initial speeds, confirming the very important fact that, depending on the configuration and the chosen parameters, there is a certain path associated with that vehicle that results in the best possible fuel economy. One more significant characteristic of these plots is the relatively low average flywheel speed, 1500 rpm (see Table 1). Notice that with the slow flywheel start the flywheel speed never reaches 2500 rpm. The flywheel studied stored 800 hp/sec at 4000 rpm. The above result indicates that a much smaller storage unit could perform as well as this large one.

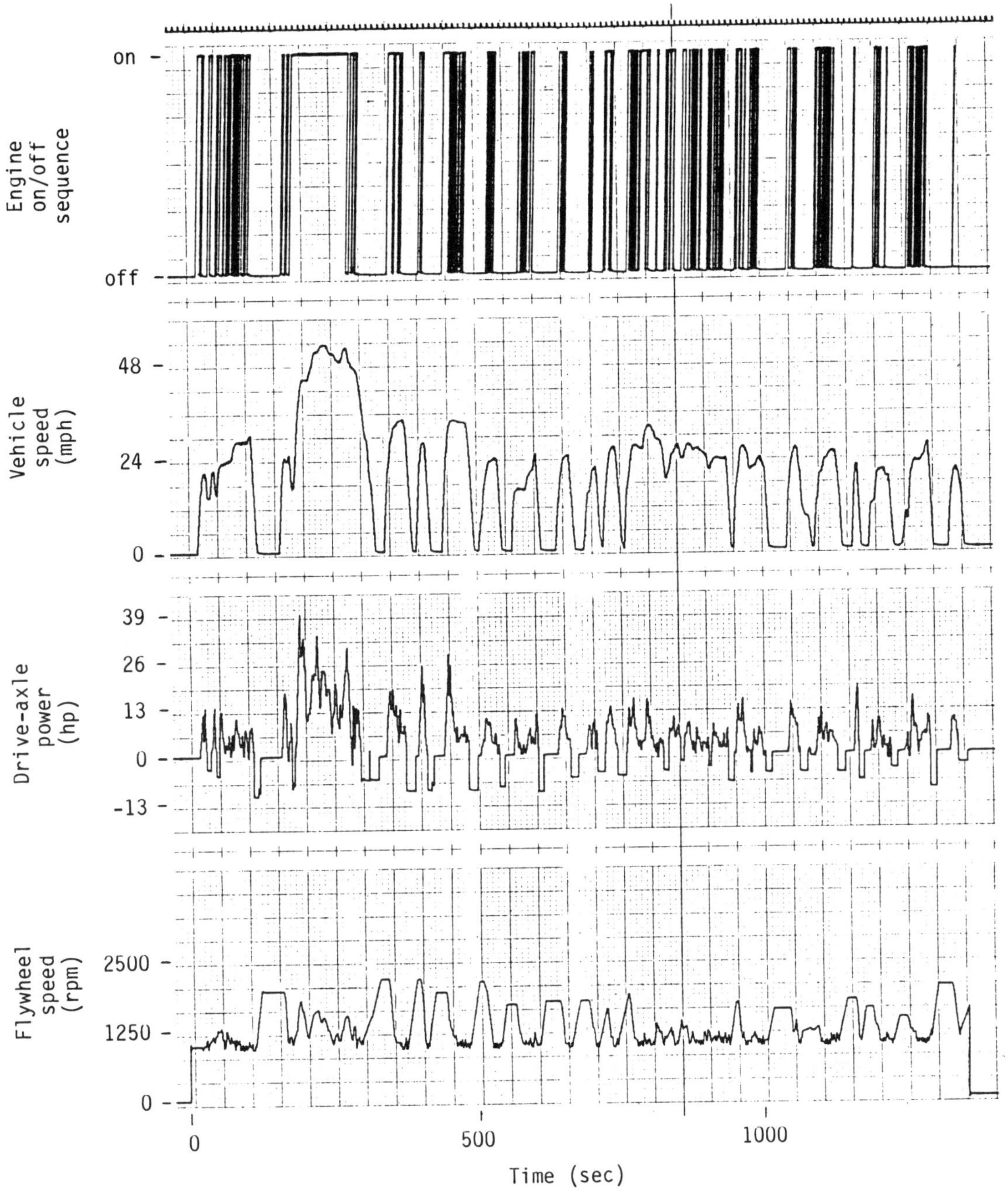

Fig. 6 - Simulation results with transmission loss and low initial flywheel charge (Run #3)

All three charts indicate frequent engine turn-on and turn-off. Our studies show that the flywheel loss overshadows the penalty considered for engine turn-on/off. Although by leaving the engine on and allowing the engine speed to rise, the BSFC drops, this improvement is not sufficient to compensate for the increased flywheel loss. Increasing the penalty for intermittent operation of the engine will reduce the number of these undesirable fast on/off states. The transmission has higher efficiency at lower speeds than at high speeds. Therefore, continuous operation of the engine, which results in high-speed operation, becomes less attractive when the transmission loss is included. Figure 6 clearly shows this.

Table 1 contains the statistical data associated with these three runs. Maximum, minimum and average values of three system variables (flywheel speed, vehicle speed, and instantaneous drive-axle power), along with the gas mileage of each run, are shown. The maximums and minimums are given with the times of their occurrence in the cycle. The values of the other two variables at that instant are also shown. These numbers can be used to reconsider some parameters of the vehicle for further simulations. Here, the strong effect of transmission loss on the gas mileage is clearly indicated. The manual-transmission Chevrolet Citation fuel economy is estimated at 24 mpg (EPA). The test inertia weight considered is 3000 lb, and the assumed

Table 1 - Data Summary of Three Different Runs

	Incident Time (sec)	Flywheel Speed (rpm)	Vehicle Speed (mph)	Drive-Axle Power (hp)
Run #1				
Slow flywheel start (994 rpm). Transmission loss not included. Gas mileage: 61.9 mpg.				
Maximum flywheel speed	319	2488	29.0	-8.30
Minimum flywheel speed	24	922	11.5	9.86
Average flywheel speed		1509		
Maximum drive-axle power	196	1519	36.2	41.73
Minimum drive-axle power	116	2102	28.6	-14.93
Maximum vehicle speed	240	1286	56.7	24.40
Run #2				
Fast flywheel start (3394 rpm). Transmission loss not included. Gas mileage: 60.2 mpg.				
Maximum flywheel speed	21	3396	3.0	0.56
Minimum flywheel speed	299	923	49.5	14.82
Average flywheel speed		1593		
Maximum drive-axle power	196	1519	36.2	41.73
Minimum drive-axle power	116	2102	28.6	-14.93
Maximum vehicle speed	240	1286	56.7	24.40
Run #3				
Slow flywheel start (994 rpm). Transmission loss included. Gas mileage: 45.2 mpg.				
Maximum flywheel speed	319	2218	29.0	-8.30
Minimum flywheel speed	372	924	36.0	15.77
Average flywheel speed		1293		
Maximum drive-axle power	196	1672	36.2	41.73
Minimum drive-axle power	116	2029	28.6	-14.93
Maximum vehicle speed	240	1321	56.7	24.40
All Runs				
Average positive power				9.29
Average negative power				-7.76
Average vehicle speed			25.4	

road load at 55 mph is 22 hp. Therefore, a projection of 80% fuel-consumption improvement is very realistic. The effect of the extra weight of the flywheel and CVT is kept in mind.

CONCLUSIONS

These studies indicate that the use of a flywheel for energy management results in substantial improvement of mileage, i.e., over 80%. This projection is based on the use of actual test data from Volvo and General Motors, and it is therefore realistic to expect that a hardware implementation will be able to verify these results. More efficient continuously variable transmissions designed for this size vehicle promise further fuel-economy improvements.

We have constructed a 1976 Pinto vehicle whose test results show a mileage improvement of 50%, operating with a non-optimum control policy and non-optimum components. The test inertia weight is 3000 lb, and the road load at 55 mph is 22 hp. The optimum schedule of operation shows that the flywheel should spin at its slowest controllable speed. This results from the flywheel loss, which increases at the 2.4 power of its speed. Inclusion of transmission loss in the study causes the flywheel to rotate at even a slower speed. The reason is poor transmission efficiency at high speeds.

The approach chosen for control of the vehicle is deterministic because the driving cycle is given. Random daily driving cannot be scheduled in the same way. However, further simulation using different driving cycles can provide general guidelines for scheduling such a system optimally. The general trend is obvious, however: it is desirable to run the flywheel at the slowest possible speed. The effect of flywheel size and sensitivity on the optimum schedule remains to be studied.

REFERENCES

1. N. H. Beachley and A. A. Frank, "Flywheel Energy Management Systems for Improving the Fuel Economy of Motor Vehicles." Final Report under Contract: DOT-OS-60177 (DOT/RSPA/DPB-50/79/1). Washington, D. C.: U. S. Dept. of Transportation, August 1979.

2. S. M. Rohde and N. A. Schilke, "The Fuel Economy Potential of Heat Engine/Flywheel Hybrid Vehicles." General Motors Research Laboratories, Research Publication GMR-3641, April 1981.

3. D. E. Koehler and W. F. Marshall, "Performance Characteristics of Automotive Engines in the United States, Third Series--Report No. 4, 1978, Pontiac, 151 CID (2.5 Liters)." Department of Energy Report #DOT-TSC-NHTSA-79-4, February 1979.

4. R. Bellman, "Dynamic Programming." Princeton, N. J.: Princeton University Press, 1957.

5. H. M. Wagner, "Principles of Operations Research." Englewood Cliffs, N. J.: Prentice-Hall, 1969.

6. Volvo Hydraulics, Data Sheet: Variable Pump/Motor V20, Ref. No. 45-8001 GB. Volvo Flygmotor AB, Trollhätan, Sweden.

System Design and Control Considerations of Automotive Continuously Variable Transmissions**

C. Chan, T. Volz, D. Breitweiser, A. Frank and F. S. Jamzadah
Department of Electrical and Computer Engineering
University of Wisconsin
Madison, WI
T. Omitsu
Aisin Seiki Co., Ltd.
Japan

**Paper 840048 presented at the International Congress and Exposition, Detroit, Michigan, February, 1984.

ABSTRACT

The Continuously Variable Transmission (CVT) concept has been available for many years. Reliability, durability, efficiency and controllability have been problems in the past. All but controllability have been greatly improved recently. This paper outlines the special control considerations necessary in CVT system design. Stressed is the concept of ideal engine operation and independent operation of engine and transmission control systems.

CVT HISTORY

When automobile development began in the late 19th century it was recognized that to achieve good efficiency and performance a transmission which varied its ratio was necessary. Early transmissions were often continuously variable [1]*. Reliability and durability were two main problems with these transmissions. Control was also rather haphazard with little regard for maximum efficiency or maximum performance. Because of these rather serious short comings CVT's gave way to a set of fixed ratio gears which could rather easily achieve the necessary reliability. Further, control was considerably simplified since gear selection could be rather independent from engine control, i.e., the two controls needed only discrete coordination rather than continuous as was apparently required by the CVT.

CVT DEVELOPMENT PROBLEMS

Obviously durability and reliability are a requisite to acceptance of a transmission in the automotive market. Recently, many manufacturers, inventors, and governments have renewed their interest in the CVT because of developments in materials and concepts as well as the motivation of better performance and fuel economy. Components in the form of V-belts with variable sheaves made of either rubber or steel [2] hydro static transmission [3] and other steel traction configurations have been designed which have reasonable efficiency and durability. It is now again important to reexamine the issues of vehicle system control.

CVT SYSTEM CONTROL

There are a number of general design factors to keep in mind. These are:

1. Minimum fuel consumption for a given power of the engine under all possible conditions.
2. With a finite CVT ratio, some vehicle starting device and control is necessary, such as a clutch or torque converter. If an infinitely variable transmission (IVT) were available such a starting device would not be necessary.
3. Some technique should be provided to limit maximum engine speed.
4. Engine idle or minimum speed needs to be controlled.
5. Rate of change of ratio must be high enough to satisfy all driving conditions including a panic stop.
6. Driveability must be similar to existing vehicles.
7. CVT losses must be minimized at all times.
8. It is desireable to have the CVT control scheme independent of the engine control system to allow manufacturing interchangeability.
9. Stable operation must be achieved under all conditions.
10. Performance under all conditions should be equal or better than that achievable with a standard transmission.

*Numbers in parentheses designate References at end of paper.

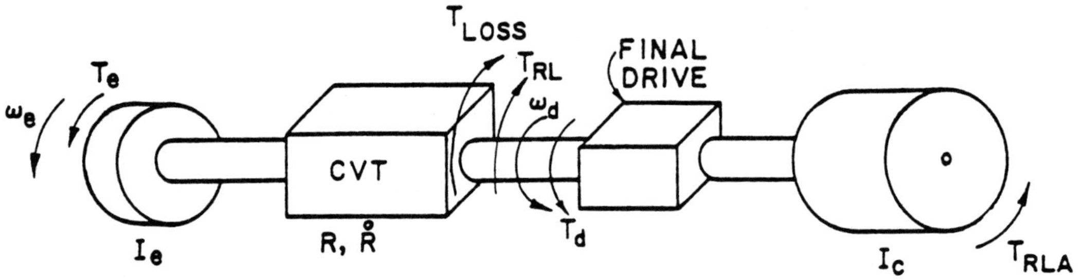

EQUIVALENT SYSTEM MODEL

$$\dot{\omega}_d = -\dot{R}\,\frac{I_e\,\omega_e}{(I_e R^2 + I_{c_d})} + \frac{T_e R - T_{RL} - T_{Loss}}{(I_e R^2 + I_{c_d})} \qquad \text{EQTN. 1}$$

Fig. 1 - Drive-train schematic and dynamic equation

There are many solutions to implement these objectives, discussed here will be one set of possible control schemes.

CVT DYANAMICS

To begin the discussion of design and control it is important first to understand that the dynamics of an engine-CVT vehicle need a more careful approach than that of a conventional vehicle. It's as if the conventional vehicle were continually being shifted from gear to gear. The configuration of a CVT vehicle is shown in Fig. 1. This figure shows that the ratio and the rate of change of ratio (ratio rate) of the CVT must be considered in the system. While this is true also in conventional transmissions during shifting in CVT's the ratio is constantly changing and therefore the full equation is necessary at all times. That is to say, since the inertias are coupled together through a continually variable element its rate of change is an equally important factor in the equation of motion. It is seen from Equation 1 that the vehicle acceleration is dependent upon ratio rate with a negative sign as well as upon the torque developed by the engine with a positive sign. For example, if acceleration is desired, R should be increased, thus $\dot{R}$ is positive. Then if the transmission ratio is changed too fast the result on the vehicle could be an initial deceleration rather than the acceleration as desired. Understanding this equation and consequences of the various terms it is necessary for the design and control of a CVT system. This problem is further complicated by the engine if it produces nearly a constant torque as its speed increases since the engine cannot then be used to counteract the effect of $\dot{R}$.

The definition of the terms in this particular equation is provided in the nomenclature list. The most critical term to define is the ratio R. We have chosen to define the ratio R as the engine speed divided by the drive shaft speed. Thus, the final drive ratio has been left out of the definition of R. Of course, R may be defined in any fashion one desires and since the final drive ratio is constant it could be included as part of R. The former is chosen because the sensitivity of R and $\dot{R}$ is reduced by the final ratio if it is included.

ENGINE CONTROL CONCEPTS

The dynamics of the system have been defined above, let us now examine the objective of the CVT with respect to a vehicle. Fig. 2 illustrates the fuel efficiency in terms of brake specific fuel consumption "BSFC" contour map for a typical internal combustion engine. On this curve is shown maximum power, maximum torque, idle speed, and maximum engine speed. It is desirable to operate this engine by varying its power. As the vehicle proceeds, the power must be variable by the driver. In a conventional automobile, vehicle output power is varied by adjusting engine torque at a given speed. Instantanously the vehicle and engine speed does not change, thus, if the driver wishes to change the acceleration characteristic of his vehicle he will vary the engine torque by changing the throttle of the engine. This is possible because the ratio between the engine and the final drive is fixed assuming he does not change gears. Unfortunately, this generally causes the operating condition of the engine to go toward a higher BSFC configuration since the minimum BSFC is only at approximately one throttle setting (70-80%).

Now with a continuously variable transmission it is possible to operate in the best BSFC area for a given power. Looking at Fig. 2 it is obvious that for any power demanded by the driver, the engine can operate along the line shown in the figure as the "ideal operating line" if the CVT has the appropriate capability as will be explained in the next section. It is immediately clear that at a constant vehicle speed it is the engine speed that should be controlled by the accelerator pedal rather than the engine torque. That is, it is necessary to

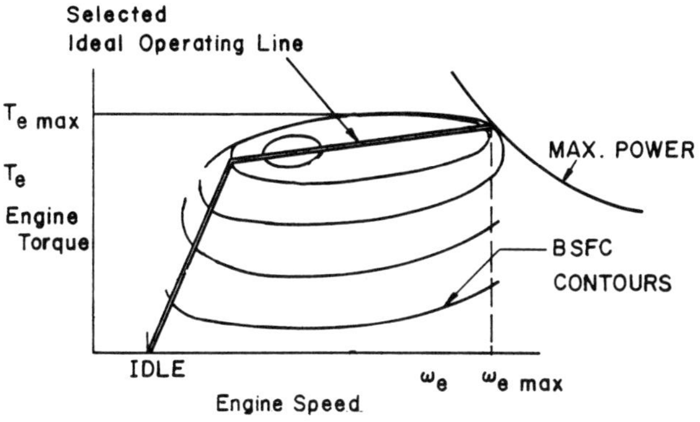

Fig. 2 - Engine operation map

change the transmission ratio to modulate the vehicle output torque. The engine throttle can now be governed solely by engine speed and this means that an engine throttle governor can be used for engine control.

Ideally, the range of control should vary from idle in which the engine is producing zero power to maximum power in a continuous fashion. Thus, as the accelerator pedal is moved fore and aft, the engine speed and torque should respond by moving up and down along the ideal operating line of Fig. 2.

Now that the general engine control requirement has been discussed, consider next the transmission.

CVT RATIO RANGE AND FINAL DRIVE SELECTION

A ratio selection of CVT and axle truly depends upon how the engine will be controlled. As the authors stressed, to operate the engine on its ideal operating line, it is necessary to have fairly wide ratio range such as about 8:1. (Maximum ratio between the engine and the axle is 15:1 and minimum is 2:1).

A 15:1 ratio is required to provide comparable vehicle acceleration performance to the conventional transmission-engine vehicle because the engine is operated on ideal operating line, the torque of the engine at each engine speed tends to be sacrificed (except maximum engine power area).

Figure 3 is an engine speed versus vehicle axle speed map. In this map, lines A and B are ratio of 15:1 and 2:1 respectively. Line C is a level road load curve using the engine ideal operating line. It is necessary not to have any intersections between line C and minimum ratio line B in order to keep the engine operating on its ideal operating line. It is also important to evaluate the engine's characteristics to find its ideal operating line first and then select the minimum ratio of CVT and axle. But as a rule of thumb, a minimum overall ratio of 2:1 is low enough.

In Fig. 3, the hatched area between D and E represents a narrow ratio range CVT-axle such as 4:1 (maximum ratio is 12:1 and minimum is 3:1). Clearly, line E has intersections with line C. This means the engine must be throttled off the ideal line.

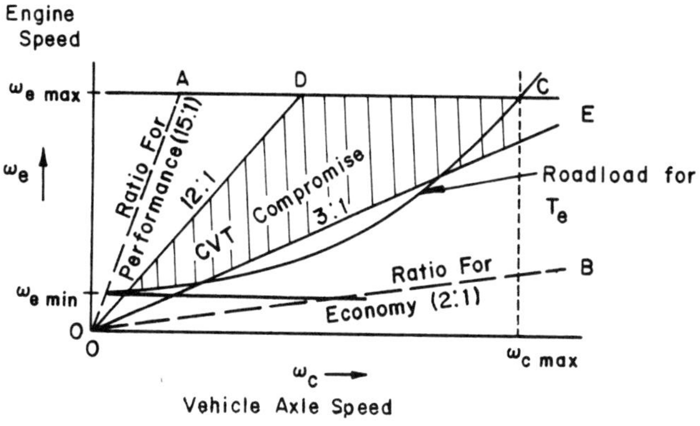

Fig. 3 - CVT operating range

If an infinitely variable transmission (IVT) rather than a continuously variable transmission is available, the problem is simplified. Since the maximum ratio is infinity, selection of the final drive ratio for best fuel economy is only necessary. This is shown in Fig. 4 as a relation minimum ratio line and line C of the level road load (same as Fig. 3). Now that the general control range has been discussed let us examine in detail the regions and boundaries of control.

There are other reasons for selecting a particular control range and final drive such as maximum output torque, efficiency, average operating point durability, etc., but these will not be discussed here.

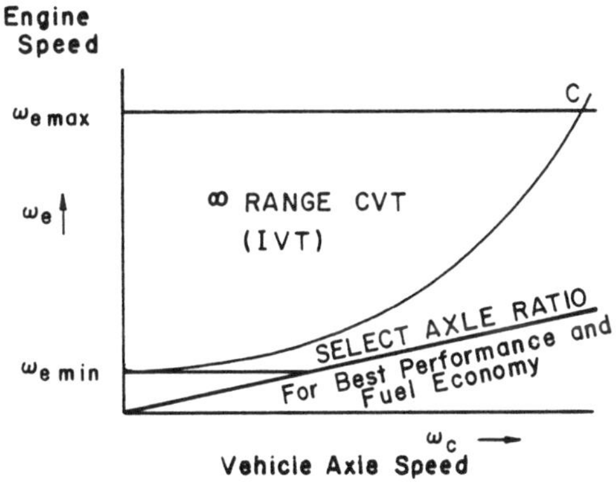

Fig. 4 - IVT operating range

CONTROL REGIONS AND BOUNDARIES

The regions and boundaries of control that are important to consider are shown in Fig. 5. Here again is plotted engine speed versus vehicle axle speed. This graph is divided into regions: I and II; and boundaries: III, IV, and V. The hatched area Region I is for starting the vehicle with a CVT and not an IVT. With a CVT this area is covered by slipping a clutch or using a torque converter or possibly a fluid coupling. Obviously in Region I, the ratio is held at the maximum available. The control of the system in this area must then necessarily be different than the other areas.

The next area of control is defined as the

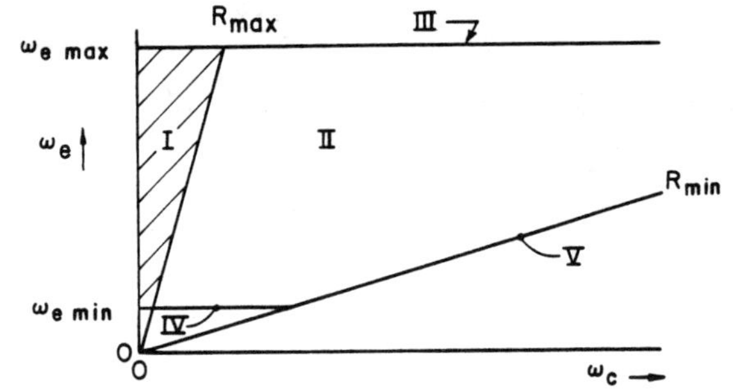

Fig. 5 - CVT control regions and boundaries

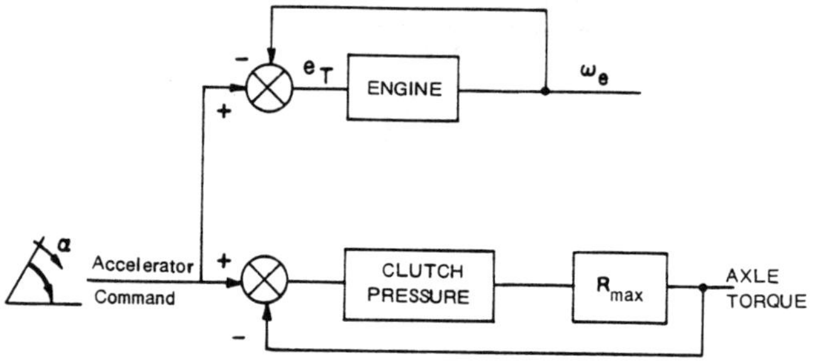

Fig. 6 - A clutch slip control system

CVT control region and labelled II. This is the region of control in which the CVT will vary the engine speed by changing its ratio. Boundary III is the maximum engine speed. Some control must be provided to prevent engine overspeed. This is also engine maximum power because if we are staying on the ideal engine line then maximum engine speed implies maximum engine power. The only possible exception of this at Boundary III is the clutch slip mode, perhaps the engine speed should not be maximum to provide maximum power during clutch slip.

Boundary IV is the minimum engine speed. This boundary is necessary to prevent the engine from going below the idle speed since its torque would again rise and stall the engine.

Below Boundary V is unreachable since minimum ratio of the transmission will not allow this area to occur. Operation along the minimum ratio line results in the minimum amount of output torque. If an output torque less than this value is required by the driver then some means of reducing torque other than control of the transmission must be resorted to, such as throttling the engine or cutting it's fuel. It is desireable to locate this minimum ratio line so that, for example, the engine is operating at a very low speed with the vehicle at a cruising speed of perhaps, 100 kmph. It would be ideal to be able to operate the engine at idle with the vehicle at cruise speed. Now that regions and boundaries of control have been introduced, techniques of control for these various regions and boundaries will next be explored.

REGION I: CLUTCH SLIP REGION

The objective for the control system in this region is to provide torque to the drive shaft regardless of vehicle speed up to the point where vehicle speed approaches the maximum ratio limitation of the CVT. Of course, the CVT ratio is held at its maximum during this period of operation. Figure 6 shows a possible block diagram for this controller. Here this block diagram takes the accelerator command, measures the output axle torque, compares the difference between the accelerator pedal command and the axle torque and generates, then from the error signal, a clutch pressure. This clutch pressure is then translated via the clutch mechanism to output torque which when multiplied by the maximum ratio provides axle torque. There are many ways to control the engine during this mode of operation, one way is to control the engine according to the indicated block diagram, that is, the engine is controlled via another control loop by the accelerator pedal in which the speed of the engine is set dependent upon the accelerator pedal position. This will provide an engine speed which is above engine idle. Thus when the vehicle is shifted into the CVT control range the engine speed is always above engine idle and the CVT control system then can take over.

REGION II: CVT CONTROL REGION

In this region the CVT is controlling the engine and the vehicle. The accelerator pedal thus is commanding power output from the engine-CVT system. Since the engine is to be operating on the ideal operating line according to Fig. 2 there is a maximum power output available from the engine. There is also a maximum torque available from the output of the transmission. The junction of the maximum output torque and the maximum engine power curve as shown on Fig. 7 indicates a corner. This corner is a natural junction for which the control system can be tailored. It is proposed that below this corner vehicle speed, that the system provide output torque control. That is the output torque would be proportional to the accelerator pedal position. Above this corner speed the vehicle would be controlled with respect to power. That is

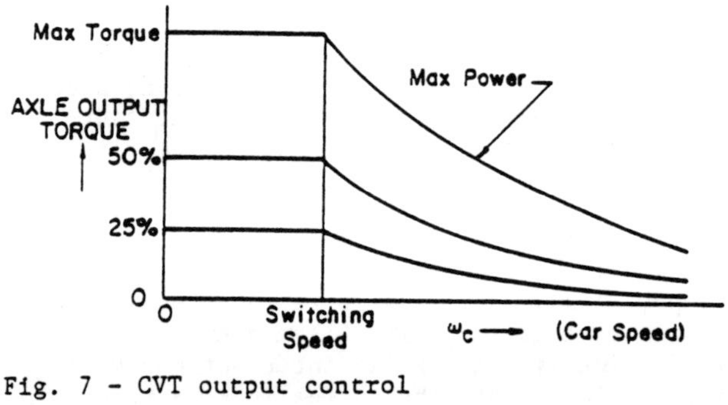

Fig. 7 - CVT output control

the accelerator pedal would provide the available power from the engine-transmission system. Thus when the vehicle is proceeding at very high speeds moving the accelerator pedal to the floor results in maximum engine power but the actual output torque is reduced. With this kind of control a host of problems involving sensitivity of the accelerator pedal can be solved. The basis for this seems sensible when a standard automatic transmission is considered since a kick down switch or downshifting does discretely what is being done continuously by the CVT. Furthermore this control scheme provides a smooth transition from low speed to high speed operation. A block diagram for this system is shown in Fig. 8.

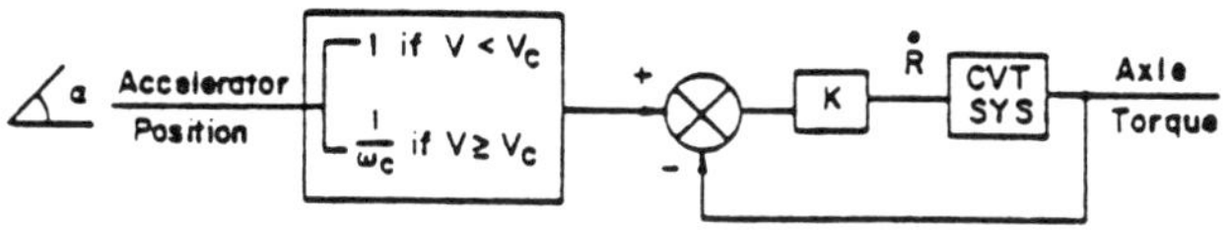

Fig. 8 - Torque/power control system block diagram

In Region II the main CVT control equation (Eqtn 1) must also be satisfied. Examing the equation the only variables that can be controlled are $\dot{R}$ and Te. But since the engine torque Te is fixed to an engine speed ωe by the ideal operating line, Te is not free. This implies the CVT must be one in which $\dot{R}$ can be varied. Fortunately all CVT's will have some means of varying $\dot{R}$. Explicitly obtaining $\dot{R}$ for a particular mechanism may be difficult. However, $\dot{R}$ could be implemented approximately in many cases. For example, $\dot{R}$ may be obtained from sheave pressure for a V-belt CVT, or stroke position pressure for a hydrostatic CVT.

According to Fig. 8 output axle torque should be measured and compared with the desired output commanded by the driver. The difference then is used to change $\dot{R}$ in the CVT system.

It should be pointed out that this control scheme separates the engine controls from the transmission controls. Engine control is solely a function of engine speed and transmission control is done only with ratio rate.

This concept should be contrasted with systems designed by other researchers. One such system is the one designed by British Leyland and Peter Stubbs [4]. This system is depicted in Fig. 9. It requires a coordinated system between the engine and the transmission for control and it cannot stay on the ideal engine operating line at all times. This control system, however, attempts to always approach the ideal operating line. It changes the engine throttle and ratio simultaneously to keep the engine on its ideal operating line for the required vehicle power.

The simple approach discussed above implies that the system could be constructed totally from mechanical components reducing the cost greatly.

BOUNDARIES III AND IV: MAXIMUM AND MINIMUM ENGINE SPEED CONTROL

In Fig. 10 shows a block diagram which

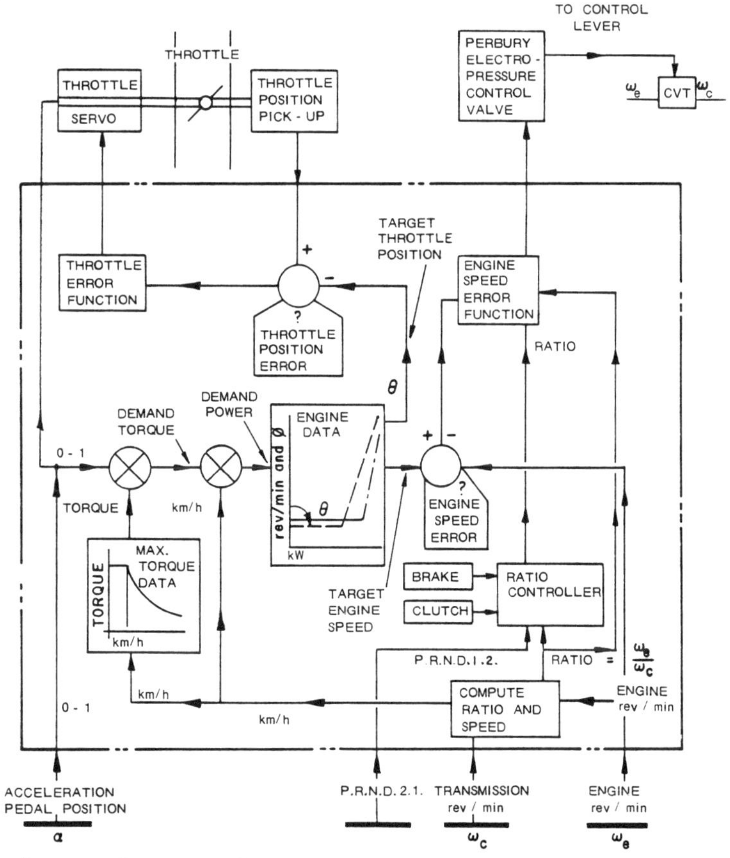

Fig. 9 - The British Leyland control system for the Perbury transmission by Peter Stubbs (ref. 4)

indicates the maximum and minimum engine speed control for Boundaries III and IV. This diagram indicates that when the engine speed is below the allowable, no limitation is necessary but once the engine tends to exceed the predetermined maximum speed the ratio rate of the transmission should be changed such that the engine stays at its maximum speed. The same holds conversely for the minimum speed, that is, the engine speed is regulated by changing the ratio rate of the transmission. Since the engine is operating on its ideal operating line at all times then the ratio rate is the only parameter which should be controlling the vehicle.

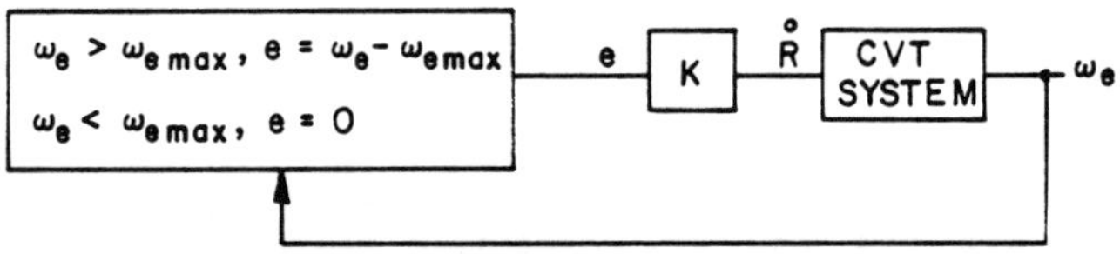

Fig. 10 - Engine maximum speed regulator block diagram

BOUNDARY V: MINIMUM RATIO CONTROL

When the CVT or IVT [5] is at its minimum ratio, vehicle control has to be done by varying the engine output torque. In conventional vehicles, when the ratio is fixed the vehicle output torque is varied by changing the engine throttle. In a CVT vehicle such torque variations must also be accomplished. Depending on the minimum ratio one may find alternatives to throttling the engine. For example, if the minimum ratio is low enough so that the engine could be close to idling at a vehicle cruise speed of 100 km per hour then torque can be varied simply by turning the engine "off" and "on" since the engine is not providing much positive power. Many electronically fuel injected engine systems today have such a fuel cut control for deceleration purposes above a certain vehicle speed that could be incorporated in a CVT control system very nicely. Since throttling the engine means lower engine efficiency this alternative means of power modulation may save fuel.

SIMULATION STUDY RESULTS

Using these control system ideas, a study has been completed indicating the engine and vehicle dynamic relationships. The control system block diagram is shown in Fig. 11. This block diagram is implemented for a CVT with a ratio range of 10:1. On the block diagram is indicated a number of system gains (K_1 thru K_5). These gains must be adjusted to provide smooth and consistant vehicle response for all values of driver input and vehicle speed.

Figure 12 shows the vehicle response for various values of driver input and one set of optimized gains.

Figure 13 shows the result of a set of unoptimized gains. Here the vehicle driveability and controllability becomes unacceptable.

CONCLUSION

The authors have attempted to present in this paper the essence of control considerations for a CVT or IVT system. A discussion of the dynamics and control parameters involved was presented. There are separate controls in regions and boundaries which must be considered for most CVTs. These controls must, of course, be coordinated in such a way that transition from one region/boundary to the other is smooth and without any bothersome phenomenon. With the CVT it is very clear that once an ideal or optimum engine operating line is selected for an engine, the CVT controls the vehicle. Thus if a manufacturer wishes interchangeability between engines with the same interia I_e but with different operating characteristics the CVT really doesn't care because it is operating only on available power and regardless of the specific lines selected by the engine manufacturer or engine builder, the CVT will do its job. Thus great flexibility in vehicle design

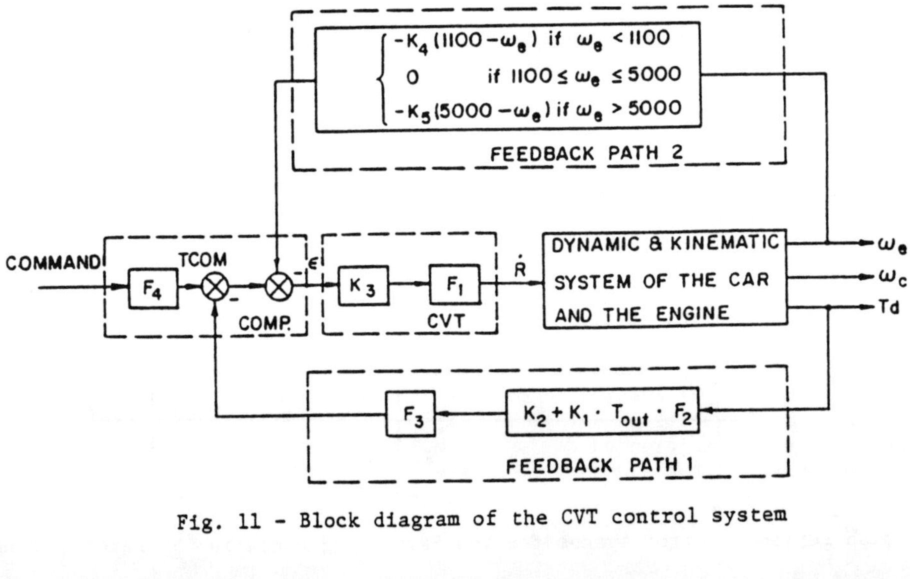

Fig. 11 - Block diagram of the CVT control system

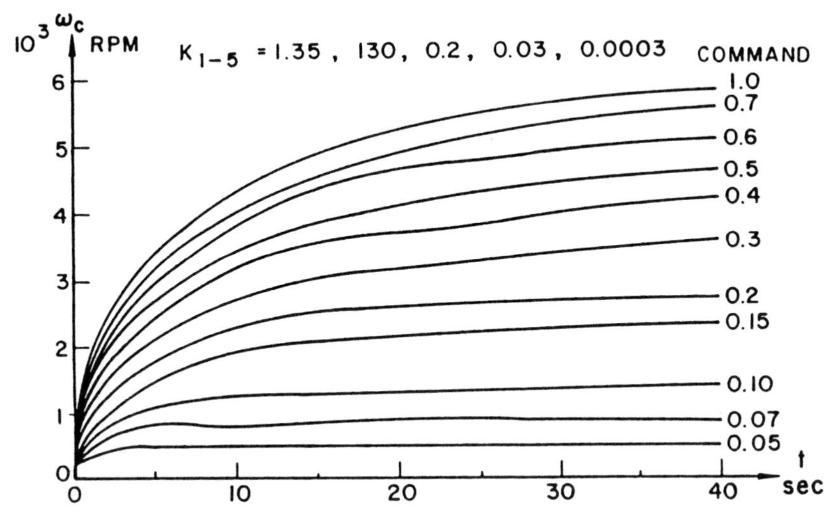

Fig. 12 - Speed of the vehicle vs. t with optimized system gains

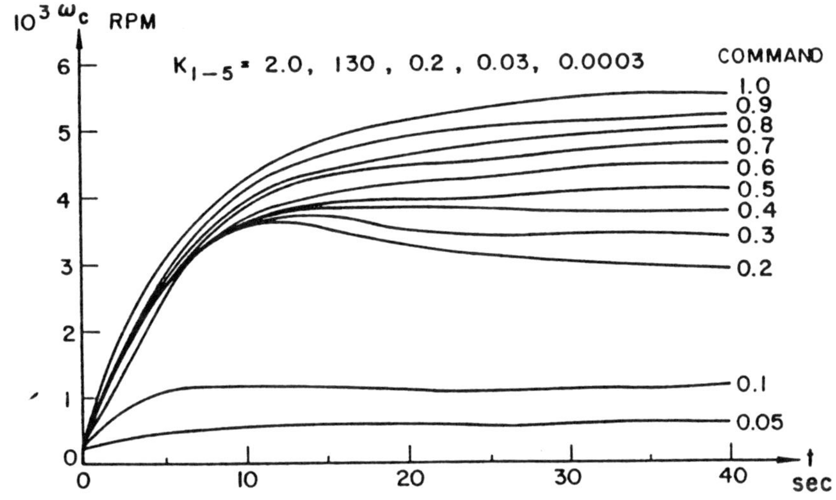

Fig. 13 - Speed of the vehicle vs. t with unoptimized system gains

can be achieved with CVTs. Engine possibilities which were not suitable in the past, such as two cycle engines, may now become possible with the CVT. Turbo charged engines, of course, are also attractive with the possibility of eliminating the effect of turbo lag. The CVT provides a new and exciting design tool if its system limitations are kept in mind and treated in manners as discussed in this paper. Of course, what has been discussed are some specific solutions to these problems. These solutions are by no means unique and many other possible techniques and control block diagrams can be used to achieve the same objectives [6,7,8]. The authors have presented a series of techniques of control which maintains maximum possible engine efficiency at all times, i.e., the engine will never come off the ideal operating line. Simulations results show that these control concepts can be implemented to provide a vehicle with good driveability and response.

NOMENCLATURE

F_{1-4}	= Control functions
I_c	= Vehicle inertia at the axle
Icd	= Vehicle inertia + CVT output inertia reflected at drive shaft
I_e	= Engine inertia + CVT input inertia reflected at input shaft to CVT
K_{1-5}	= System gains
R	= CVT ratio (ω_e/ω_d)
$\dot{R}$	= Ratio rate
Td	= Drive shaft torque
Te	= Torque generated in engine
Tloss	= Torque loss in CVT evaluated at drive shaft
T_{RL}	= Road load torque evaluated at drive shaft
ω_c	= Vehicle axle speed
ω_d	= Drive shaft speed
ω_e	= Engine speed

ACKNOWLEDGEMENT

We wish to thank Aisin Seiki Co., LTD. for their monetary and technical support of the research leading to this paper.

REFERENCES

1. U.S. Patent 2,150,456. Variable Speed Drive filed 1933, issued 1939, L.E. Perrine.

2. R.F. Stieg and W.S. Worley, "A Rubber Belt CVT for Front-Wheel-Drive Cars." SAE 820746.

3. Peter Huntley, Borg Warner, Chicago, "Design and Development of a Continuously Variable Ratio Transmission for an Automotive Vehicle," Orshansky Transmission Corporation, November, 1975.

4. P.W.R. Stubbs, "The Development of a Perbury Traction Transmission for Motor Car Applications," ASME 80-C2/DET-59.

5. N.H. Beachley and A.A. Frank, "Principles and Definitions for Continuously Variable Transmission, with Emphasis on Automotive Applications," ASME 80-C2/DET-95.

6. K. Srinivasan, D.R. Houser, G.L. Kinzel, D.L. Mattern, "Microprocessor Controlled Continuously Variable Rubber V-belt Transmissions for an Automobile," SAE 820745.

7. M.H. Cuypers, "Numberical Evaluation of the Torque Capacities of Mechanical Drives, Especially for Continuously Variable Transmissions," ASME 80-C2/Det-120.

8. B.G. Gerbert, "Some Notes on V-belt Drives," ASME 80-C2/DET-91.

On the Use of Engine Modulation for Deceleration Control of Continuously Variable Transmission Vehicles**

D. Yang and A. Frank
Dept. of Electrical and Computer Engineering
University of Wisconsin
Madison, WI

**Paper 850490 presented at the International Congress and Exposition, Detroit, Michigan, February, 1985.

ABSTRACT

The design and evaluation of a control system for a continuously variable transmission (CVT) vehicle is presented in this paper. Studied and simulated are the following three topics: design and evaluation of the control system with a CVT vehicle in acceleration, deceleration and to the sensitivity of the system with respect to changing characteristics of the engine (hot, cold, age). It is shown that the control system proposed here could insure good stability and controllability not only for acceleration of the system but also for deceleration operation, provided that a pulse width modulated (PWM) fuel control scheme for the engine is employed in the system. It seems that the control system designed in this paper is not complicated and does not have large sensitivity with respect to variation of the characteristics of the components of the system and provides the best possible fuel efficiency. In addition, a number of alternate decelerating schemes are presented at the end of the paper for comparison.

THE CONTINUOUSLY VARIABLE TRANSMISSION concept has been available for many years. Reliability, durability, efficiency and controllability have been problems in the past. All but controllability have been greatly improved recently. Some special control considerations and techniques on the design of a CVT control system have been proposed [1,2]*. This paper will present simulation results for a CVT control system for both acceleration and deceleration operation. All evaluations presented here are based on a typical vehicle with a wide range belt drive CVT device, partial data of which are listed in Table I.

----

*Numbers in parentheses designate references at the end of paper.

The general requirements that the control system designed here must meet are also listed as follows:

1. Stable operation must be achieved under all conditions of acceleration and deceleration operation.
2. Some technique should be provided to automatically limit engine speed within an allowable range as listed in Table I.
3. It is desirable to have the CVT control scheme independent of the engine control system to allow manufacturing interchangeability.
4. The engine is to be operated at the best possible efficiency at all times.
5. Performance under all conditions should be equal to or better than that achievable with a standard transmission and standard engine controls.
6. A start up procedure (not discussed here).

There are many solutions to meet the requirements listed above, the system proposed in this paper will be one set of all possible control schemes. Four parameters of this system would first be selected by using an optimization technique with a quadratic cost function, based on a designed structure of the CVT control system (block diagram of Fig. 5). The system is able to insure good controllability and stability for acceleration operation over all conditions. Then "good" performance would be achieved for deceleration operation by employing a PWM engine fuel control scheme along with the designed control system.

This scheme provides the best possible fuel efficiency of the engine. It makes a gasoline engine operate in principal like a diesel, i.e., without a throttle. Finally, the sensitivity of the system with respect to a change of characteristics of the engine operating line such as would occur when the engine is cold or when the engine ages would be studied.

Before introducing the control system, it is necessary to describe some key components including the hardware and software of the system.

Table I - Example system parameters

		Unit
Moment of inertia of the engine I_e	= 0.0027	Ft-lb sec/RPM
Effective moment of inertia of the vehicle, I_c (Transformed onto the output shaft of the CVT)	= 0.1638	Ft-lb sec/RPM
Maximum ratio of the CVT, R_{max}	= 4.69	Input/output
Minimum ratio of the CVT, R_{min}	= 0.44	Input/output
Fixed ratio of the final gear	= 4.6	Driveshaft/axle
Maximum engine speed $W_{e,max}$	= 5500	RPM
Maximum engine torque $T_{E,max}$	= 37	Ft-lb
Minimum engine speed $W_{e,min}$	= 1000	RPM

DESCRIPTION OF THE SYSTEM

The control system used here is divided into coupled parts of "hardware" and "software", each part of which will be presented, respectively.

HARDWARE OF THE SYSTEM - The hardware structure of the CVT control system is shown in Fig. 1.

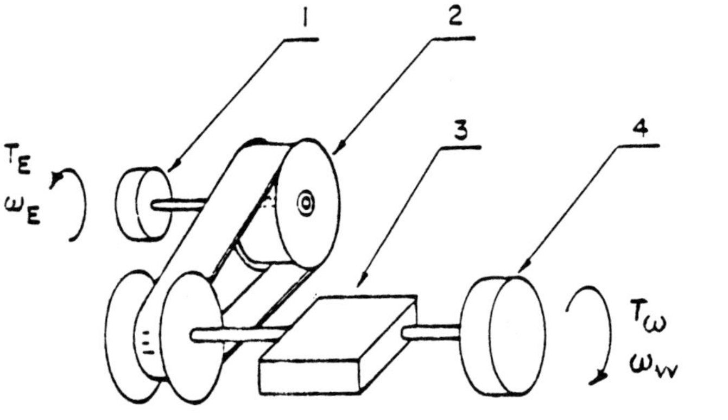

1. Engine and inertia
2. Belt or other CVT
3. Fixed gear box
4. Wheels of the car

Fig. 1 Hardware structure of a typical CVT control system

Three key components of the hardware in the control system are the engine, the CVT concept, and the final drive ratio. These will each be discussed below.

Engine - In order to obtain the most effective characteristics for fuel consumption and emissions the engine is forced to work along a special operating line referred to as the "Optimum Operating Line" (OOL) [1,2], labeled $f_1(\omega_e)$ and shown in Fig. 2. This line is chosen by the engine designer to produce the best operating characteristics for a given power. It may be a curve in general, but here it is represented as a set of straight lines.

In addition to the positive torque function $f_1(\omega_e)$ associated with OOL, there is a negative function labeled $f_2(\omega_e)$ to be used along with the positive torque $f_1(\omega_e)$, alternatively, for deceleration operation as will be discussed later. This negative characterististic is obtained from spinning the engine without fuel when the throttle is opened according to $f_1(\omega_e)$. Below 1000 RPM the engine torque again increases as the engine operates on idle fuel.

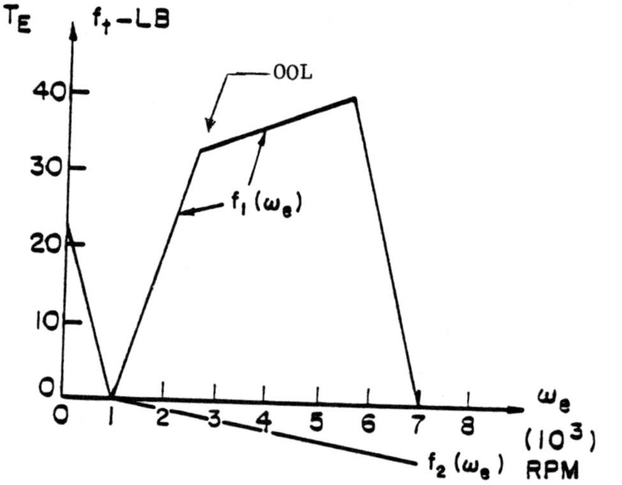

Fig. 2 Characteristics of the engine

It should be noted that the curve (OOL) shown in Fig. 2 (also in general) has a positive slope, i.e., the derivation of the OOL with respect to the engine speed, ω_e, is greater than zero in the main operating range of engine speed (1000, 5500). That is to say the engine, if it is forced to operate always with such an OOL, is an unstable element [2], which could not be used by itself in any conventional vehicle system unless a suitable control system is provided.

The CVT - The most important element in the system, the CVT, is a fixed center rubber V-belt device. There are two forces, FDR, FDN (or TCOM), and an input torque T_{in}, which act on the driver and driven shaft of the CVT, respectively as shown in Fig. 3.

The forces and torque shown in Fig. 3 along with the ratio, R, of the CVT are four key vari-

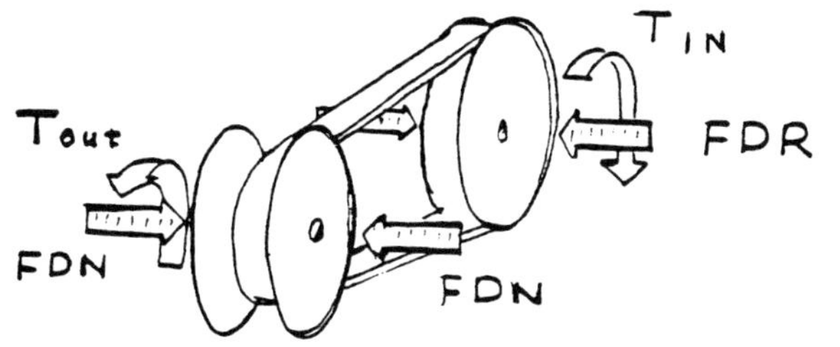

Fig. 3 Force balance of the V-belt device

ables which can be used to control the drive. The force balance between both FDR and FDN will determine the actual speed ratio of the CVT in a running drive. It is essential to know what axial forces will be required to hold the ratio constant at any given condition of the CVT, and how these axial forces must change as the other conditions are changed. The first characterization that must be performed so as to design the control system is the establishment of steady state values of these forces, torques and ratio of the CVT [3] for maximum operating efficiency. The data map of the best steady states for a typical CVT device is shown in Fig. 4, which must be obtained experimentally.

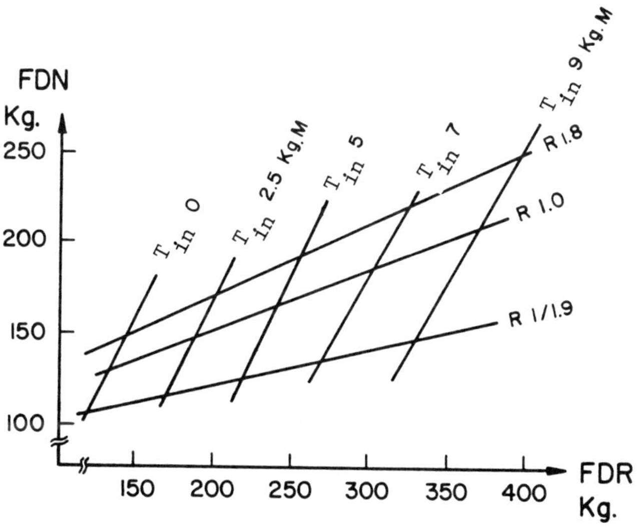

Fig. 4 Data map of a typical V-belt device for steady state

It should be noted that the four variables, FDR, FDN, T_{in}, R, must satisfy the data map of the typical CVT shown in Fig. 4 at the steady state. Any two of the four variables could be determined if the other two are given. Usually the force FDR is expressed by the formula FDR = $F_1'(T_{in},R)$ and the force FDN by the formula FDN = $F_2(FDR,R)$. The two functions, $F_1'(T_{in},R)$, $F_2(FDR,R)$, satisfy the data map in Fig. 4 and are very useful to the system design as will be seen later.

Besides the characteristic of the CVT in steady state, it is also important to know how the rate of the change of ratio, $\dot{R}$, varies when there is an unbalance between FDN and FDR.

Final Drive Ratio - The required fixed ratio in addition to the V-belt ratio needed was selected as a constant of 4.6 to obtain the satisfactory speed ratio between the engine and the vehicle.

SOFTWARE OF THE SYSTEM - Software included in the control system are referred to as the functions in the system, each of which provides a kind of computational function denoted by F_0 thru F_4 shown in Fig. 5.

From Fig. 5, it is noted that there are only two measured variables, ω_e (engine speed) and ω_c (vehicle speed or output shaft speed of the CVT) in the system. The software of the system has to provide the five computational functions for the control system as derived from these two variables. These functions are discussed as follows:

F_0 is simply a division function, which is expressed as follows:

$$R = \begin{cases} \dfrac{\omega_e}{\omega_c} & \text{if } R_{min} \le \dfrac{\omega_e}{\omega} \le R_{max} \\ R_{max} & \text{if } \dfrac{\omega_e}{\omega_c} > R_{max} \\ R_{min} & \text{if } \dfrac{\omega_e}{\omega_c} < R_{min} \end{cases}$$

where R is the ratio between engine speed and output shaft of the CVT. R_{max} and R_{min} are the limits of the ratio of the CVT.

F_1 is a function generator which gives the value of the driver force FDR as a function of R and ω_e. In fact, as was mentioned above, the value of the driver force FDR should be a function of R and T_{in} for any steady state condition of the CVT. Since the engine torque is a function of ω_e,$[f_1(\omega_e)$ or $f_2(\omega_e)]$ and the torque loss of the engine inertia can be neglected compared with the engine torque, the driver force FDR could be expressed by a function of R and ω_e. The advantage of taking a function of R and ω_e rather than a function of R and T_{in} is that one can eliminate the measurement of torque to simplify the implementation of the real system. While the information may be incorrect when the engine is cold for example, the objective here is to study the system operation even under these conditions.

F_2 expresses the force FDN as a function of FDR and R, which is shown in Fig. 4 from the data map of the CVT and could be used to determine the force FDN which is required to keep the CVT de-

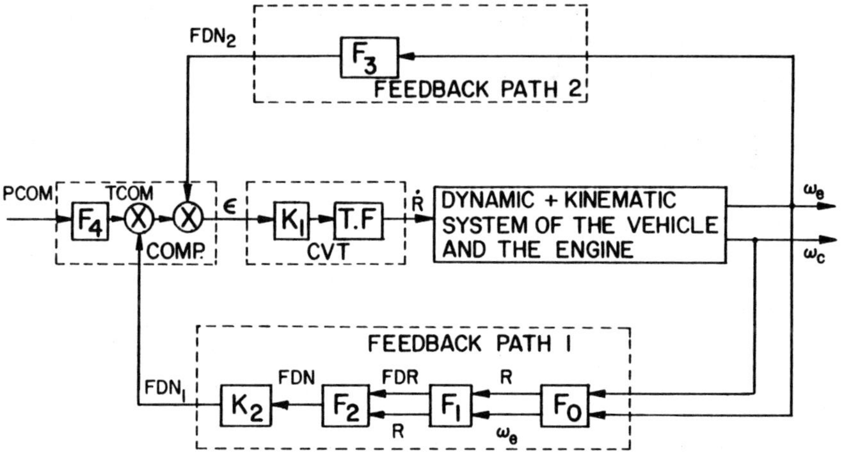

Fig. 5 Block diagram of a CVT control system

vice at a steady state along with the force FDR at a given point of R and T_{in}.

K_1 and K_2 are two gains which are to be selected so as to get good controllability and stability of the system.

F_3 is used to keep the engine speed within a permissible range (1000, 5500) RPM and expressed as follows:

$$FDN2 = \begin{cases} -K_3(1100-\omega_e) & \text{if } \omega_e < 1100 \text{ RPM} \\ 0 & \text{if } 1100 \le \omega_e \le 5000 \\ -K_4(5000-\omega_e) & \text{if } \omega_e > 5000 \end{cases}$$

where FDN2 is a feedback force acting on the driven shaft of the CVT. The two gains K_3 and K_4 are also designed in this paper.

F_4 is a switch function which can be expressed as follows:

$$TCOM = PCOM \qquad \text{if } \omega_c < \omega_{sw}$$

$$TCOM = \frac{1}{\omega_c} PCOM \qquad \text{if } \omega_c \ge \omega_{sw}$$

where PCOM, power command, is the system input signal provided by the accelerator pedal.

TCOM is the torque command to the system.

ω_{sw} is the switching speed, above which the vehicle would be controlled with respect to power, below which it is controlled with respect to torque.

ω_c is the car speed.

In addition to the five function blocks and two gains in the block diagram of the control system, the block named $T.F_{CVT}$ expresses the transfer function of the CVT, which is used to determine how to change the ratio rate, $\dot{R}$, of the CVT, and could be described by an experimental result as follows:

$$T.F_{CVT} = \frac{\dot{R}}{\varepsilon} = \begin{cases} (0.01138-1.3(E-3)\cdot T_E)\cdot\omega_e & \text{if } R < 1.46 \\ (0.01138-7.3(E-4)\cdot\frac{T_E}{7.2}) * \omega_e & \text{if } R > 1.46 \end{cases}$$

where ε is the error force signal, it is the sum of the dimensionless forces TCOM, FDN1 and FDN2. R = 1.46 is a CVT shift point. The block named "dynamic and kinematic system of the vehicle and engine" consists of all dynamic and kinematic equations of both the vehicle and the engine. The most important equation, in which the inertias of both engine and vehicle are coupled together through the CVT device is as follows:

$$\dot{\omega}_c = -\frac{I_e\omega_e}{I_c + I_eR^2}\dot{R} + \frac{T_ER - T_{L.C} - T_{L.R}}{I_c + I_eR^2} \qquad (1)$$

where

$\dot{\omega}_c$ = the acceleration of the output shaft of the CVT

$\dot{\omega}_e$ = the acceleration of the engine

I_e = the moment of inertia of the engine

I_c = the effective moment of inertia of the vehicle, which was transformed onto the output shaft of the CVT.

T_E, $T_{L.C}$, $T_{L.R}$ are the engine torque and the torque losses in CVT and on the road load, respectively.

DESIGN AND EVALUATION OF THE SYSTEM FOR THE ACCELERATION CASE

As was mentioned above, there are four gains K_1 through K_4, which could be selected to get good drivability. These gains have been selected by using an optimization technique with a quadratic cost function, which was presented in details in another paper [2]. We now choose one set of acceptable gains [2], that is, K_{1-4} = 0.2, 1.30, 0.0003, 0.03. With this set of gains, all variables of the CVT-vehicle system was computed. A set of results of two vehicle variables, ω_c and ω_e, is presented in Fig. 6 and 7. The responses of ω_c and ω_e caused by 10 discrete accelerator step signals with equal space, 0.1 through 1.0 are shown.

From this set of evaluations, it is shown that with the given gains K_1 through K_4, 1) the system could be operated with good stability and controllability in all conditions of acceleration. 2) the engine speed could be limited to within a desired range (1000, 5500) over all operating conditions. Additionally, the sensitivity of the system with respect to characteristic variations of the engine is acceptable as will be discussed later.

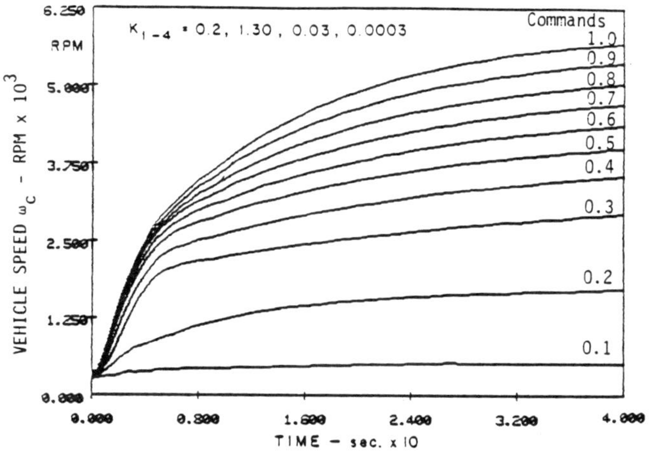

Fig. 6 Speed of the vehicle vs. t for different commands, PCOM

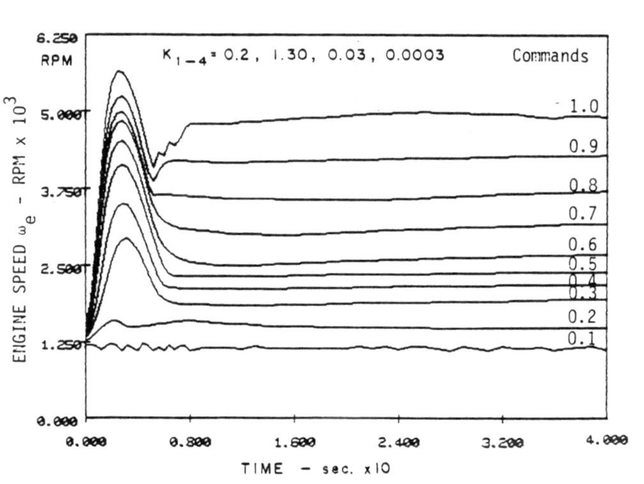

Fig. 7 Speed of the vehicle vs. t for different commands, PCOM

DESIGN AND EVALUATION OF THE SYSTEM FOR THE DECELERATION CASE

DESIGN OF THE SYSTEM FOR DECELERATION - The system shown in Fig. 5 is able to be used for the deceleration control of the vehicle, provided that a PWM #1 scheme (pulse width modulated scheme #1) is employed to the engine fuel delivery system. Without some techniques of reducing engine torque the vehicle would "overshoot". That is in the transient when the accelerator pedal is lifted the vehicle would continue to accelerate sometime before deceleration actually occurs. This is especially noticeable at low vehicle speeds. The conventional technique is to close the throttle valve, but this causes inefficient engine operation. The technique proposed here provides the best possible engine efficiency to accomplish the desired deceleration characteristic. Using this scheme the engine provides a positive and a negative torque alternately during the deceleration operation. Both the positive and negative torque are functions of engine speed as was mentioned above. Average torque could be controlled by a PWM signal which has the pattern shown in Fig. 8.

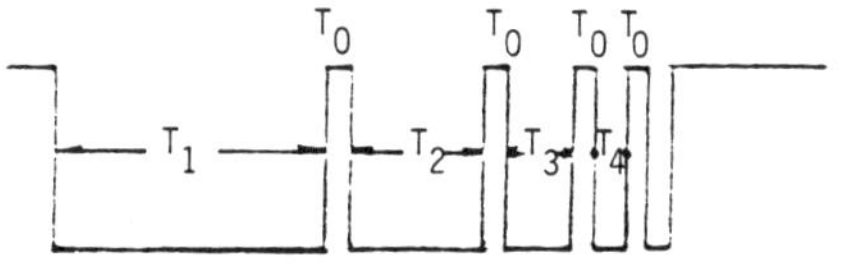

Fig. 8 Pattern of a PWM signal

While the PWM signal changes between high and low levels, the engine torque changes from positive to negative in the same manner as the PWM does. That is to say, while the PWM is on a low level, the engine torque will be a negative value depending on $f_2(\omega_E)$. On the contrary, while the PWM is on a high level, the engine torque is positive and depends on $f_1(\omega_E)$. The the parameters of the PWM signal, T_1,T_2, etc., are mainly determined by the degree of deceleration operation, DPCOM (change or derivative of the power command). The following algorithm for determining the time parameters are used here.

T_0 = 0.3 sec. to insure a number of engine firings before the fuel is shut off. This may be a function of engine speed. It is kept constant for this study.

$$T_1 = \begin{cases} 0.1 + 2 \cdot \text{DPCOM} & \text{if PCOMN} \geq 0.4 \\ 1.10 + 7 \cdot (0.4\text{-PCOMN}) & \text{if PCOMN} < 0.4 \end{cases}$$

DPCOM = PCOM - PCOMN

Then

$$Ti + 1 = Ti \cdot 0.5 \qquad \text{if } Ti + 1 \geq 0.3 \text{ sec.}$$

$$Ti + 1 = 0.3 \text{ (sec)} \qquad \text{if } Ti + 1 < 0.3 \text{ sec.}$$

$$i = 1, 2, \ldots .$$

Where PCOM represents the old accelerator (or initial) power command, PCOMN the new one. If Ti + 1 is equal to 0.3, that means the PWM signal will be kept at the high level or it will terminate in the next 0.3 sec.

For the PWM #1 scheme, it is necessary to point out that while the engine torque changes strongly from $f_1(\omega_e)$ with a positive sign to $f_2(\omega_e)$ with a negative sign, the feedback signal FDR changes smoothly since the function $F_1(\omega_e,R)$ is dependent on the function $f_1(\omega_e)$ rather than both $f_1(\omega_e)$ and $f_2(\omega_e)$ no matter what the real engine output torque is.

The PWM #1 scheme used here can insure good drivability for the deceleration operation over all conditions. This is presented in the next section.

EVALUATION OF THE SYSTEM FOR DECELERATION - Evaluation of the system is studied by dynamic simulation, i.e., integration of (1). The deceleration case is presented in Fig. 9 through Fig. 14, over different conditions. Fig. 9 through Fig. 11 show three groups of the responses of the system corresponding to 10 different step reductions of PCOM, from 1.0 to 0.0, 0.1, 0.2, ...0.9. Each of these three groups consists of the responses of both vehicle speed and engine speed. Each group has the same initial power command of 1.0, and different time points at which the decelerating operation starts. The time points starting the decelerating operation are chosen to be the 2nd sec., 5th sec., and 10th sec. from the beginning of a 100% vehicle acceleration. The reason for chosing these points is that the CVT is at different reduction conditions when the deceleration begins. The contents shown in Fig. 12 through Fig. 14 are the same as that in Fig. 9 through Fig. 11 with the exception that the initial PCOM is equal to 0.5. These conditions represent the worst cases expected. Other conditions should respond in a similar manner.

A number of schemes for deceleration operation have been studied and are partially listed in the Appendices A, B, and C at the end of this paper. They are:

A. The deceleration characteristics of a CVT control without PWM control in Appendix A. Here, the overshoot in the vehicle velocity after the acceleration command change should be noted and represents the major problem in CVT vehicle deceleration.

B. A deceleration scheme (PWM #2), in which the function $F_1(\omega_e,R)$ depends on both $f_1(\omega_e)$ if the engine output torque is positive, and $f_2(\omega_e)$ if the engine output torque is negative, listed in Appendix B. Here, the response is unacceptable.

C. A deceleration scheme (PWM #3), basically the same as PWM #1 with the exception that the engine inertia torque is considered to obtain the driver force FDR, listed in Appendix C.

Comparing the simulation results listed in Appendices A, B, and C, the scheme PWM #1 has obvious advantages, which can be listed below.

1. The response of the system, ω_e and ω_c for instance, are more consistant and smooth during the deceleration operation. The engine speed is able to go down from the old or current state into a new steady state quickly without overshoot and oscillation while the vehicle speed enters the new state in an acceptable manner.

2. The scheme for deceleration operation does not require complicated hardware or software. The construction of the control system for the decleration case is exactly the same as the one in the acceleration case provided that a fuel cutting device is available for the engine which is able to produce an alternating torque according to the PWM control signal.

3. The performance under all conditions are almost equal to that of PWM scheme #3, which has the disadvantage that the input torque of the CVT has to be measured. Thus simple speed sensors can be used to monitor the vehicle condition.

SENSITIVITY WITH RESPECT TO THE ENGINE

As was discussed above, the software of the system requires programming the function $F_1(\omega_e,R)$ which is dependent on the normal engine operating line. It is natural to speculate what would be the change in response of the vehicle and engine if the characteristics of the engine changes around the normal operating line (OOL). This may occur due to engine tolerance, cold or warm and age. The programmed function $F_1(\omega_e,R)$ based on the normal OOL does not change. In order to answer this question, the OOL characteristics are intentionally changed by modifying the function $f_1(\omega_e)$. A coefficient CT, an engine scale factor, is inserted. The simulation results of the system with a varying engine scale factor are presented for both cases of acceleration and deceleration in Fig. 15 through Fig. 18.

The results show that the sensitivity of the system with respect to the variation of the engine operating line is not large, if CT is within the limit (80%, 120%). On the other hand, if CT is out of this range, then a suitable scheme has to be selected in order to maintain good drivability. For instance, the function $F_1(\omega_e,R)$ may be reprogrammed based on the new OOL or the system gains K_1 or K_2 may be readjusted.

CONCLUSIONS

The control system of a continuously variable transmission vehicle has been simulated in the cases of both acceleration and deceleration operations. The parameters of the control system, K_1 through K_4, were selected by using an optimum technique with a quadratic cost function to insure good controllability and stability not only

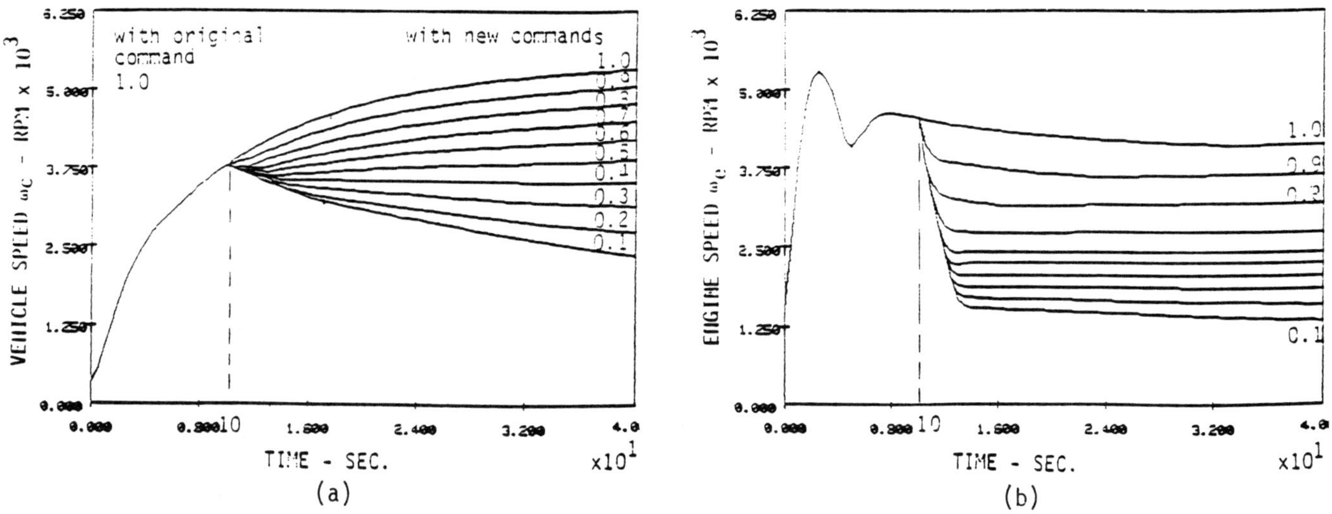

Fig. 9 Deceleration characteristics of the vehicle at the 10th sec. with PCOM = 1.0 using PWM #1

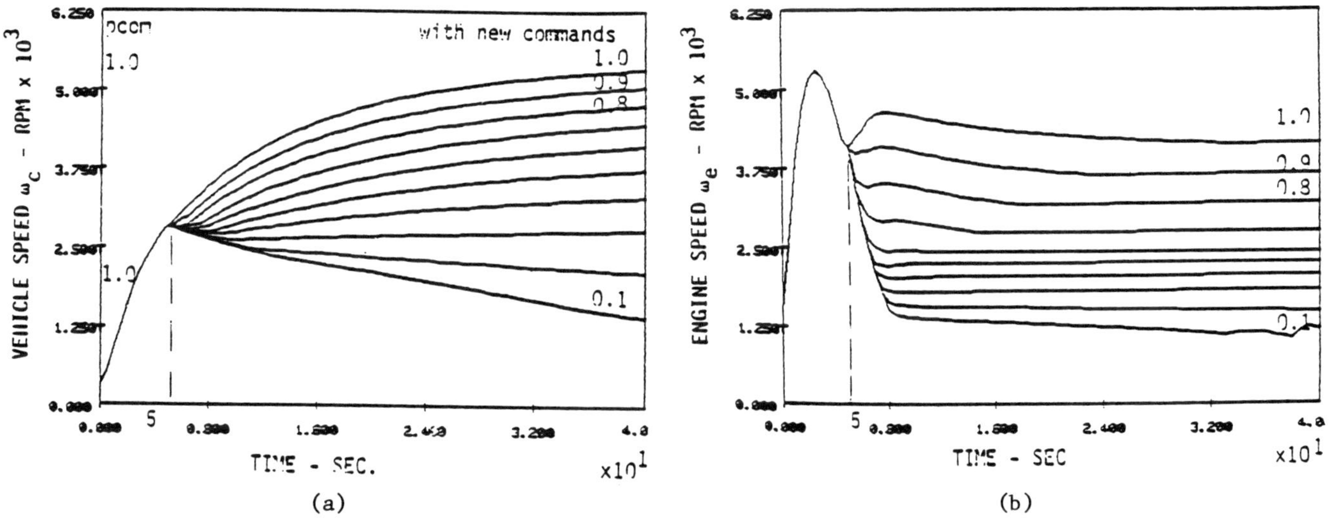

Fig. 10 Deceleration characteristics of the vehicle at the 5th sec. with PCOM = 1.0 using PWM #1

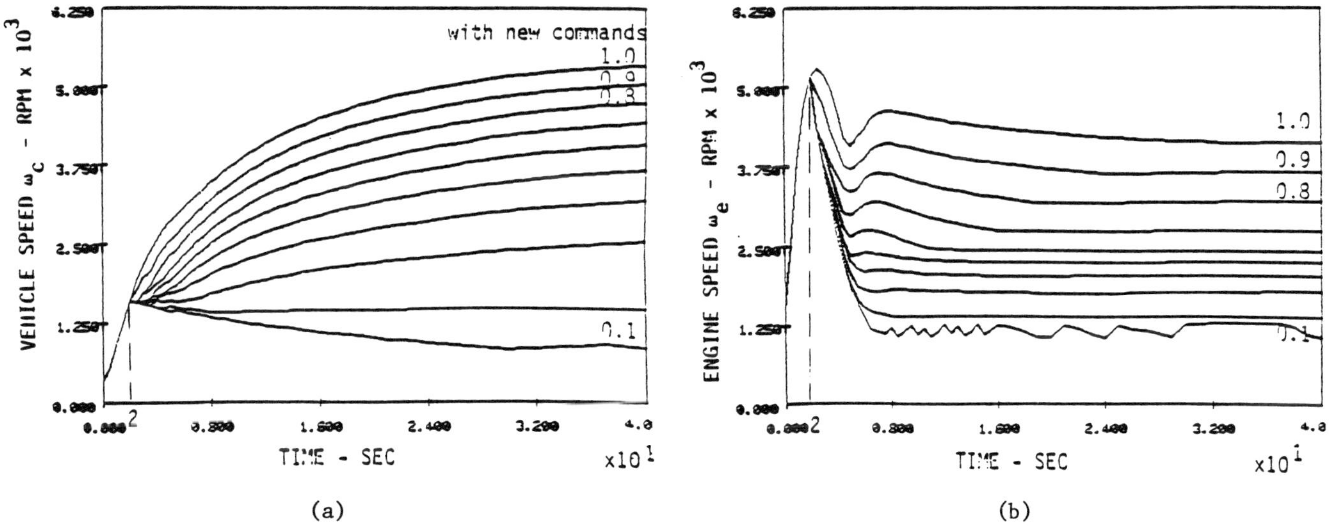

Fig. 11 Deceleration characteristics of the vehicle at the 2nd sec. with PCOM = 1.0 using PWM #1

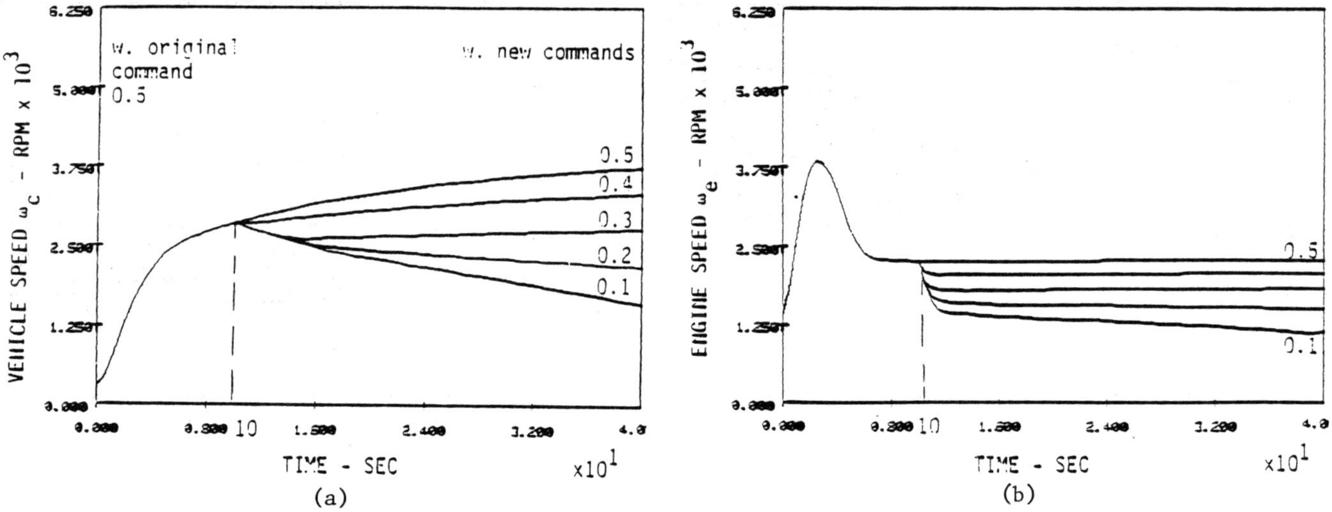

Fig. 12 Deceleration characteristics of the vehicle at the 10th sec. with PCOM = 0.5 using PWM #1

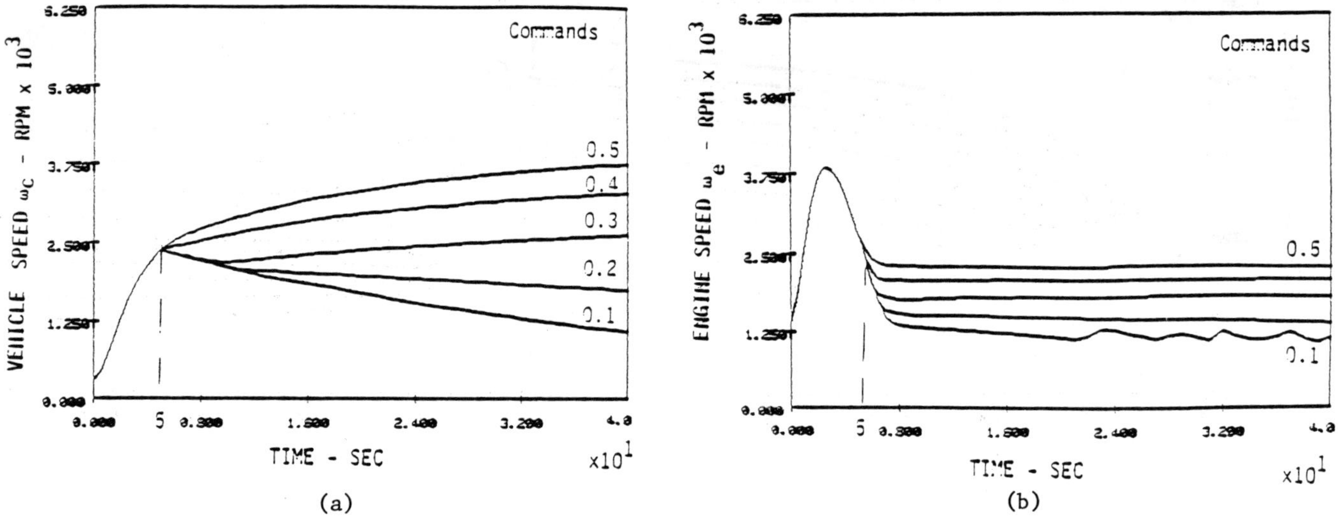

Fig. 13 Deceleration characteristics of the vehicle at the 5th sec. with PCOM = 0.5 using PWM #1

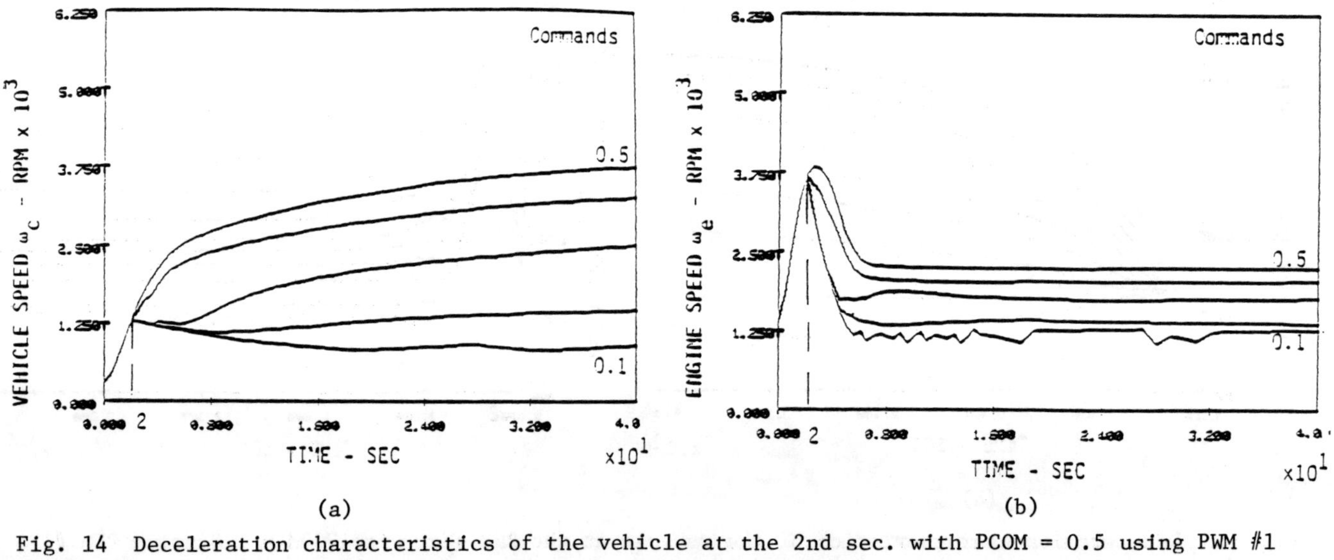

Fig. 14 Deceleration characteristics of the vehicle at the 2nd sec. with PCOM = 0.5 using PWM #1

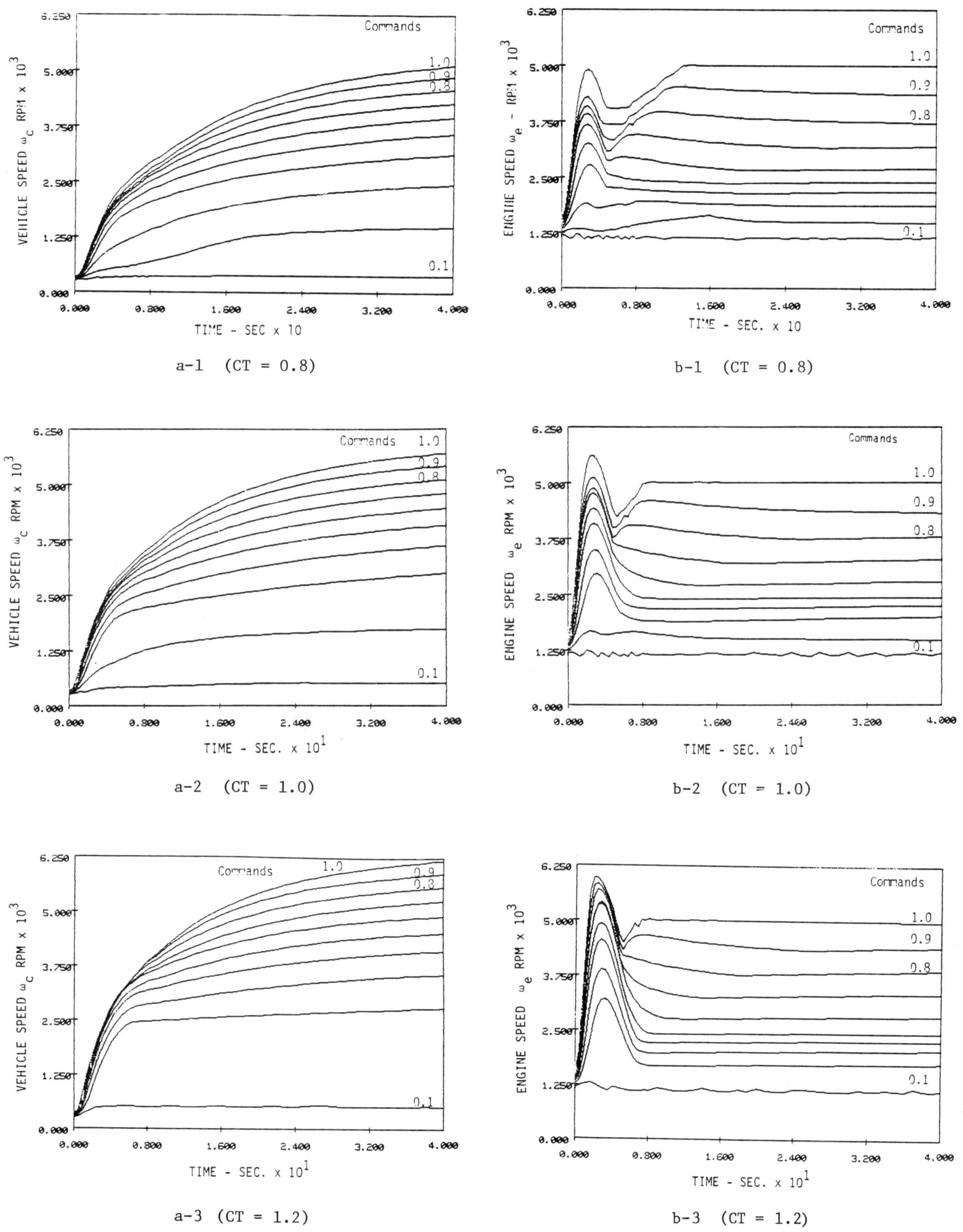

Fig. 15 Sensitivity of the system W.R.T. engine characteristics for acceleration operation

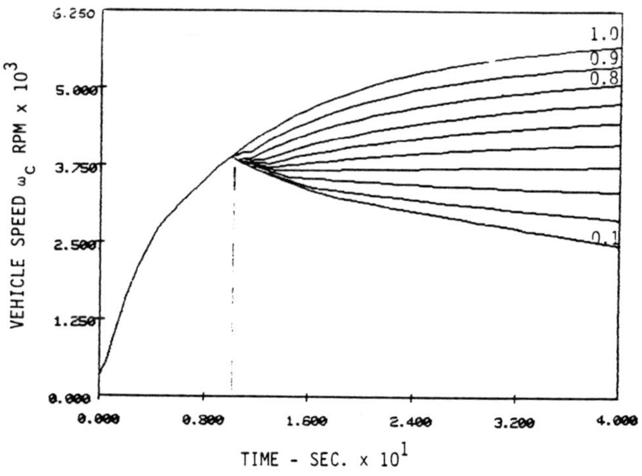

a-1 (CT = 0.8)

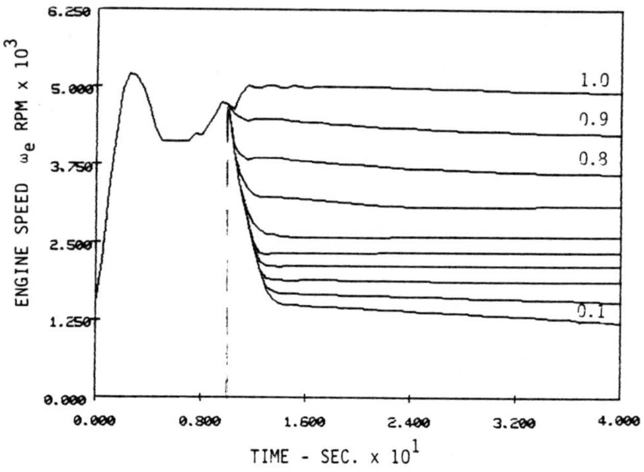

b-1 (CT = 0.8)

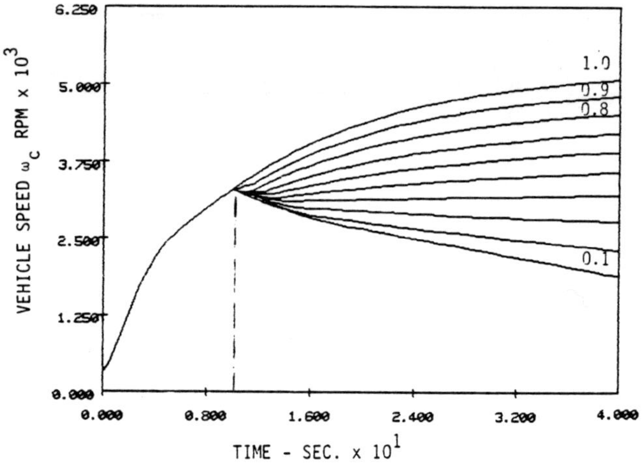

a-2 (CT = 1.0)

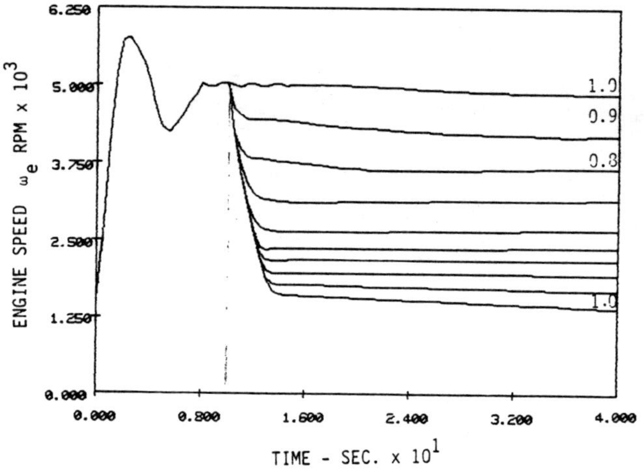

b-2 (CT = 1.0)

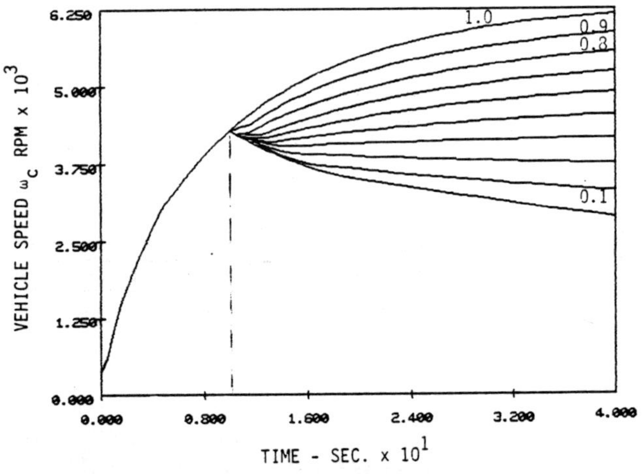

a-3 (CT = 1.2)

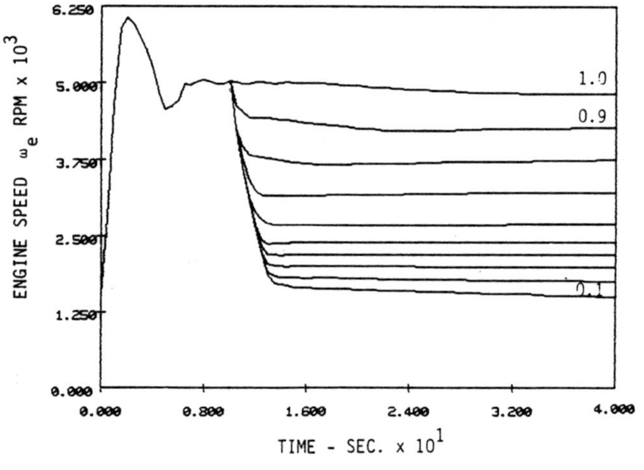

b-3 (CT = 1.2)

Fig. 16 Sensitivity of the system W.R.T. engine characteristics for deceleration operation at the 10th sec. with PCOM = 1.0

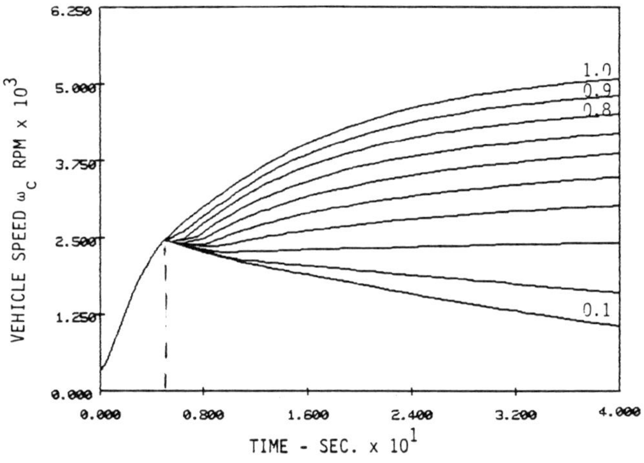

a-1 (CT = 0.8)

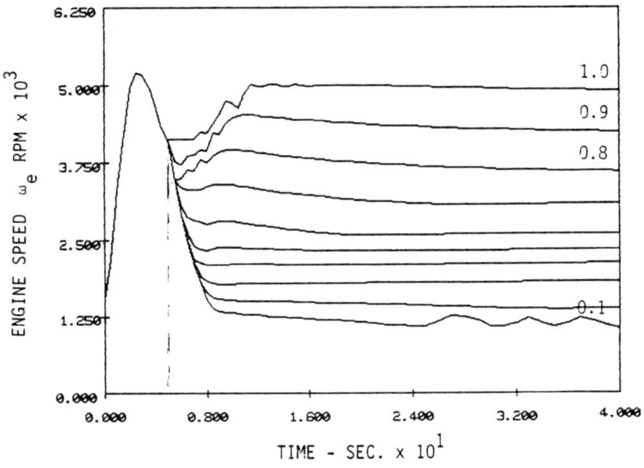

b-1 (CT = 0.8)

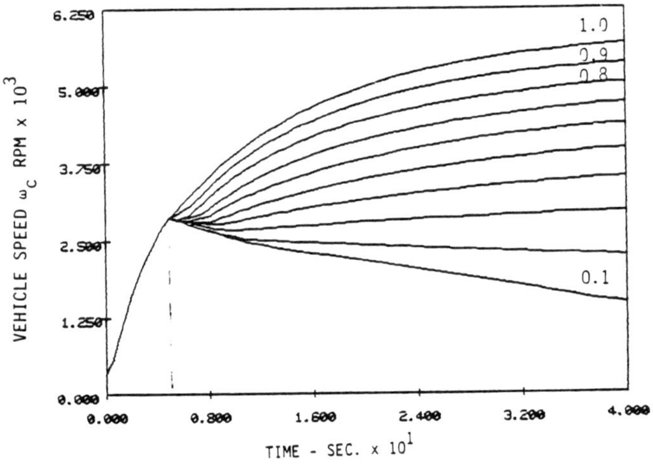

a-2 (CT = 1.0)

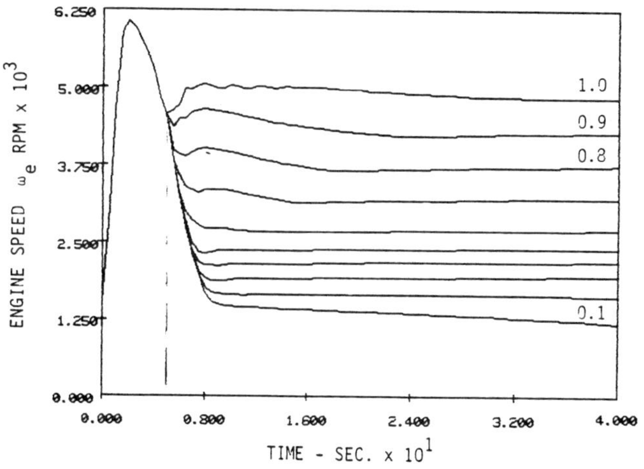

b-2 (CT = 1.0)

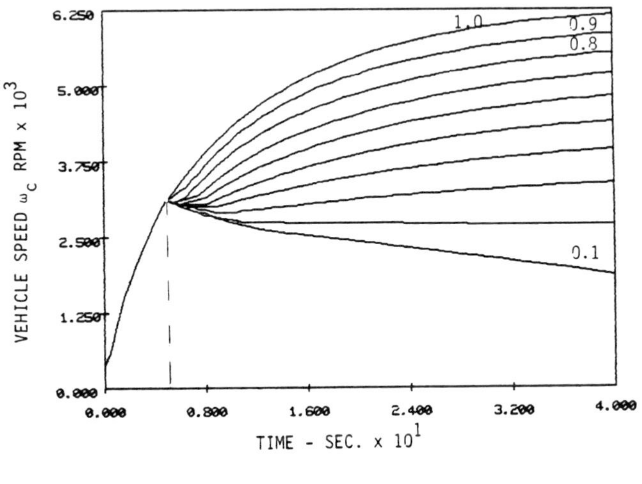

a-3 (CT = 1.2)

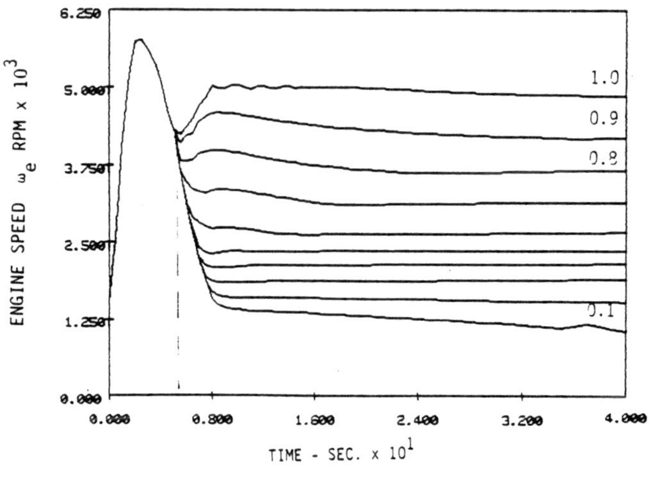

b-3 (CT = 1.2)

Fig. 17 Sensitivity of the system W.R.T. engine characteristics for deceleration operation at the 5th sec. with PCOM = 1.0

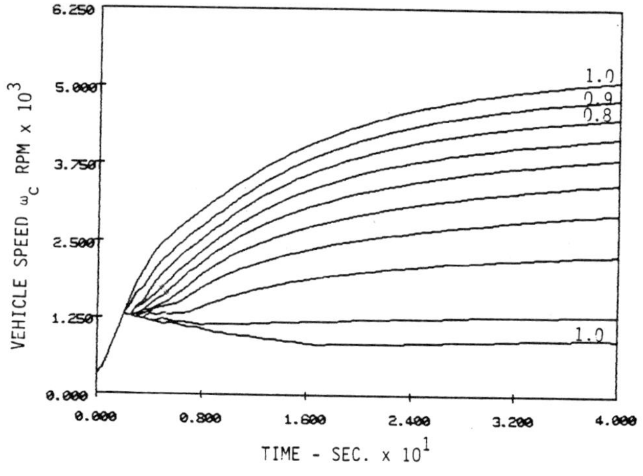

a-1 (CT = 0.8)

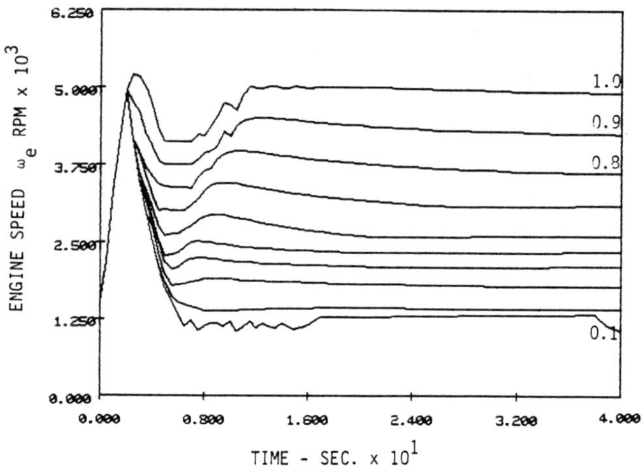

b-1 (CT = 0.8)

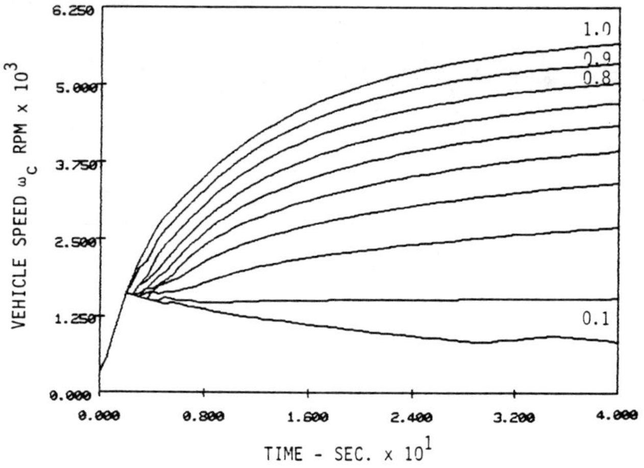

a-2 (CT = 1.0)

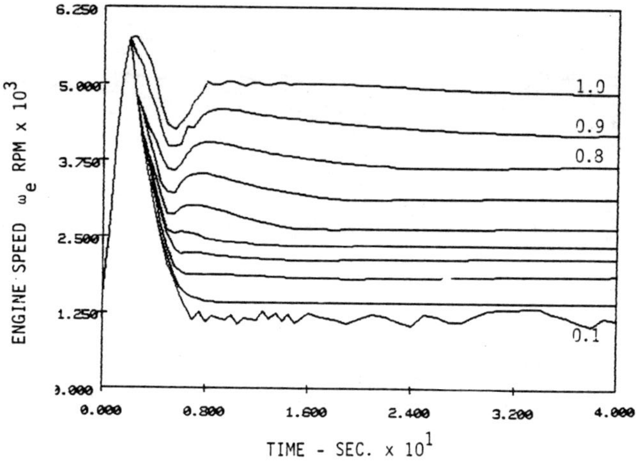

b-2 (CT = 1.0)

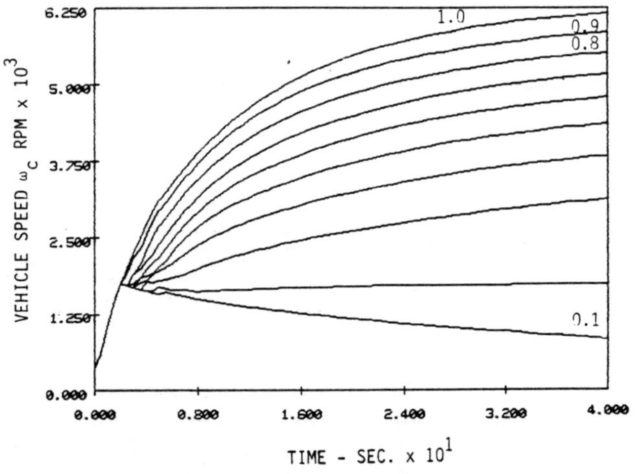

a-3 (CT = 1.2)

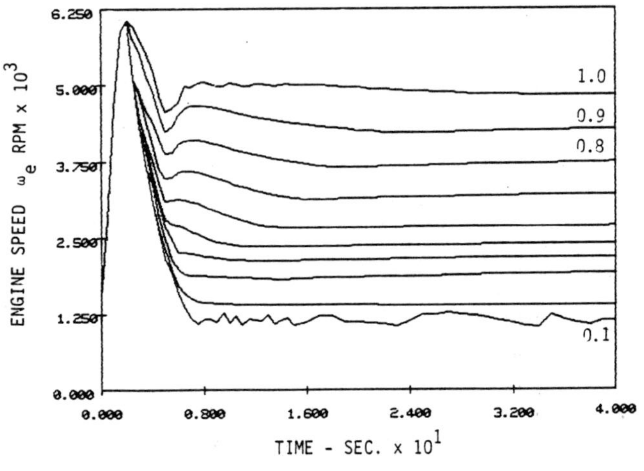

b-3 (CT = 1.2)

Fig. 18 Sensitivity of the system W.R.T. engine characteristics for deceleration operation at the 2nd sec. with PCOM = 1.0

for the acceleration operation but also for the deceleration operation where an appropriate pulse-width-modulated control scheme is used. The candidate scheme PWM #1 is one set of many possible deceleration schemes which can take the system from any current state into a new state in an acceptable way. The concept of fuel shut off to reduce torque works because we are using the vehicle inertia to provide a smooth ride. Also the engine inertia is as small as possible to further aid in the transition. Finally, the engine's power is reduced without throttling making the engine the most efficient possible.

PWM #1 is able to provide a good performance with a simple implementation under all possible operating conditions. For the PWM #1 scheme, it is necessary to point out again that while the engine torque changes strongly from $f_1(\omega_e)$ with a positive sign to $f_2(\omega_e)$ with a negative sign, the feedback signal FDR changes smoothly since the function $F_1(\omega_e,R)$ is dependent on the function $f_1(\omega_e)$ no matter what the real engine output torque is.

There seems to be a shortcoming due to the software dependency on the characteristics of the true engine operating line. But it is not serious and can be solved by simple design patches.

NOMENCLATURE

CT	= Engine operating line scale factor to simulate different engine performance such as cold to hot engine.
$f_1(\omega_e)$	= Positive engine torque-speed characteristcs to provide optimum engine performance
$f_2(\omega_e)$	= Negative engine torque-speed characteristics when fuel is shut off but throttle is in accordance with $f_1(\omega_e)$
$F_1'(T_{in},R)$	= Engine torque and ratio characteristics of belt sheeve system (FDR)
$F_1(\omega_e,R)$	= Engine speed and ratio dependent loading effects on the driver pulley
$F_2(FDR,R)$	= Belt-sheeve characteristics
FDN	= Driven sheeve force
FDN_2	= Driven sheeve command for regulating engine speed below maximum or above minimum
FDR	= Driver sheeve force
I_c	= Equivalent car moment of inertia reflected to the drive shaft
I_e	= Engine moment of inertia
K_1	= Control gain of the TCOM
K_2	= Feedback gain
K_3	= Minimum engine speed regulation gain
K_4	= Maximum engine speed regulation gain
OOL	= Optimum operating line $f_1(\omega_e)$
PCOM	= Power command
PWM #i	= The i th pulse width modulation scheme
R	= CVT ratio input/output
$\dot{R}$	= Rate of change of CVT ratio
R_{max}	= Maximum ratio of the CVT (reduction)
R_{min}	= Minimum ratio of the CVT (overdrive)
T_0	= Minimum engine "on" time to insure a even number of engine firings
T_1	= First engine "off" time
TCOM	= Torque command
T_E	= Engine torque
$T.F_{CVT}$	= Transfer characteristics of CVT
T_{i+1}	= Succeeding engine "off" times
$T_{L.C}$	= Torque losses in the transmission and drive line
$T_{L.R}$	= Road load losses at steady state speeds
T_{in}	= Input torque of CVT
T_w	= Torque of the wheels
ω_c	= Speed of the car at the driveshaft
ω_e	= Engine speed
ω_{sw}	= Switching speed
ω_w	= Speed of the wheels

ACKNOWLEDGGEMENT

We wish to thank AISIN SEIKI CO., LTD. for their support of this research.

REFERENCES

1. Chan, C., Yang, D., Volz, T., Breitweiser, D., Jamzadeh, F.S., Frank, A., and Omitsu, T., "System Design and Control Considerations of Automotive Continuously Variable Transmissions," SAE 840048

2. Yang, D., and Frank, A., "An Optimization Technique for the Design of a Continuously Variable Transmission Control System for Automobiles," in review at ASME.

3. Bents, David J., "Axial Force and Efficiency Test of the Fixed Center Variable Speed Belt Drive," NASA Lewis Research, SAE 810103.

APPENDIX A - DECELERATING A VEHICLE WITH THIS CVT CONCENT WITHOUT ANY PWM CONTROL

The block diagram of the central system shown in Fig. 5 was directly used for deceleration operation without any secondary control in this case. Whenever the driver wants to decelerate the vehicle, he need only reduce the power command, PCOM, by moving the accelerator pedal. Partial simulation results of this scheme are shown in Fig. 19-24. The contents shown in these figures are based on the same conditions as those shown in Fig. 9-14 with the exception that both come from different control schemes. From these results, the following conclusions can be made:

1. The desired feature of this kind of control scheme is that whenever deceleration is required, the engine speed goes into a new steady state quickly. This feature is independent of both old and new power commands.

2. The problem of this kind of control is that there is an overshoot of the vehicle speed, after the operator command is made. Particularly it is most serious in the case that the starting point of the deceleration is close to the beginning of the original acceleration, i.e., the 2nd or 5th sec. These results are shown in Fig. 20 and Fig. 21, Fig. 23 and Fig. 24.

3. In order to solve the overshoot problem and retain the desired features, a new scheme has to be found.

APPENDIX B - DECELERATING OPERATION OF SCHEME PWM #2

The control scheme, PWM #2 is the same as PWM #1 with the main exception that the function $F_1(\omega_e,R)$ for control scheme PWM #1, was dependent on $f_1(\omega_e)$ rather than $f_2(\omega_e)$ regardless of the engine output torque. However, the function $F_1(\omega_e,R)$ was chosen to depend on both $f_1(\omega_e)$ and $f_2(\omega_e)$ according to the sign of the engine output torque for the scheme PWM #2. That means the function $F_1(\omega_e,R)$ has two forms corresponding to both positive engine torque and negative engine torque, respectively. Partial simulation results of the scheme PWM #2 are shown in Fig. 25-27.

From these results some conclusions are listed as follows:

1. PWM #2 was able to control the vehicle speed significantly. For some cases, it would overcome the overshoot problem associated with the scheme shown in Appendix A.

2. But the engine does not go to a new state quickly for some cases which results in bad performance of the CVT-vehicle system.

3. It seems that this scheme is not acceptable because the feedback signal FDR changes incorrectly due to the form of the function $F_1(\omega_e,R)$ during deceleration.

APPENDIX C - DECELERATING OPERATION SCHEME PWM #3

The difference among three kinds of PWM control schemes is in how the function $F_1(\omega_e,R)$ is defined. The definition of the functions $F_1(\omega_e,R)$ for both PWM #1 and #3, are almost the same. The slight difference between these two is simply the consideration of the torque loss of engine inertia in establishing the function $F_1(\omega_e,R)$ based on the data map shown in Fig. 4 The scheme PWM #1 does not consider loss but PWM #3 does. Partial simulation results of the scheme PWM #3 are shown in Fig. 28-30.

It is very clear that the performance of both PWM #1 and PWM #3 are almost the same when comparing the results shown in Fig. 9-14 for PWM #1 and Fig. 28-30 for PWM #3. This is because the torque loss of the engine inertia is small compared with the engine torque in most cases. The scheme PWM #1 is perferred rather than PWM #3 because a real input torque of the CVT must be measured for the scheme PWM #3. This is not necessary for the scheme PWM #1, thus PWM #1 is the preferred scheme.

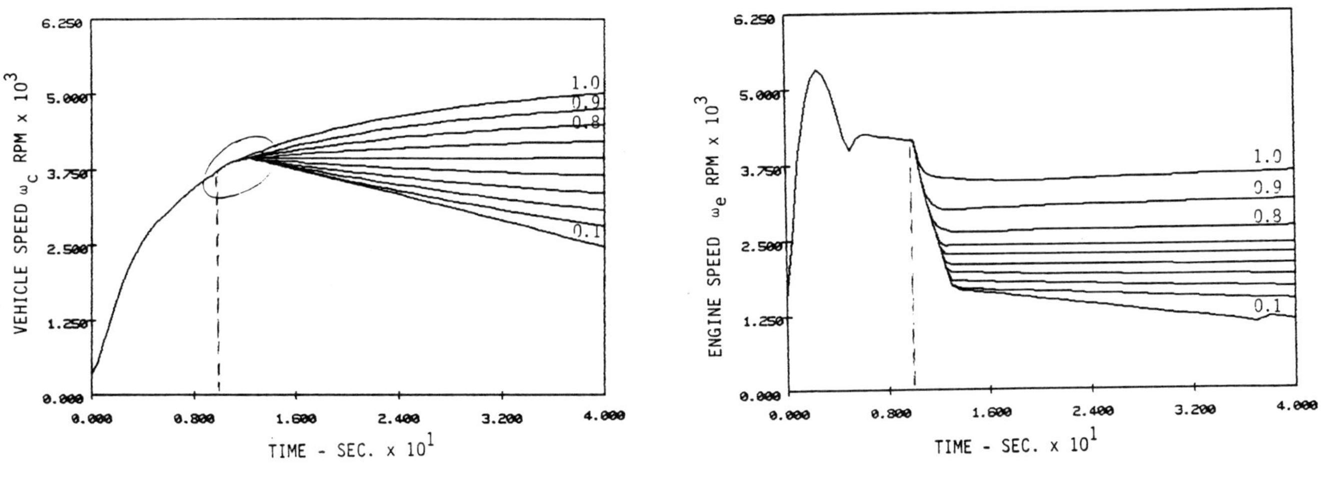

(a) (b)

Fig. 19 Deceleration characteristics of the vehicle at the 10th sec. with PCOM = 1.0 w/o PWM

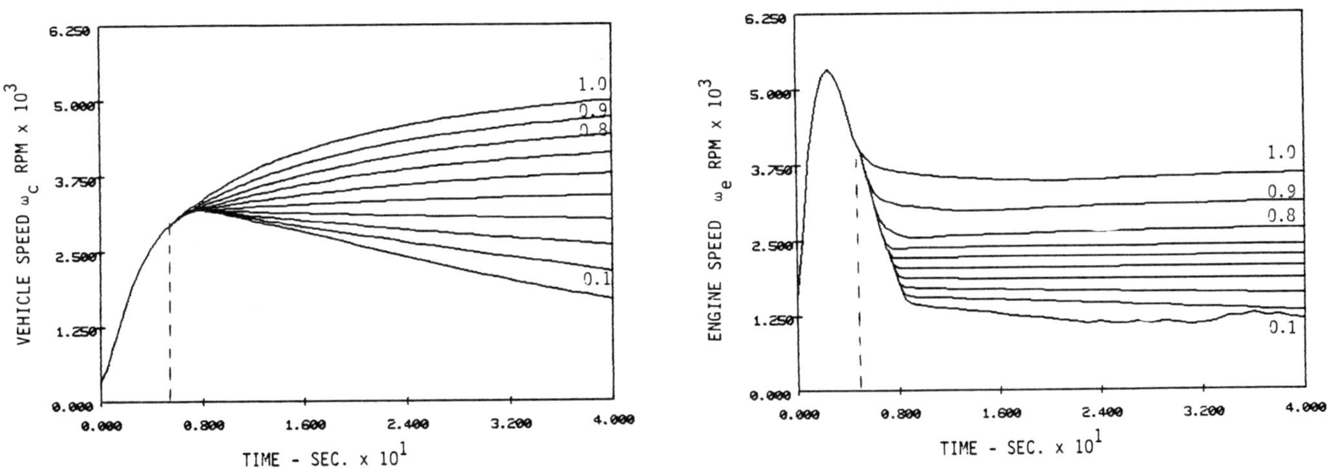

(a) (b)

Fig. 20 Deceleration characteristics of the vehicle at the 5th sec. with PCOM = 1.0 w/o PWM

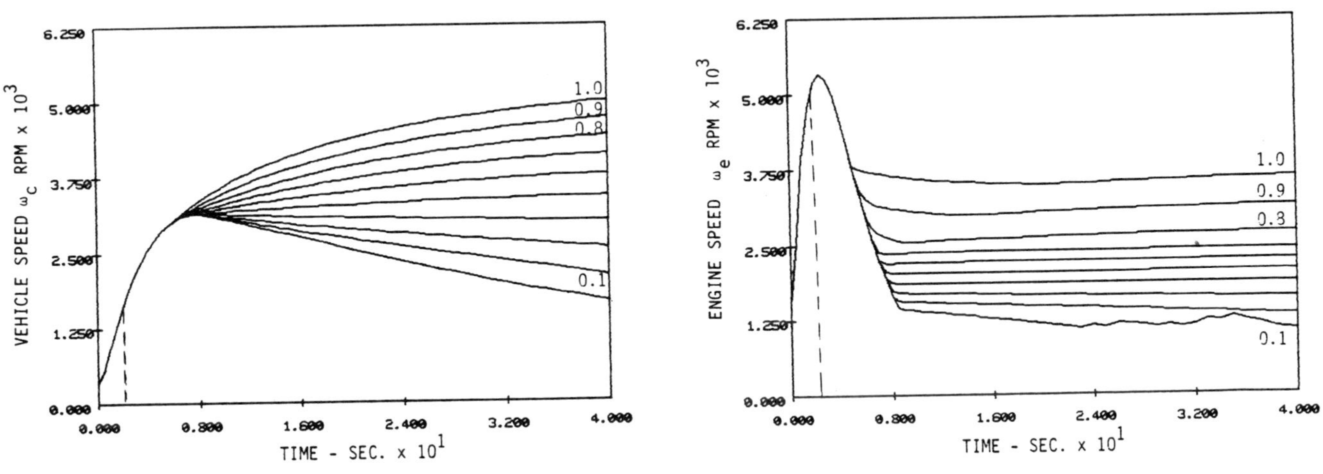

(a) (b)

Fig. 21 Deceleration characteristics of the vehicle at the 2nd sec. with PCOM = 1.0 w/o PWM

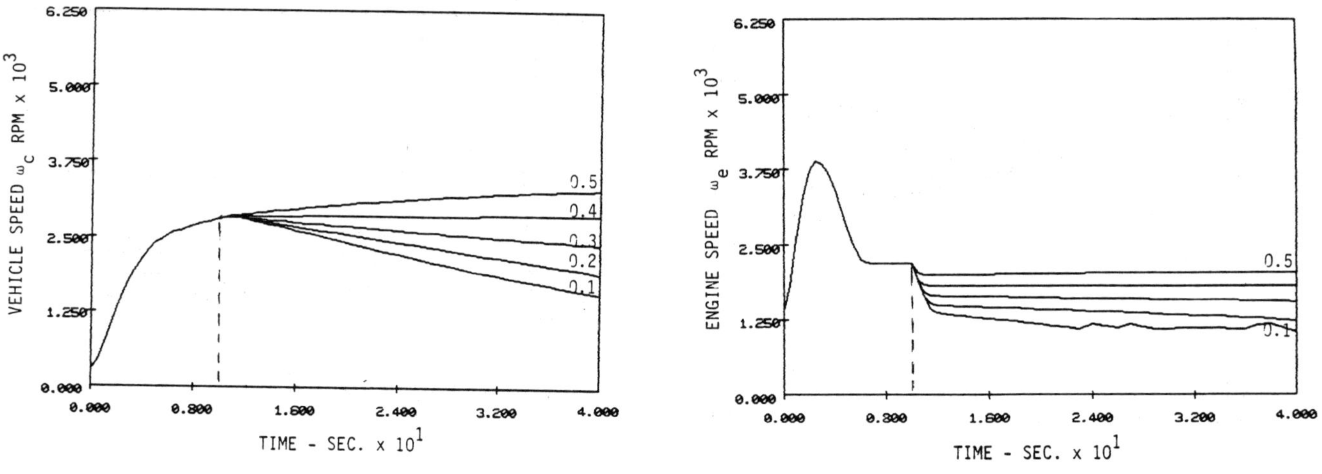

(a) (b)

Fig. 22 Deceleration characteristics of the vehicle at the 10th sec. with PCOM = 0.5 w/O PWM

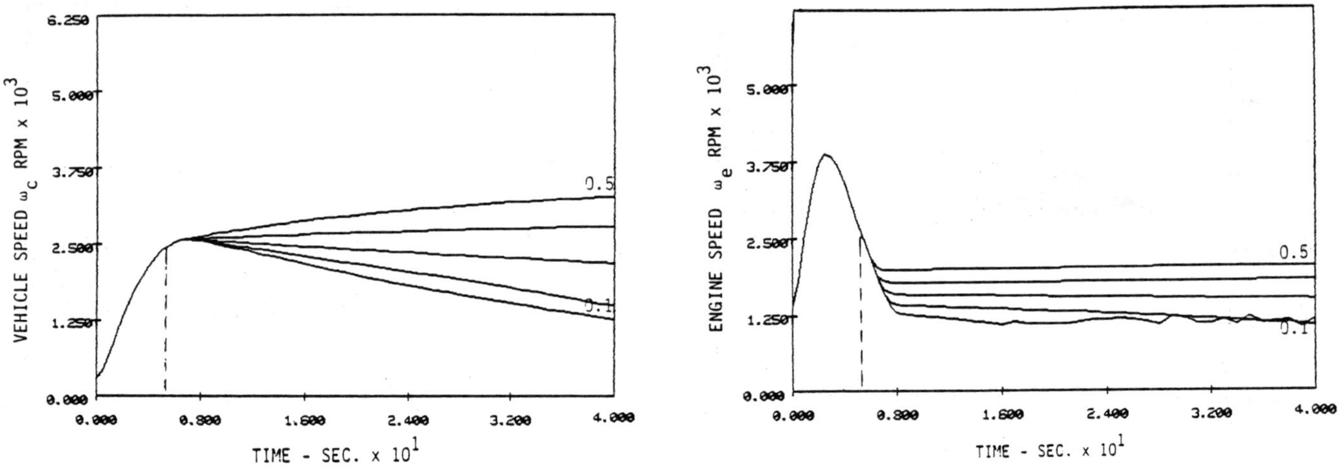

(a) (b)

Fig. 23 Deceleration characteristics of the vehicle at the 5th sec. with PCOM = 0.5 w/o PWM

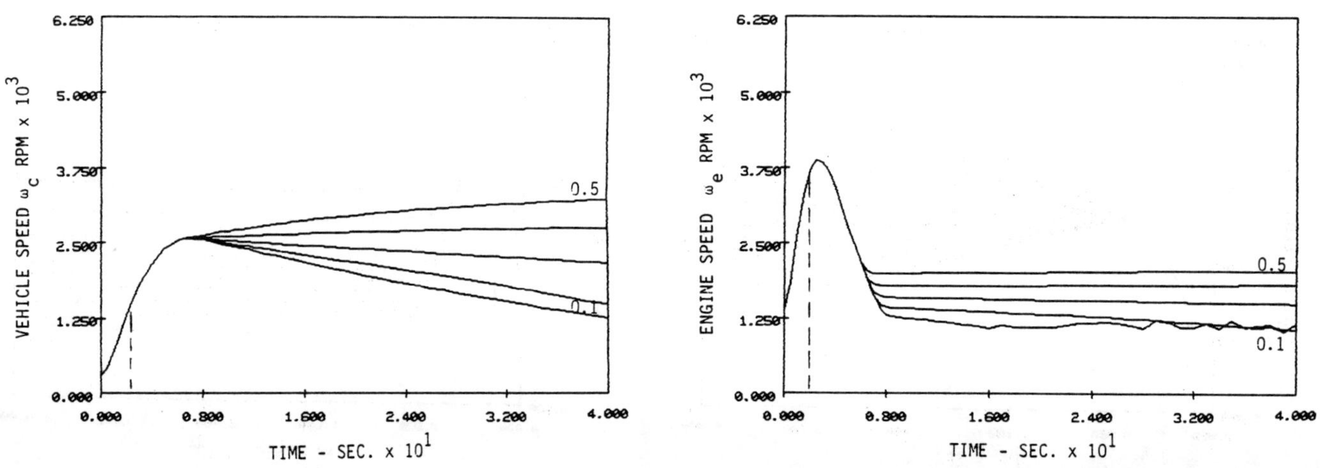

(a) (b)

Fig. 24 Deceleration characteristics of the vehicle at the 2nd sec. with PCOM = 0.5 w/o PWM

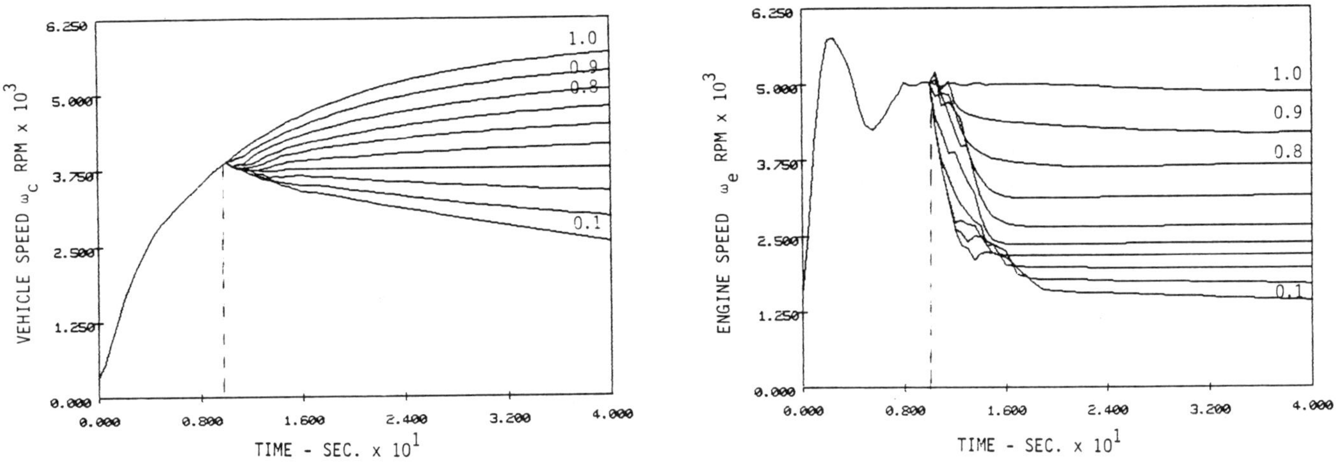

(a) (b)

Fig. 25 Deceleration characteristics of the vehicle at the 10th sec. with PCOM = 1.0 using PWM #2

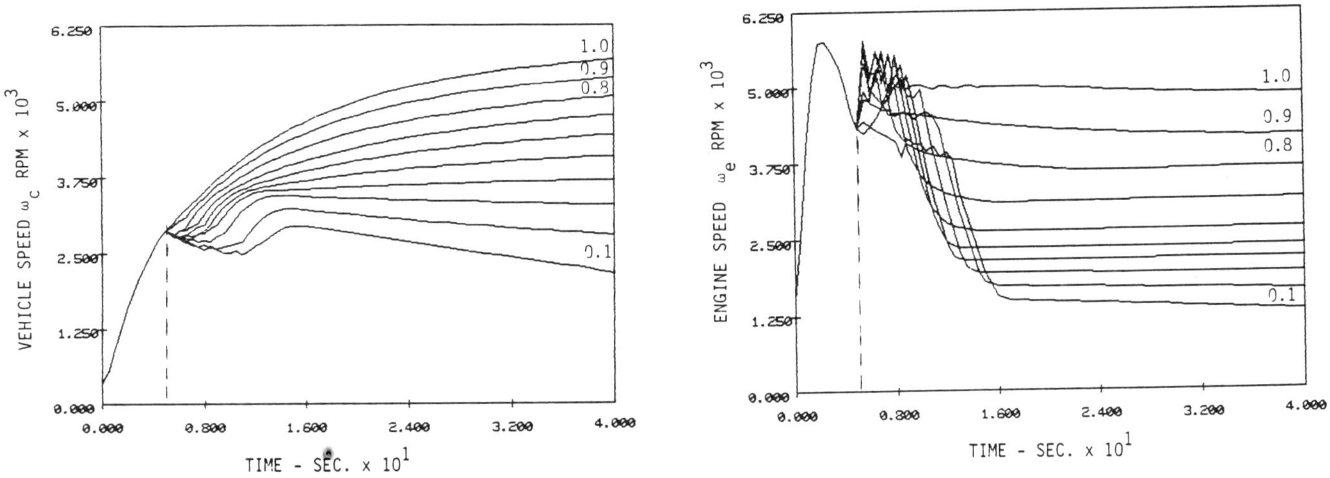

(a) (b)

Fig. 26 Deceleration characteristics of the vehicle at the 5th sec. with PCOM = 1.0 using PWM #2

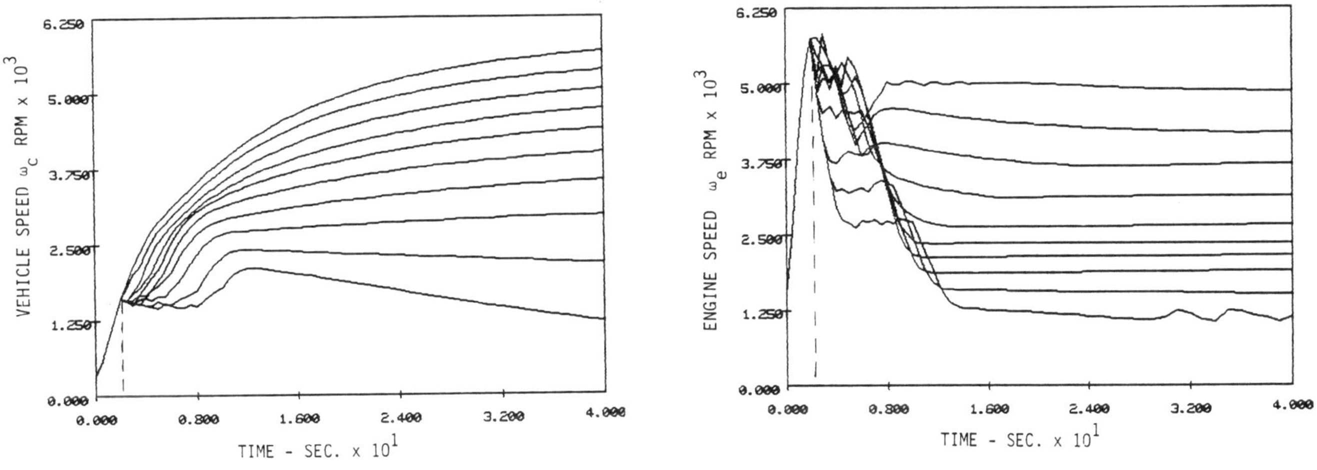

(a) (b)

Fig. 27 Deceleration characteristics of the vehicle at the 2nd sec. with PCOM = 1.0 using PWM #2

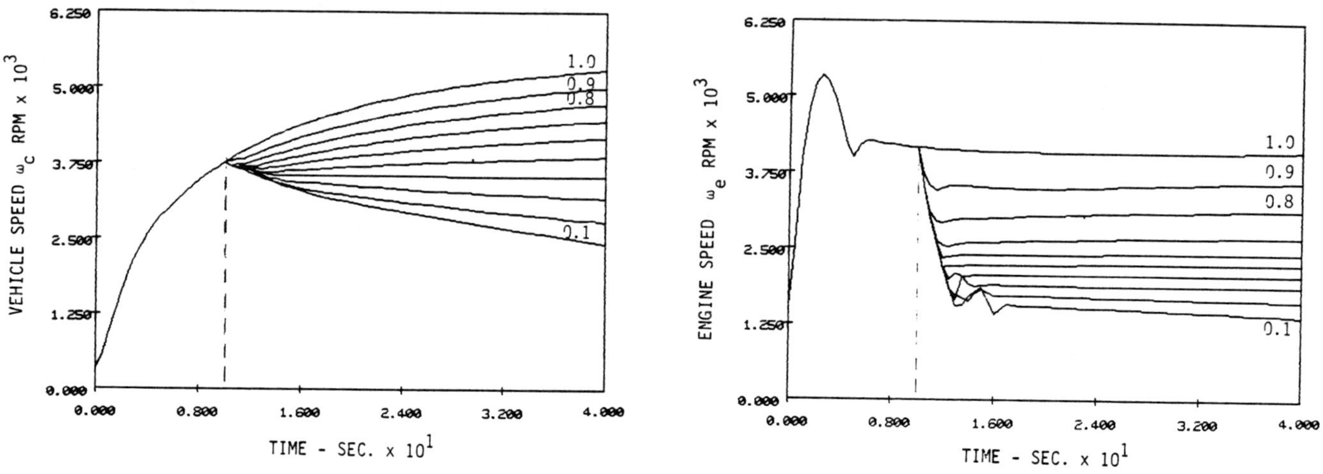

(a) (b)

Fig. 28 Deceleration characteristics of the vehicle at the 10th sec. with PCOM = 1.0 using PWM #3

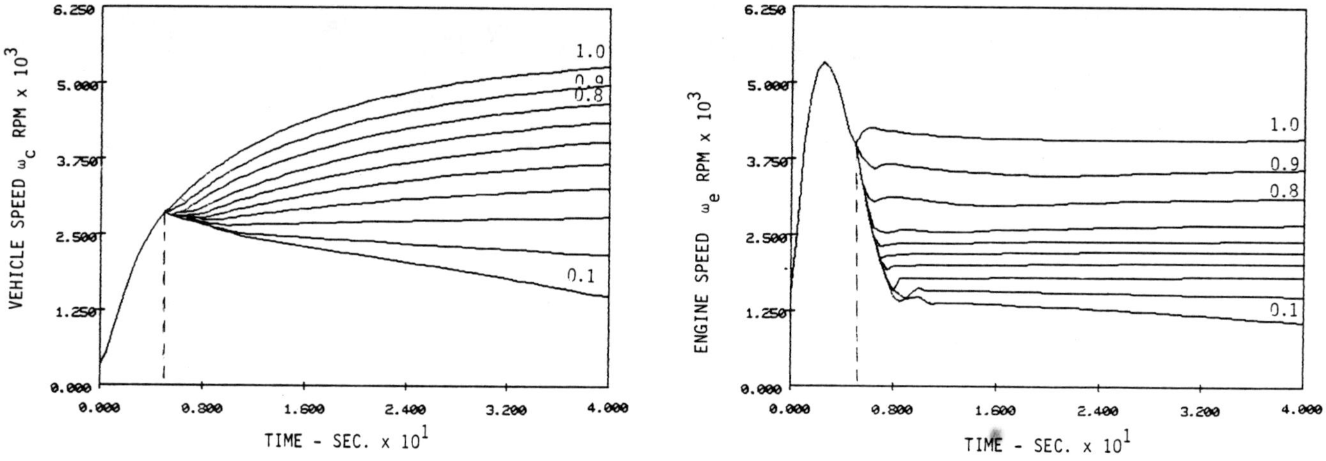

(a) (b)

Fig. 29 Deceleration characteristics of the vehicle at the 5th sec. with PCOM = 1.0 using PWM #3

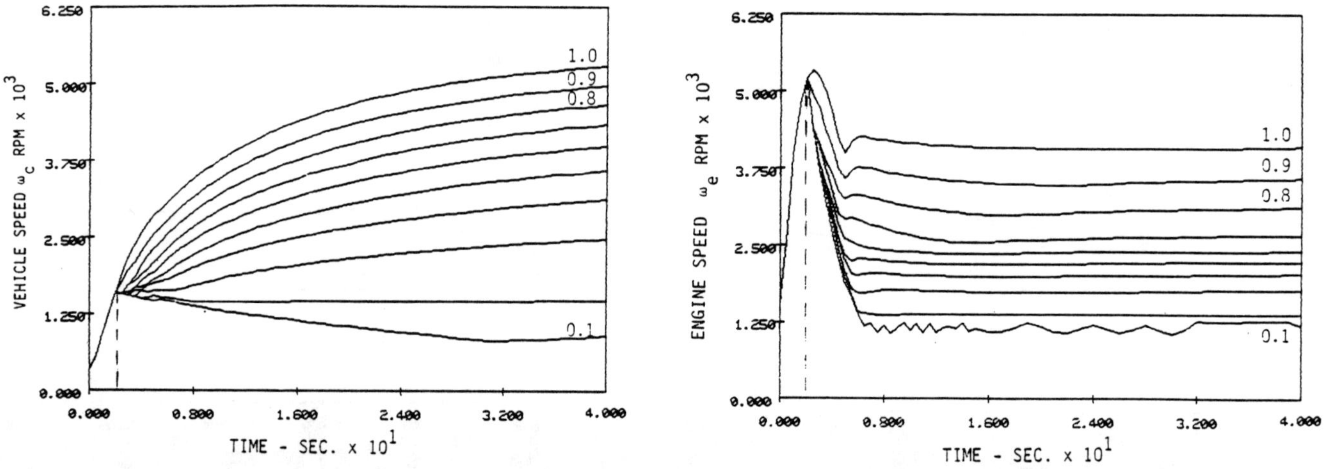

(a) (b)

Fig. 30 Deceleration characteristics of the vehicle at the 2nd sec. with PCOM = 1.0 using PWM #3

Metal Belt or Chain Drives

The Ford Research Dual Mode Continuously Variable Transmission*

T. R. Stockton
Ford Motor Co.
Dearborn, MI, USA

**Paper 841305 presented at the International Conference on Fuel Efficient Power Trains and Vehicles, London, England, October, 1984.

ABSTRACT

The familiar continuously variable steel belt transmissions being introduced for automotive use are "straight" CVT's, where total speed ratio span is limited to capability of the variable speed belt and pulley system. Meeting start-up performance and gradeability requirements penalizes the final drive or axle ratio resulting in the transmission reaching the maximum overdrive ratio prematurely.

The Ford Research front wheel drive Dual Mode CVT employs a torque converter for start-up in addition to a full range of CVT operation as a means of improving both smoothness and fuel economy.

The torque converter and a separate chain drive power path from the engine to the variable speed unit are arranged in parallel. This combination provides an approximate 30% effective increase in ratio coverage to 7:1. Vehicle fuel economy is thus improved an estimated 5-9% during highway driving because the final drive ratio can be reduced 30%. Vehicle start-up through the torque converter is inherently smooth.

A 4 x 4 system is also derived from the base FWD, having a bevel gear power take-off near the vehicle center.

INTRODUCTION

This paper describes the application of a torque converter to an automotive CVT as a means of extending overall ratio range for improving fuel economy, while also providing excellent start-up feel. The limitations of speed ratio span of candidate variable speed belt and pulley systems without some scheme of ratio augmentation causes an area of vehicle operation where the transmission ratio remains in the max. overdrive position and consequently the engine operates at speeds above the optimum for fuel economy.

Transmission design and control strategies, with vehicle test results of the Dual Mode CVT concept have contributed to the understanding of trade-offs between fuel economy improvement and design factors.

SYSTEM CONCEPT AND KEY FEATURES

The Dual Mode CVT employs a torque converter for start-up in addition to a full range of CVT operation as a means of maximizing both smoothness and fuel economy. This is a two-axis system, shown schematically on Fig. 1. The torque converter and variable speed drive pulley are located on the engine centerline with the driven pulley, the planetary forward-reverse reduction gearing and differential on the drive wheel axis.

The design cross section (Fig. 2) is sized for 136 N-M torque capacity. The prototype powertrain assembly shown on Fig. 3 has been installed in a 1.6L Ford Escort for test and evaluation. Specifications for the dual mode CVT are listed below.

Ratios (N_I/N_O)	
Chain	2.241
Belt	
Max. Underdrive	2.42
Max. Overdrive	.41
Range	5.90
Final Drive	
Forward	3.871
Reverse	3.452
N/V In Escort (Engine Rev./Min./Vehicle Speed Km/Hr)	
Converter Mode	80.03
Belt Mode	
Max Underdrive	86.44
Max Overdrive	14.76
Torque Capacity (Engine)	136 N.M.
Dimensions	
Pulley Radii	33.27-80.52 mm
Center Distance	226.82 mm
Torque Converter	
Stall Ratio	2.1
Stall K $\frac{\text{Engine Rev/Min.}}{\sqrt{\text{Engine Torque N.M.}}}$	202

The steel "V" belt is an experimental development provided by the Morse Chain Division of Borg Warner Corp.

The torque converter chain drive power path is arranged in parallel with the variable speed unit. This combination provides an approximate 30% effective increase in ratio coverage to 7:1 overall for the transmission. The final drive ratio can be accordingly reduced 30% without compromising performance and gradeability.

Vehicle start-up through the torque converter is inherently smooth. Near the converter coupling point a power transfer clutch is engaged during acceleration, transferring power flow from the converter to the variable speed belt drive. The chain transfer drive ratio is set approximately equal to the pulley underdrive ratio so that at heavy-to-full throttle the transfer clutch is approximately synchronous with the input side of the variable ratio pulley drive, which is in its maximum underdrive ratio position. At light-to-medium throttle, the reduced converter speed ratio results in a small engine speed drop after the shift.

As vehicle speed increases above the power transfer point, the pulley system strokes from maximum underdrive towards maximum overdrive position. An overrunning clutch unloads the converter in this mode. It also permits a conventional forced downshift and a rapid change in ratio for passing maneuvers. Smoothness of the upshifts and downshifts is assured by the overrunning clutch on the chain transfer drive sprocket.

Hydraulic pressure to activate the three clutches, to provide the pulley clamping force and to charge the converter and lubrication system is provided by a variable displacement vane type oil pump, which is engine driven through a quill shaft.

CONTROL STRATEGY

The primary objective of the control system is to force the engine to operate at the most fuel efficient combination of RPM and torque for any power level. Figure 4 is a map of the 1.6L engine for the vehicle, showing a family of constant RPM lines plotted aginst B.S.F.C. and horsepower. The lowest B.S.F.C. point for each of these fish hook curves is identified by the corresponding throttle opening. Thus for any power demand level the control must seek the optimum RPM. The vehicle speed can vary at constant engine RPM to accommodate hills or accelerations. At closed throttle the engine speed drops to the max. overdrive value, and continues to drop with vehicle speed to idle RPM. The transmission then downshifts into converter mode.

Figure 5 shows the relationship between the 1.6L engine and vehicle speeds in an Escort for various throttle positions. N/V lines for the Ford ATX high gear and equivalent straight CVT in maximum overdrive are included for reference. Assuming a minimum engine speed of 1200 RPM, it can be seen that at road load condition, the dual mode CVT upshifts to maximum overdrive at 80 km/h. As vehicle speed increases the CVT gradually downshifts to meet road load power requirements. Note that the straight CVT reaches limiting overdrive ratio positon at a much lower vehicle speed.

During a typical start-up acceleration at 20° throttle, engine speed will rise to 2100 RPM and fall to 1700 RPM at the power transfer point. It will then hold constant RPM and power to 103 km/h. At larger throttle angles the engine speed change at the power transfer point diminishes toward zero, consistent with increasing converter speed ratio. At full throttle, engine speed in the CVT mode holds constant to maximum vehicle speed. At this point the variable pulley system has shifted from maximum underdrive to approximately 1:1 ratio.

It can be seen that the CVT is under full variable control at all vehicle speeds but that ratio range beyond that available would be of little value for this particular engine and vehicle combination.

In Fig. 6 the desired engine RPM vs. throttle opening is conveniently approximated by a straight line. Referring back to Fig. 4 the engine RPM is not overly critical for best economy and precise control of ratio in the belt mode is not necessary. This also simplifies control of the transfer clutch. As shown on Fig. 7 the clutch apply and release schedules are straight line functions of belt output pulley RPM, or road speed, and throttle angle. Above 70° throttle the forced downshift is made easily available for performance. The corresponding upshift line must be located slightly higher to maintain hysteresis between shifts. A broader hysteresis band exists at lower throttle angles, to minimize upshift-downshift business.

BELT LOADING

Belt clamping load is provided by hydraulic pressure acting upon each of the two moving pulley cones. The required belt loading to insure efficient operation and to prevent slip is directly controlled only on the lower, or output pulley. Clamping on the upper or input pulley is a resultant equilibrium force, which varies as a function of ratio and torque. These characteristics are shown on Fig. 8 for max. input torque, along with operating radii of the pulleys. Note that clamping load of the output pulley is higher than the input pullley in underdrive, but lower in overdrive. Ratio is downshifted by bleeding oil from the input pressure cavity. Conversely for upshifting oil is forced into the cavity until the desired ratio is reached. The oil quantity in the cavity is trapped during constant ratio. System line pressure must be held sufficiently above the ratio control pressure to provide adequate ratio change response.

CONTROL SYSTEM

Overall function of the transmission is electronically controlled. An Ultra PAC 500 computer was used to permit rapid reprogramming for developmental changes. The interface between electronics and the hydraulic control is comprised of four linear actuators with position feed back potentiometers, and four pressure transducers, with magnetic sensors for engine and output pulley speeds.

The hydraulic control schematic is shown on Fig. 9 and the electronic-hydraulic diagram on Fig. 10. System line pressure ranges from 552 kPa to 1930 kPa for the 1.6L engine. The linear actuators consist of a 12 v. DC motor which drives a ball screw to compress a spring to the height required. Spring load on the spool valves establishes pressure on the main regulator, belt load, and transfer clutch. The ratio control actuator serves to control flow in and out of the input pulley cavity, see Fig. 11. This system was selected at the outset because no suitable commercial actuators were available. Although numerous development problems were corrected, this type of linear actuator would not be considered for production because of high cost.

FUEL ECONOMY

At the time of writing this paper fuel economy testing was commencing and no results were available. Computer projectons indicated a 25% Metro and Highway improvement over the base Escort, with ATX automatic transmission. Compared with a straight CVT with no extension of ratio range beyond the variable speed belt and pully system, there was no Metro improvement but up to 9% Highway improvement. Due to uncertainty of the projections the level of improvement is not yet established.

CVT 4x4 DERIVATIVE

Historically, 4x4's have been RWD adaptations, but such systems penalize the complexity of a base FWD.

A modest cost 4x4 drive line is derivable from some transverse FWD powertrains. However, the "U" drive transaxles, such as the dual model CVT are much more favorable for packaging because the rear drive shaft can be conveniently located near the vehicle centerline.

A bevel gear set is integrated with the FWD final drive gearing to provide the longitudinal drive shaft take-off for driving the rear axle.

On the premise that all new FWD powertrain designs should consider 4x4 option, an "either-or" modification to the dual mode CVT was derived (Fig. 12). As a result the base transaxle would serve either 2x4 or 4x4 installations.

In the interest of minimizing parasitic drag losses, spiral bevel gearing operating in ATF is employed. This not only eliminates the gear tooth sliding losses inherent in hypoids, but avoids the need for heavy E.P. lubes.

The above system is potentially a cost/weight effective means of providing a 4x4 option for small transverse FWD powered passenger cars, vans, or pick-up trucks, and is compatible with various rear axle drive schemes.

A vehicle test bed installation is being evaluated.

CONCLUSIONS

1. The torque convertor is an effective means of extending the overall ratio range of a CVT.
2. A potential improvement in EPA Highway fuel economy results from reducing the final drive ratio of a CVT.
3. Extending CVT ratio range beyond the optimum value for a particular vehicle offers no benefit in fuel economy.
4. A modest cost and favorably packageable power take-off to drive the rear axle for a 4x4 option is deriveable from the Dual Mode CVT.

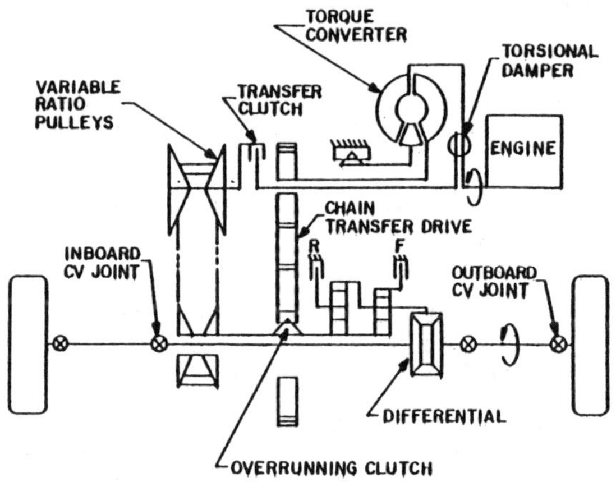

Fig 1 Dual mode CVT schematic

Fig 3 Dual mode CVT prototype with 1.6ℓ engine

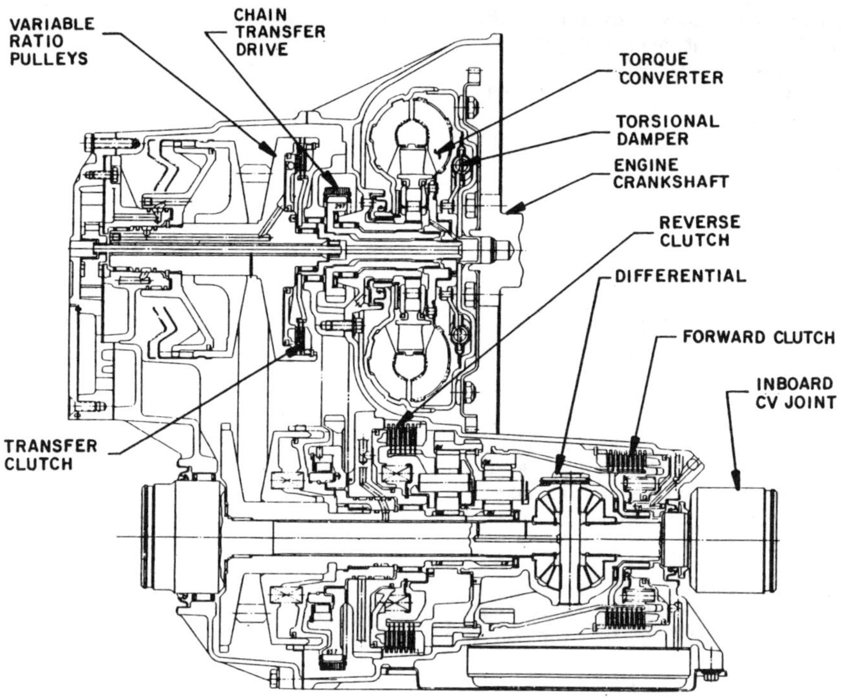

Fig 2 Dual mode CVT

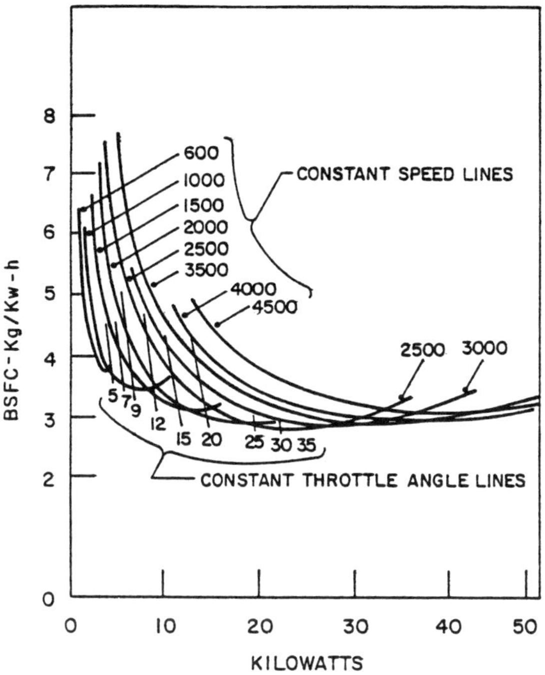

Fig 4 1.6ℓ engine BSFC versus power

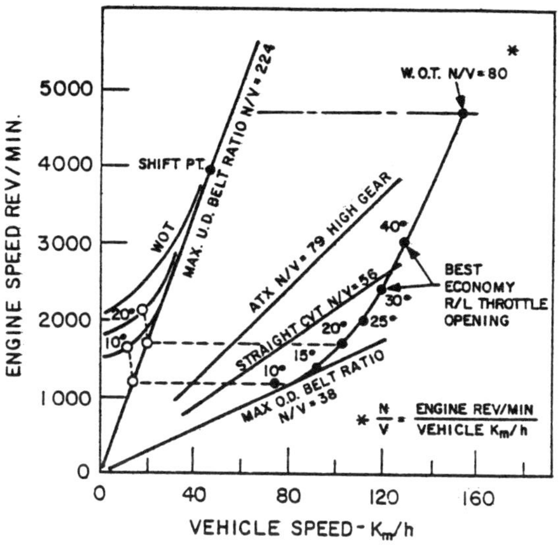

Fig 5 Engine speed versus vehicle speed

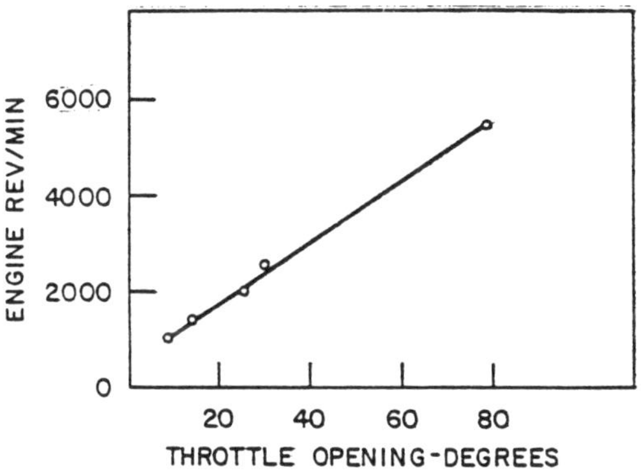

Fig 6 Engine speed versus throttle opening – CVT range

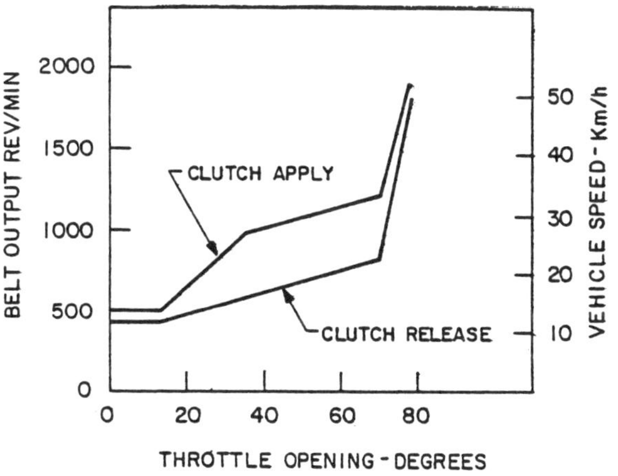

Fig 7 Transfer clutch control schedules

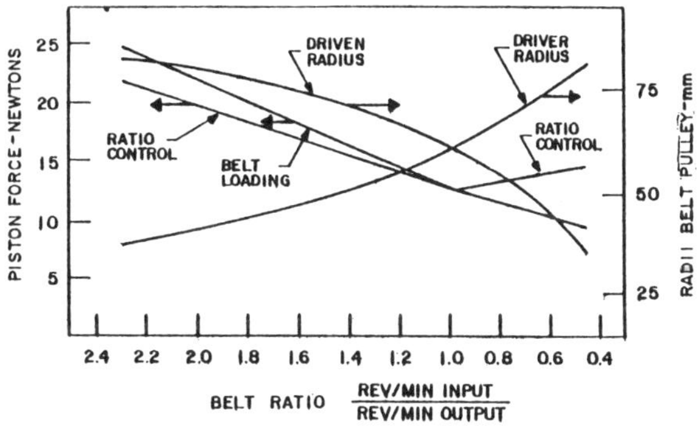

Fig 8 Belt and clamping forces – 136 Nm engine

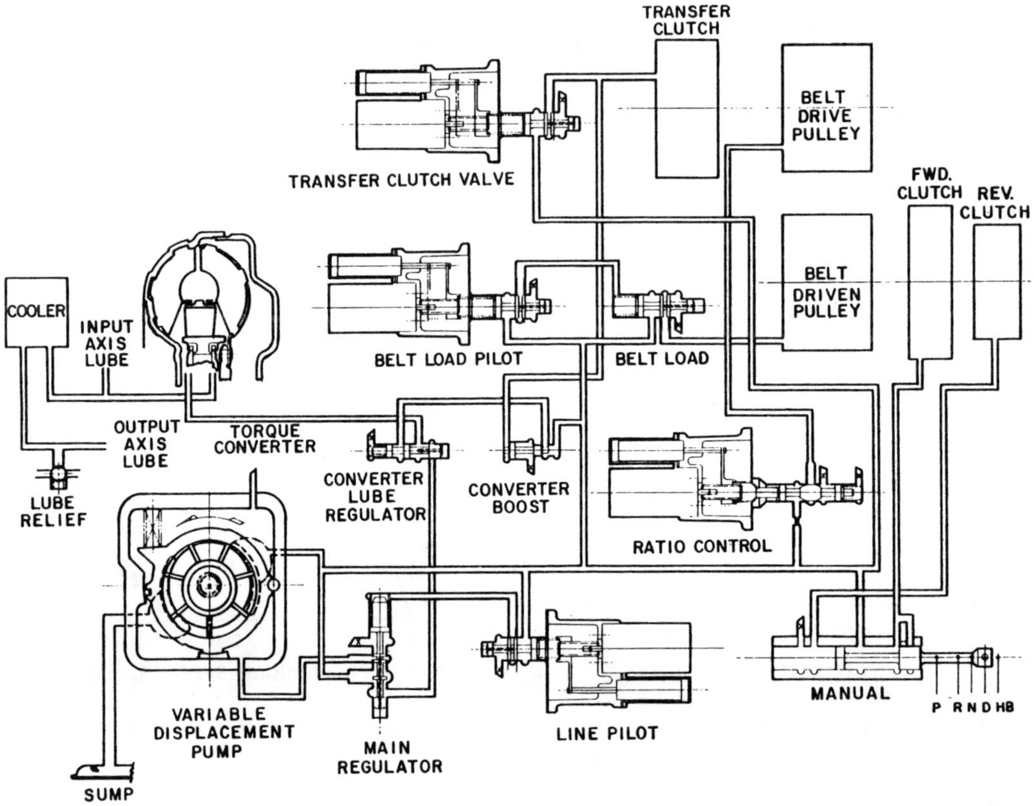

Fig 9 Hydraulic control schematic

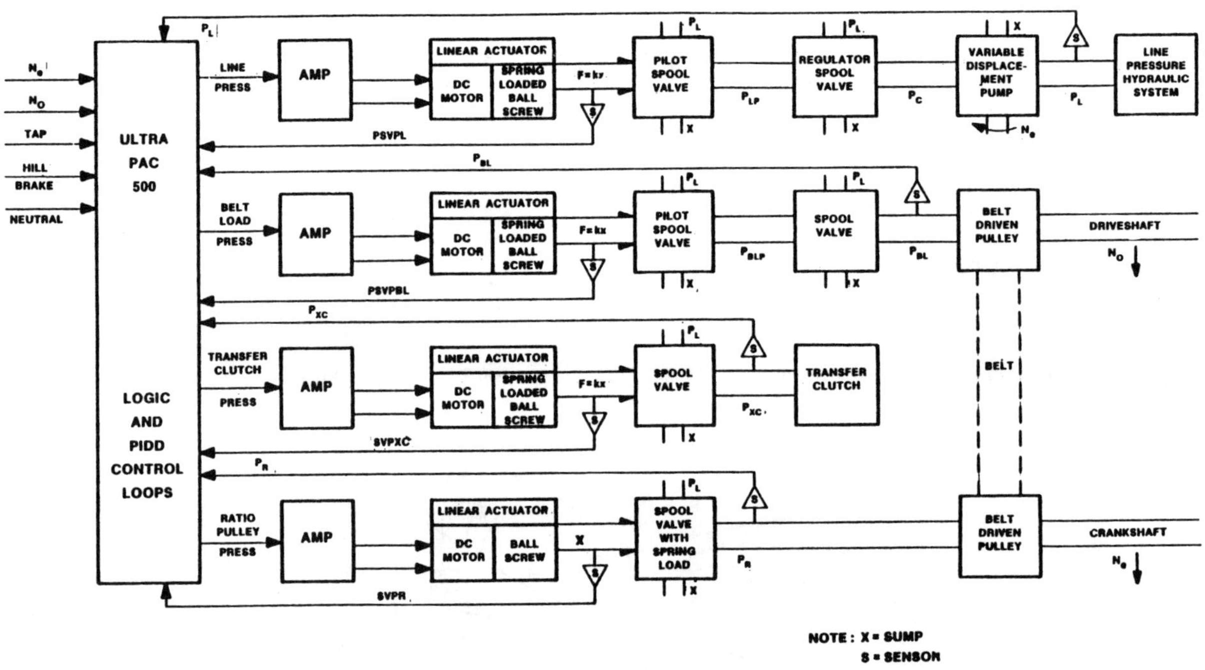

Fig 10 Electronic–hydraulic control diagram

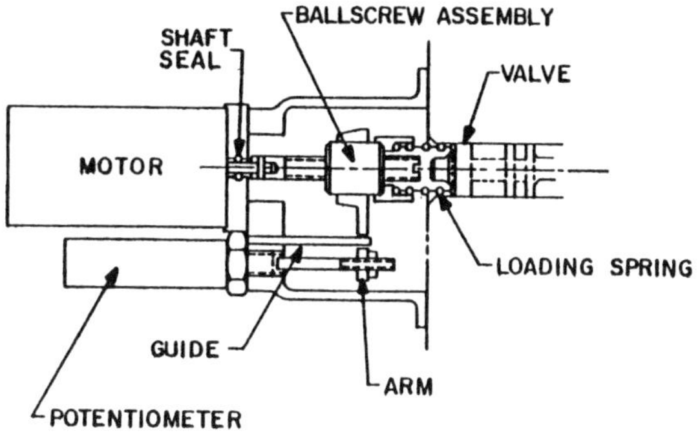

Fig 11 Actuator assembly

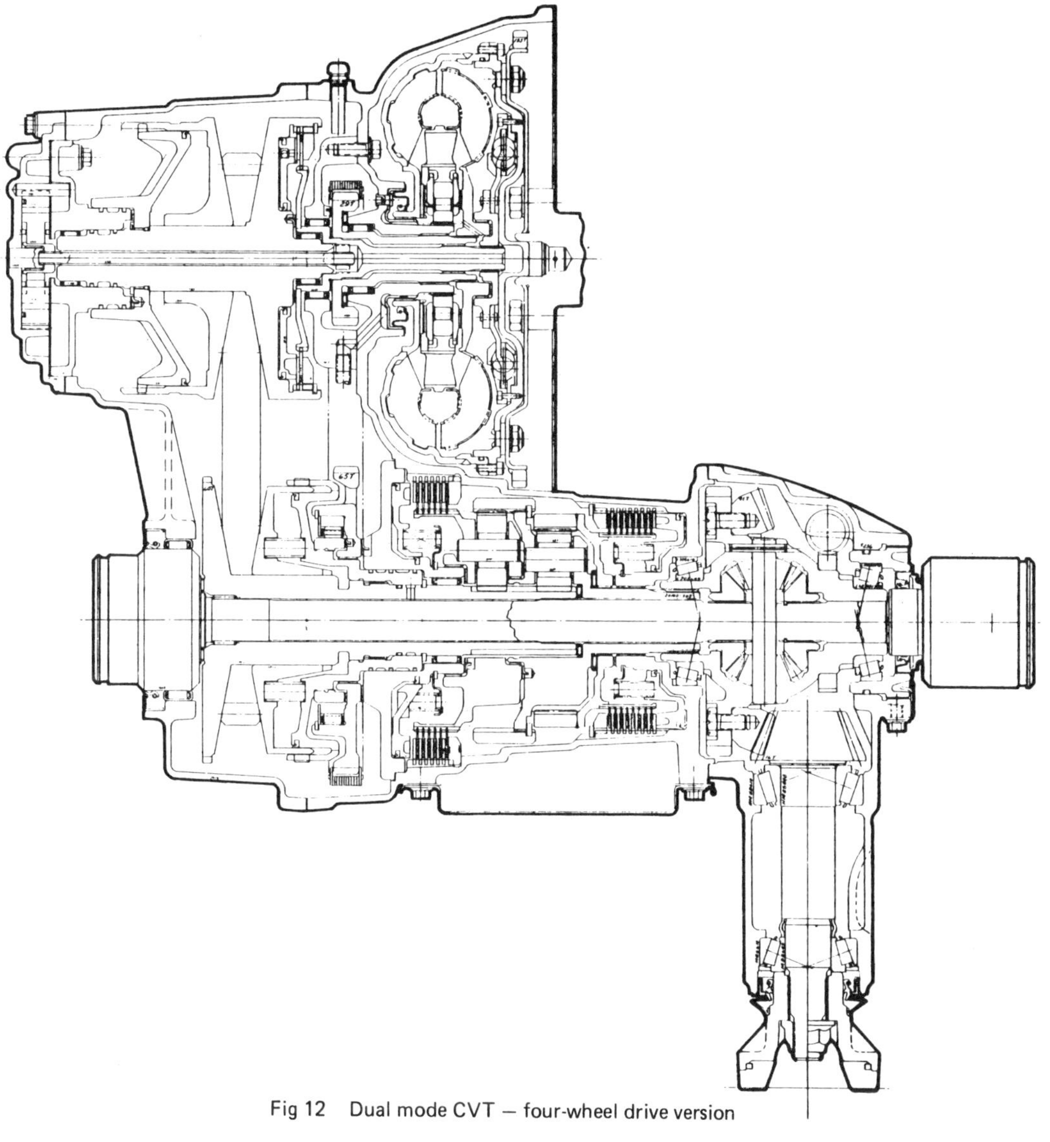

Fig 12 Dual mode CVT – four-wheel drive version

Non-Metallic Belt Drives

Adjustable Speed V-Belt Drives— Mechanical Properties and Design**

B. G. Gerbert
Lund Technical University

**Paper 740747 presented at the Off-Highway Vehicle Meeting and Exposition, Milwaukee, Wisconsin, September, 1974

ADJUSTABLE SPEED V-belt drives, both manually and automatically regulated, are used in many applications in different fields such as agricultural equipment, snow tracks, automobiles, etc. In the design and calculation of the mechanical properties of the drives, Worley (1),* Oliver and Henderson (2), and others used a combination of theoretical and empirical rules because they lacked a full knowledge of the mechanism of a V-belt working in the pulley groove.

In previous works by the author (3-7), different aspects of the mechanics of V-belt drives have been examined both theoretically and experimentally. The axial forces exerted by the belt on the pulley for different belt, pulley, and driving properties are specially treated in Ref. 7. In adjustable speed V-belt drives, the pulleys are divided and the axial forces are taken up in different ways by axial loading equipment outside the pulley. The combined action of the axial forces from the belt and those from the outer axial loading equipment determines the mechanical properties of the drive with regard to transmitted torque, velocity ratio, slip, efficiency, etc.

A discussion of various axial loading equipment and the properties attained by the adjustable speed V-belt drive when different axial loading equipments are combined is included in this paper.

*Numbers in parentheses designate References at end of paper.

AXIAL FORCE FROM BELT

The basic theory for the mechanism of V-belt drive is given in Ref. 3, and in Ref. 7 the axial forces that the belt exerts on the pulley are calculated for wide V-belts, that is, those V-belts commonly used in adjustable speed V-belt drives.

Figs. 1 and 2 are taken from Ref. 7. Fig. 1 shows the axial force F_z as a function of the coefficient of traction $\lambda = (F_2 - F_1)/(F_2 + F_1)$. F_1 and F_2 are the belt forces in the slack and tight part, respectively. The axial force is made nondimensional by the variable $F_{0z} = F_z/(F_2 + F_1)$. The force difference $F_2 - F_1$ is given by the transmitted torque, and therefore λ is a measure of the loading of the drive. Fig. 1 is valid for representative values of friction (see Ref. 2), elasticity of the belt, wedge angle, and center distance in relation to the length of the belt.

The maximum force ratio is obtained from the well-known equation

$$\frac{F_2}{F_1} = e^{\mu\alpha/\sin\beta} \tag{1}$$

Data used in Fig. 1 ($\mu = 0.4$ and $\beta = 14$ deg) and a contact angle of $\alpha = \pi$ gives $F_2/F_1 = 182$, which corresponds to $\lambda = 0.989$.

ABSTRACT

This paper is a theoretical examination of both the manually and automatically regulated adjustable V-belt drive. Simple relations for external axial loads are assumed to show the analytical treatment of various axial loading equipment and the properties attained by the V-belt drive when different equipment are combined. It is demonstrated that some combinations give stable automatic drives while other combinations give unstable drives.

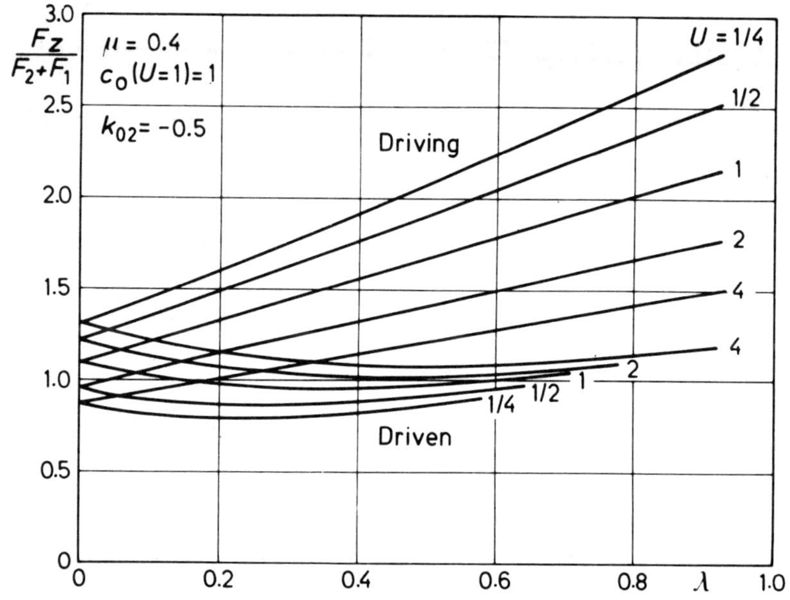

Fig. 1 - Axial force $F_z/(F_2 + F_1)$ as function of coefficient of traction $\lambda = (F_2 - F_1)/(F_2 + F_1)$. Velocity ratio $U = R_{dn}/R_{dg}$, groove angle β = 14 deg, c_0 and k_{02} are measures of elasticity of belt, L/A = 3.75, L = length of belt, and A = center distance (Ref. 7)

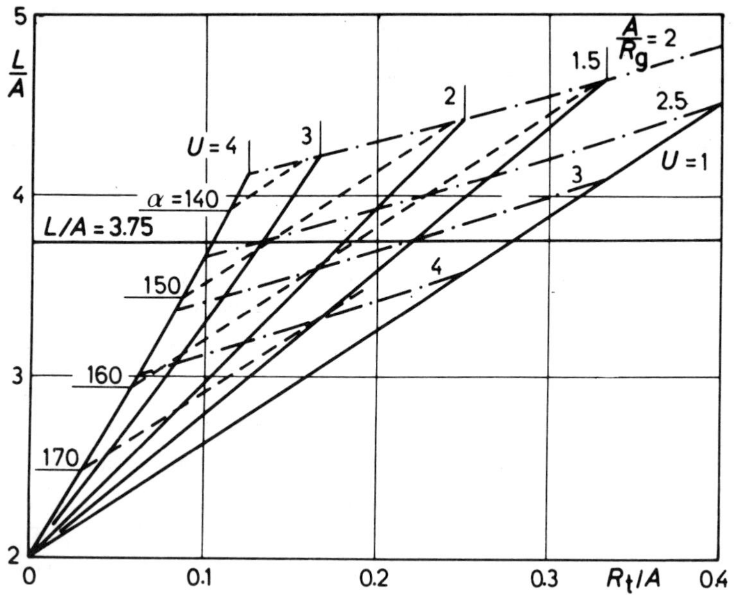

Fig. 2 - Relation between length of belt L, center distance A, and radii of pulleys R_t and R_g. R_t = pitch radius of small pulley and R_g = pitch radius of large pulley. α = contact angle of small pulley (Ref. 7)

Calculations of V-belt drives are commonly based on the force ratio $F_2/F_1 = 5$ giving $\lambda = 2/3$.

It is easy to calculate the length of the belt for given values of the center distance and pitch radii of the pulleys. Adjustable speed V-belt drives operate usually at locked center distance. In this case, the values of the length of the belt and the center distance are given, and the pitch radii for a certain velocity ratio are to be calculated. The solution of this problem is a little more complicated than the first one, and as an easy tool, the diagrams in Fig. 2 can be used. Known values of L and A give a horizontal line L/A in Fig. 2. In the intersections between the horizontal line and the curves for different velocity ratios U, the pitch radius of the small pulley is obtained on the horizontal axis R/A.

The diagram in Fig. 1 is valid for L/A = 3.75, and this line is especially marked in Fig. 2. Pitch radii for different velocity ratios are shown in Table 1.

AXIAL LOADING EQUIPMENT

The axial loading equipment commonly used to take up the axial force from the belt are: manually regulated equipment, spring-loaded regulators, torque sensing regulators, and velocity sensing regulators.

MANUALLY REGULATED EQUIPMENT - Examples of manually regulated equipment are screw mechanisms, cam curves, and hydraulic and pneumatic devices. All of this equipment is adjusted in such a way that the V-belt drive runs at a definite velocity ratio independent of transmitted torque and velocity.

Table 1 - Pitch Radii for Different Velocity Ratios U*

U	$\frac{R_t}{A}$	Small pulley, $\frac{R_t}{R_1}$	Large pulley, $\frac{UR_t}{R_1}$
1	0.279	1	1
2	0.182	0.652	1.304
3	0.133	0.477	1.431
4	0.105	0.376	1.504

*$L/A = 3.75$, $R_1 = R(U = 1)$

SPRING-LOADED REGULATORS - Spring-loaded regulators are the most commonly used in adjustable speed V-belt drives because of their simple design and application. The basic element is an axially mounted spring which compresses the movable disc against the fixed disc. The spring-load determines the tensioning of the belt. The principle is shown in the following example.

Example 1 - A spring-loaded regulator is used on the driving pulley. The spring-load $F_z = F_s$ = constant for all positions of the movable disc, which means that the spring-constant is very small. The torque of the pulley M = constant for velocity ratios U in the interval $1/2 \leqslant U \leqslant 2$. The pitch radius of the pulley is R_1 when U = 1. Chose maximum coefficient of traction to $\lambda = 0.6$ with respect to gross slip. Determine the spring-load and the characteristics of the pulley.

$$F_2 - F_1 = \frac{M}{R} \tag{2}$$

The definition of λ gives

$$F_2 + F_1 = \frac{F_2 - F_1}{\lambda} \tag{3}$$

The nondimensional axial force now can be written

$$F_{0z} = \frac{F_z}{F_2 + F_1} = \frac{F_z}{F_2 - F_1} \cdot \frac{F_2 - F_1}{F_2 + F_1} = \frac{F_z}{\frac{M}{R}} \lambda \tag{4}$$

Here $F_z = F_3$, M, and R are known for different velocity ratios U, and the equation appears as a straight line in Fig. 1. The smallest inclination of the line gives the largest risk for gross slip, so the axial force in this case is determined for U = 2 and $\lambda = 0.6$ in Fig. 1, that is, $F_{0z} = 1.50$. Table 1 gives the pitch radius $R = 0.652\ R_1$ and Eq. 4 the axial force

$$F_s = \frac{F_{0z}M}{\lambda R} = \frac{1.50\ M}{0.6 \cdot 0.652\ R_1} = 3.83\ \frac{M}{R_1}$$

For other velocity ratios Eq. 4 gives

$$F_{0z} = 3.83 \frac{R}{R_1} \lambda$$

where the pitch radius R is obtained from Table 1.

for U = 1

$$\frac{R}{R_1} = 1 \text{ and } F_{0z} = 3.83\ \lambda$$

for U = 1/2

$$\frac{R}{R_1} = 1.304 \text{ and } F_{0z} = 4.99\ \lambda$$

Draw these two lines in Fig. 1 and the intersections between these lines and the curves for the actual velocity ratios give the operating points. The solution is shown in Fig. 3.

TORQUE SENSING REGULATORS - A torque sensing regulator or a torque ramp is shown in Fig. 4. This regulator supplies an axial force that depends on the transmitted torque. The relation between the axial force and the transmitted torque is derived by Oliver and Henderson (2), and this relation is mainly influenced by the helical angle of the ramp. If the helical angle is varied in a certain way and the influence of the springs in Fig. 4 is neglected, this relation can be written as

$$\frac{F_z}{F_2 - F_1} = k_t \tag{5}$$

The constant k_t depends on the variation of the helical angle and the dimensions of the ramp and the pulley. Application of Eq. 5 is shown in the following example.

Example 2 - A driven pulley is equipped with a torque ramp. The velocity ratio U shall vary between 1/2 and 2. Chose maximum coefficient of traction to $\lambda = 0.6$. Determine the shape of the torque ramp and the characteristics of the pulley.

The torque ramp works according to Eq. 5. Thus the nondimensional axial force can be written as

$$F_{0z} = \frac{F_z}{F_2 + F_1} = \frac{F_z}{F_2 - F_1} \cdot \frac{F_2 - F_1}{F_2 + F_1} = k_t \lambda \tag{6}$$

Draw this line in Fig. 1 through the point $\lambda = 0.6$ and U = 2 for the driven pulley. At this point $F_{0z} = 1.04$ and Eq. 6 determines the constant k_t, that is

$$k_t = \frac{F_{0z}}{\lambda} = \frac{1.04}{0.6} = 1.73$$

Other operating points are found along the line $F_{0z} = 1.73\ \lambda$. The solution is shown in Fig. 3.

VELOCITY SENSING REGULATORS - This type of regu-

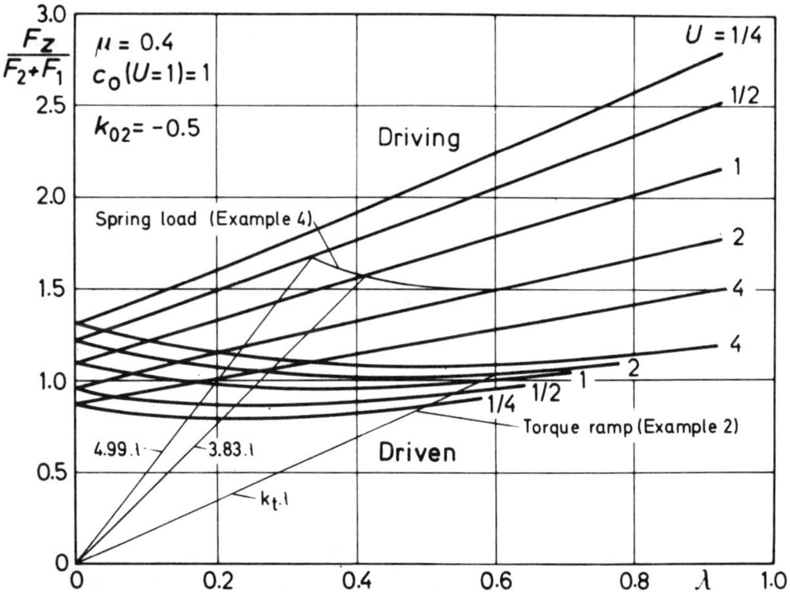

Fig. 3 - Properties of spring-loaded driving pulley and driven pulley equipped with torque ramp

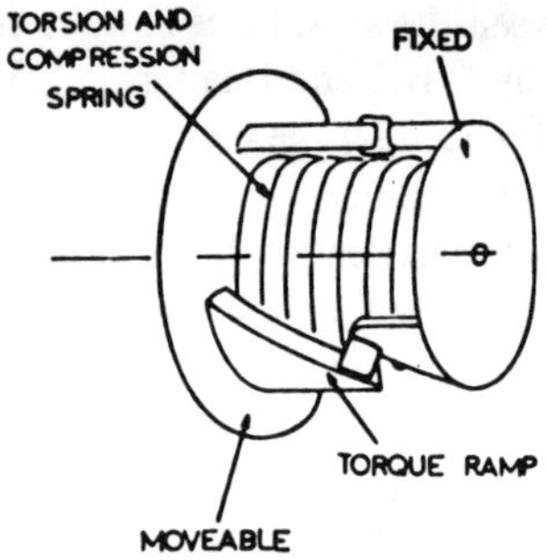

Fig. 4 - Torque ramp (Ref. 8)

lator is common in automatically regulated adjustable V-belt devices. The regulator supplies an axial force that depends on the angular velocity of the pulley. In most cases centrifugal forces are transformed to axial forces. One type of this kind of regulator is analyzed by Oliver, et al. (8). With a certain design the following relation between the axial force F_z and the angular velocity of the pulley ω can be obtained

$$F_z = k_v \omega^2 \quad (7)$$

where k_v is a constant.

Example 3 - A velocity sensing regulator is used on a driving pulley running at constant velocity. The axial force is constant according to Eq. 7. Thus the regulator behaves in the same way as the spring-loaded regulator described in Example 1 (see Fig. 3).

AUTOMATIC ADJUSTABLE SPEED V-BELT DRIVES

In the following section different drives in which different combinations of axial loading equipment are used will be studied.

SPRING LOAD-SPRING LOAD -

Example 4 - Two identical spring-loaded pulleys are used, one at the driving shaft and one at the driven shaft. What happens when torque is transmitted? This problem is studied in Ref. 7.

The solution is given in Fig. 5, which shows the velocity ratio as function of the coefficient of traction for different spring constants. The definition of the spring constant in this figure is given by the relation

$$\frac{F_z}{F_{z1}} = 1 - C\left(\frac{R}{R_1} - 1\right) \quad (8)$$

where F_{z1} and R_1 are referred to U = 1. C = 0 implies that the axial force is constant.

As mentioned before, λ is a measure of the transmitted torque, but the torque is also influenced by the radius of the pulleys. Therefore calculate the torque of the pulleys.

At the driving pulley, Eq. 4 gives

$$M_{dg} = \frac{\lambda}{F_{0z\,dg}} F_{zdg} R_{dg} = \frac{\lambda}{F_{0z\,dg}} \cdot \frac{R_{dg}}{R_1} F_{zdg} R_1 \quad (9)$$

where the radius R_1 for U = 1 is used as a reference, and the ratio R_{dg}/R_1 is found in Table 1. If $C \neq 0$, the axial load is variable as shown in Eq. 8; so use the axial force F_{z1} for U = 1 as a reference. This gives

$$M_{dg} = \frac{\lambda}{F_{0zdg}} \cdot \frac{R_{dg}}{R_1} \cdot \frac{F_{zdg}}{F_{z1}} F_{z1} R_1 \quad (10)$$

The relation between U and λ is found in Fig. 5, and then F_{0zdg} is obtained from Fig. 1. The calculations for C = 0.2 are given in Table 2, and other solutions are shown in Fig. 6.

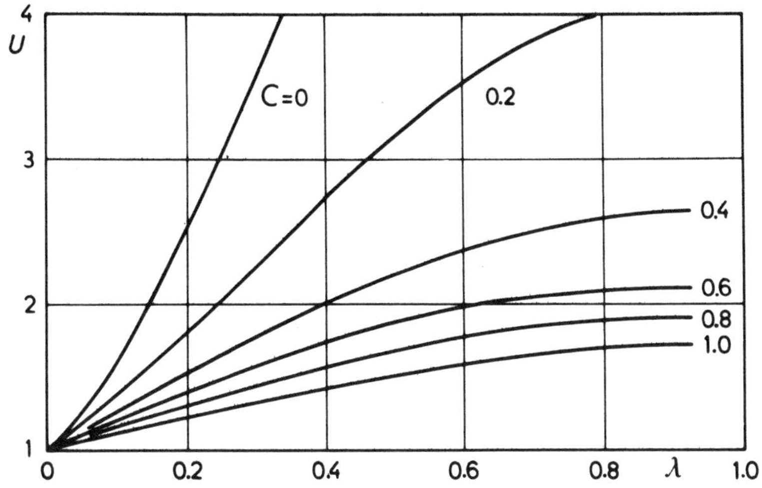

Fig. 5 - Two identical spring-loaded pulleys. Velocity ratio U as function of coefficient of traction λ for different spring-constants C (Ref. 7)

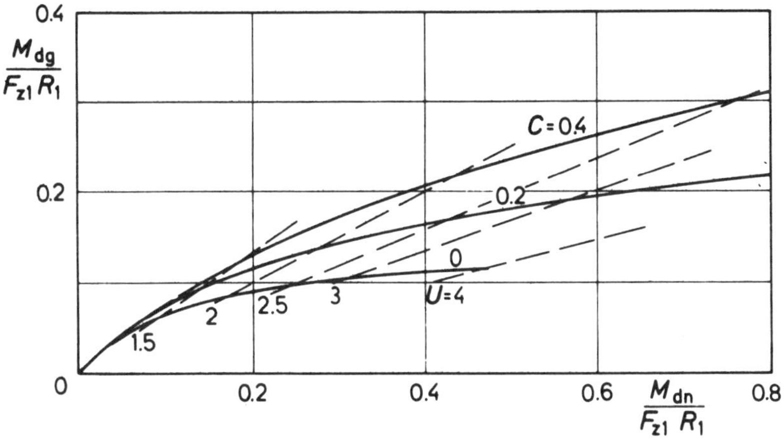

Fig. 6 - Relation between output torque M_{dn} and input torque M_{dg} for drive with two equal spring-loaded pulleys. C is spring constant

Table 2 - Transmitted Torque at Different Velocity Ratios for Two Spring-Loaded Pulleys*

U	$\frac{R_{dg}}{R_1}$	$\frac{F_{z\,dg}}{F_{z1}}$	λ	$F_{0z\,dg}$	$\frac{M_{dg}}{F_{z1} R_1}$	$\frac{M_{dn}}{F_{z1} R_1}$
1	1	1	0	1.09	0	0
2	0.652	1.062	0.250	1.19	0.145	0.290
3	0.477	1.086	0.460	1.25	0.191	0.573
4	0.376	1.110	0.790	1.41	0.234	0.936

*C = 0.2

Fig. 6 reveals that a large increase in the output torque gives only a slight increase in the input torque. This occurs at the expense of a decreasing output speed. Thus, this automatic drive works as a torque converter with positive characteristics.

SPRING LOAD-TORQUE RAMP -

Example 5 - The driving pulley is equipped with a spring-loaded regulator and the driven pulley with a torque ramp. The spring load $F_z = F_s$ = constant, and the torque ramp works according to Eq. 5. What happens when the torque is transmitted?

Use the same data as in Example 2, that is, $\lambda_{max} = 0.6$ and $1/2 \leqslant U \leqslant 2$. Then the operating points for the torque ramp are shown in Fig. 3, and from this figure the operating points for the spring-loaded regulator are also obtained because λ is the same for both the pulleys.

The transmitted torque of the driving pulley can be calculated from Eq. 9

$$M_{dg} = \frac{\lambda}{F_{0z\,dg}} \cdot \frac{R_{dg}}{R_1} F_s R_1 \qquad (11)$$

Table 3 - Transmitted Torque at Different Velocity Ratios for Spring-Loaded Driving Pulley and Torque Ramp on Driven Pulley

U	$\frac{R_{dg}}{R_1}$	λ	$F_{0z\,dg}$	$\frac{M_{dg}}{F_sR_1}$	$\frac{M_{dn}}{F_sR_1}$
2	0.652	0.600	1.50	0.261	0.522
1	1	0.575	1.76	0.327	0.327
1/2	1.304	0.535	1.96	0.356	0.178

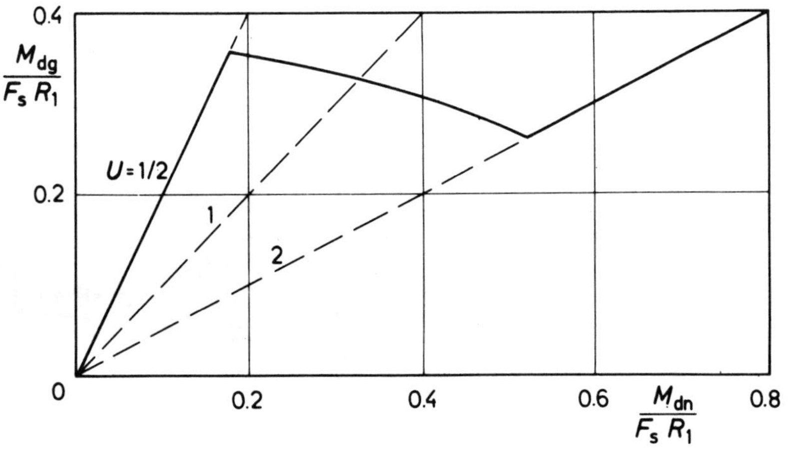

Fig. 7 - Relation between output torque M_{dn} and input torque M_{dg} for drive with spring-loaded pulley and driven pulley equipped with torque ramp

λ and $F_{0z\,dg}$ are obtained from Fig. 3 and R_{dg}/R_1 from Table 2. The calculations are given in Table 3, and the result is shown in Fig. 7.

Fig. 7 shows how the input torque is influenced by the output torque, and during the adjustment period the input torque decreases when the output torque increases, which means that this automatic drive works as a torque converter with negative characteristics. If the input torque is given, the regulator is unstable because the velocity ratio changes abruptly from U = 1/2 to U = 2 at the input torque $M_{dg} = 0.261\ F_sR_1$ for an increasing torque and from U = 2 to U = 1/2 at $M_{dg} = 0.356\ F_sR_1$ for a decreasing torque. A certain combination of spring-load and torque ramp action gives a stable regulator with positive characteristics.

VELOCITY SENSING REGULATOR-SPRING LOAD - *Example 6* - The driving pulley has a velocity-sensing regulator, and the driven pulley is spring-loaded. This combination is used in Daf's Variomatic. Here the velocity-sensing regulator behaves as stated in Eq. 7, and the spring load $F_z = F_s$ = constant. Determine the relation between the input and output velocity if the input torque M = constant.

Chose $\lambda_{max} = 0.6$ and the velocity ratio $1/2 \leqslant U \leqslant 2$. The spring load F_s is calculated from Eq. 4

$$F_s = F_z = \frac{F_{0z\,dn}}{\lambda} \cdot \frac{M}{R_{dg}} = \frac{F_{0z\,dn}}{\lambda} \frac{R_1}{R_{dg}} \frac{M}{R_1} \quad (12)$$

Here R_{dg}/R_1 is obtained from Table 1. U = 2 and $\lambda = 0.6$ give $F_{0z\,dn} = 1.04$ and

$$F_s = \frac{1.04}{0.6 \cdot 0.652} \frac{M}{R_1} = 2.66 \frac{M}{R_1}$$

Elimination of F_s in Eq. 12 gives for the other velocity ratios

$$F_{0z\,dn} = 2.66 \frac{R_{dg}}{R_1} \lambda$$

Draw these lines in Fig. 1, and the intersections with the curves for the actual velocity ratios give the operating points for the spring-loaded driven pulley. (See Example 1 for a driving pulley.) The result, both for the driven and the driving pulley, is given in Table 4.

The axial force of the driving pulley is obtained from Eq. 7. Combination with Eq. 9 gives

$$F_z = k_v \omega_{dg}^2 = \frac{F_{0z\,dg}}{\lambda} \frac{R_1}{R_{dg}} \frac{M}{R_1} \quad (13)$$

Put $\omega = \omega_1$ for U = 1. Then according to Table 4

$$k_v = \frac{1.51}{0.360} \frac{M}{R_1\,\omega_1^2} = 4.19 \frac{M}{R_1\,\omega_1^2}$$

Table 4 - Solution for Velocity-Sensing Driving Pulley and Spring-Loaded Driven Pulley*

U	$\frac{R_{dg}}{R_1}$	$F_{0z\,dn}$	λ	$F_{0z\,dg}$	$\left(\frac{\omega_{dg}}{\omega_1}\right)^2$	$\frac{\omega_{dg}}{\omega_1}$	$\frac{\omega_{dn}}{\omega_1}$
2	0.652	1.73 λ	0.600	1.50	0.915	0.957	0.479
1	1	2.66 λ	0.360	1.51	1	1	1
1/2	1.304	3.47 λ	0.250	1.55	1.135	1.065	2.130

*The input torque is constant.

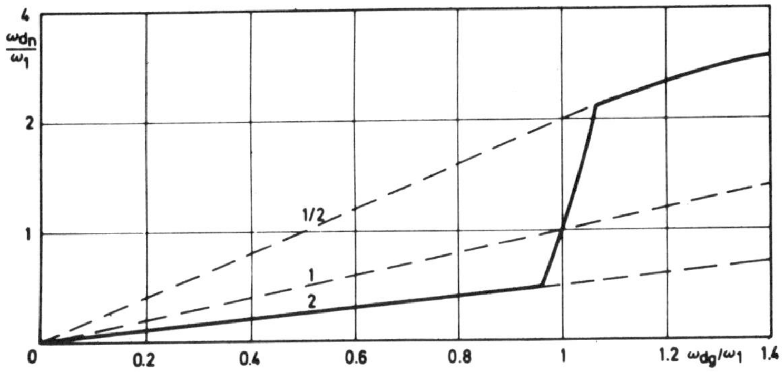

Fig. 8 - Relation between output speed ω_{dn} and input speed ω_{dg} for drive containing velocity-sensing driving pulley and spring-loaded driven pulley. Input torque is constant

Elimination of k_v in Eq. 13 gives for the other velocity ratios

$$\frac{\omega_{dg}}{\omega_1}^2 = \frac{F_{0z\,dg}}{4.19\,\lambda}\,\frac{R_1}{R_{dg}}$$

The result is given in Table 4, and Fig. 8 shows the relation between the input and output velocities.

At lower velocities the spring load is stronger than the centrifugal force, and U = 2. At $\omega_{dg}/\omega_1 = 0.957$ the regulator begins to operate, and for $\omega_{dg}/\omega_1 \geqslant 1.065$ the velocity ratio U = 1/2. Thus this drive works as an automatic gear box.

The velocity ratio of this drive also depends on the transmitted torque. At constant input velocity the axial loads of the two pulleys are constant, and the same method of solution as in Example 4 can be used. The relation between the input and output torque is similar to that in Fig. 6 for C = 0.

VELOCITY-SENSING REGULATOR-TORQUE RAMP - *Example* 7 - The drive contains a velocity-sensing driving pulley and a torque ramp at the driven pulley. The velocity-sensing regulator works according to Eq. 7 and the torque ramp according to Eq. 5. Determine the relation between the input and output velocity if the input torque M = constant.

Use the solution in Example 2 (and Example 5) for the torque ramp, that is, $\lambda_{max} = 0.6$ and $1/2 \leqslant U \leqslant 2$. Thus the operating points are given in Fig. 3 and Table 3. The torque on the driving pulley is obtained from Eq. 11 and the axial force from Eq. 7 as follows.

$$M = \frac{\lambda}{F_{0z\,dg}}\,\frac{R_{dg}}{R_1}\,k_v\,\omega_{dg}^2\,R_1 \tag{14}$$

λ, $F_{0z\,dg}$, and R_{dg}/R_1 are the same as in Table 3. If the velocity ω_1 at U = 1 is used as a reference, then the change in the input velocity $(\omega/\omega_1)^2$ can be obtained directly from Table 3 as the ratio between the input torques.

for U = 2

$$\left(\frac{\omega_{dg}}{\omega_1}\right)^2 = \frac{0.327}{0.261} = 1.253$$

$$\frac{\omega_{dg}}{\omega_1} = 1.119 \text{ and } \frac{\omega_{dn}}{\omega_1} = 0.560$$

for U = 1/2

$$\left(\frac{\omega_{dg}}{\omega_1}\right)^2 = \frac{0.327}{0.356} = 0.918$$

$$\frac{\omega_{dg}}{\omega_1} = 0.958 \text{ and } \frac{\omega_{dn}}{\omega_1} = 1.916$$

The result is shown in Fig. 9.

Low speeds give U = 2 because the ramp action is stronger than the centrifugal force. At $\omega_{dn}/\omega_1 = 0.946$ the velocity

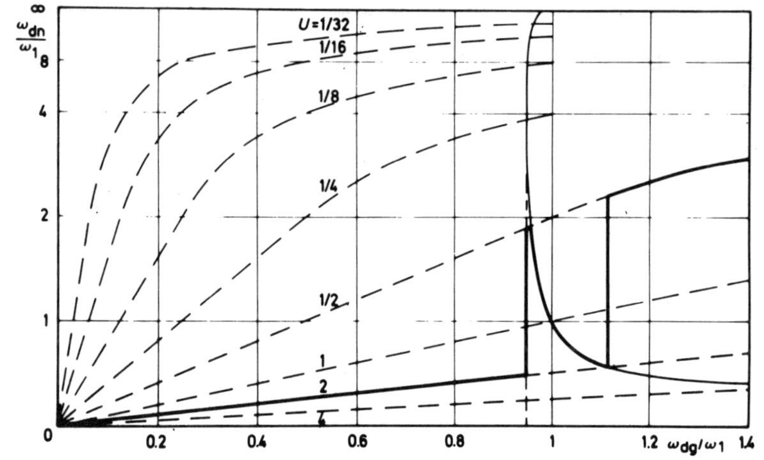

Fig. 9 - Relation between output speed ω_{dn} and input speed ω_{dg} for velocity-sensing driving pulley and torque ramp at driven pulley. Input torque is constant

ratio abruptly changes from U = 2 to U = 1/2 because this is the minimum velocity at which both the pulleys can be in equilibrium. In the interval $0.946 < \omega_{dn}/\omega_1 < 0.958$ the velocity ratio U = 1/2. A further increase in the input velocity increases the velocity ratio according to the curve for the regulator. This behavior of the drive depends on the fact that an increasing centrifugal force at the driving pulley must be compensated with an increasing axial force at the driven pulley. This is caused by increasing the torque at the driven pulley, which means that the velocity ratio must increase. At $\omega_{dn}/\omega_1 = 1.119$ the maximum velocity ratio U = 2 is reached and the torque increase on the driven pulley can not continue. Therefore the equilibrium is lost and the velocity ratio again abruptly changes from U = 2 to U = 1/2. The described behavior implies that the drive is unstable.

If the input torque is not constant the velocity ratio will vary. At constant input velocity the axial load at the driving pulley is constant and this problem is solved in Example 5.

If the output torque is constant the relation between the input and output speed is similar to that in Fig. 8, that is, the drive is stable.

COMBINATION OF SPRING LOAD AND TORQUE RAMP -

Example 8 - It is shown in Examples 6 and 7 that a spring-loaded driven pulley gives a stable drive and that a torque ramp at the driven pulley gives an unstable drive if the driving pulley is equipped with a velocity-sensing regulator. The immediate question is: What combination of spring-load and torque ramp action at the driven pulley makes the drive stable?

The stability border of the drive is obtained when the curve for the regulator is a vertical straight line in Figs. 8 and 9. This means that the input velocity is constant, and according to Eq. 7, the axial force from the velocity-sensing regulator is also constant. The case with constant axial force at the driving pulley was treated in Example 1, and the operating points are shown in Fig. 3. The result is reproduced in Table 5. (The result for U = 3 is also given.)

The combined axial load from a constant spring force F_s and torque ramp action according to Eq. 5 is

$$F_z = F_s + k_t (F_2 - F_1) \tag{15}$$

Table 5 - Combination of Spring Load and Torque Ramp at Driven Pulley Giving Constant Axial Load at Driving Pulley

U	$\frac{R_{dg}}{R_1}$	$F_{0z\,dg}$	λ	$F_{0z\,dn}$	$\frac{F_{0z\,dn}}{\lambda}$	$\frac{R_1}{R_{dg}}$
3	0.477	1.83 λ	0.830	1.15	1.39	2.10
2	0.652	2.50 λ	0.600	1.04	1.73	1.53
1	1	3.83 λ	0.410	0.96	2.34	1
1/2	1.304	4.99 λ	0.335	0.88	2.63	0.767

The input torque is M, which gives the force difference $F_2 - F_1 = M/R_{dg}$. The axial force at the driven pulley is obtained from Eq. 12. Combination with Eq. 15 gives

$$\frac{F_s}{M/R_1} = \frac{R_1}{R_{dg}}\left(\frac{F_{0z\,dn}}{\lambda} - k_t\right) \tag{16}$$

For a given velocity ratio, this relation between F_s and k_t is a straight line. Table 5 gives the following for U = 2

$$\frac{F_s}{M/R_1} = \frac{1.73 - k_t}{0.652} \tag{17}$$

What combination of F_s and k_t should be chosen in order to get a stable drive?

$F_s = 2.66\ M/R_1$ and $k_t = 0$ gave a stable drive in Example 6, and $F_s = 0$ and $k_t = 1.73$ gave an unstable drive in Example 7. Differentiate Eq. 16 and keep F_s and k_t constant.

$$0 = \frac{F_{0z\,dn}}{\lambda} - k_t + \frac{R_1}{R_{dg}} \frac{d(F_{0z\,dn}/\lambda)}{d(R_1/R_{dg})}$$

$$k_t = \frac{F_{0z\,dn}}{\lambda} + \frac{R_1}{R_{dg}} \frac{d(F_{0z\,dn}/\lambda)}{d(R_1/R_{dg})} \tag{18}$$

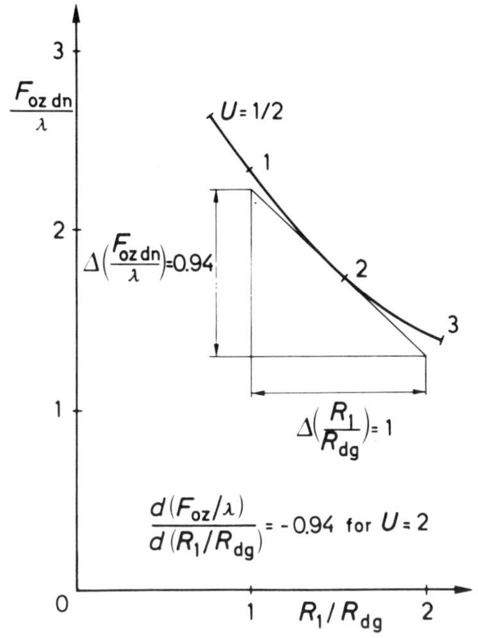

Fig. 10 - Relation between $F_{0z\,dn}/\lambda$ and R_1/R_{dg} at stability border

In Table 5 $F_{0z\,dn}/\lambda$ and R_1/R_{dg} at the stability border are given for various values of U, and Fig. 10 shows the relation between these variables. A graphic construction gives the derivative needed in Eq. 18. According to Fig. 10, the derivative is

$$\frac{d(F_{0z\,dn}/\lambda)}{d(R_1/R_{dg})} = 0.94 \quad \text{for } U = 2.$$

Put this value into Eq. 18

$$k_t = 1.73 - 1.53 \cdot 0.94 = 0.29$$

Now Eq. 17 gives the spring load

$$\frac{F_s}{M/R_1} = \frac{1.44}{0.652} = 2.21$$

Thus, combinations of F_s and k_t according to Eq. 17 give a stable drive provided $k_t \leqslant 0.29$ and $F_s \geqslant 2.21\ M/R_1$. This means that a stable drive requires strong spring-load action and weak torque-ramp action.

The same conclusion is made by Oliver and Henderson (2) who write: "Increasing tension is the sole purpose of the cam angle for the applications discussed in this paper. Specifically, it is not considered as a means of changing speed ratio, which is accomplished by some other operator-controlled shifting mechanism incorporated into the companion sheave or into an adjustable idler." The torque ramp takes care of the peak loads.

CONCLUSION

The mechanical properties, such as transmitted torque and velocity ratio, of an adjustable speed V-belt drive depend on the combined action of the internal axial forces exerted on the pulley from the belt and the external forces from some axial loading equipment. The axial forces from the belt have been treated in earlier reports. In this paper different types of axial loading equipment are treated. Simple relations for the external axial loads are assumed in order to show how different equipment are treated analytically and what properties are given to the V-belt drive.

Automatic adjustable speed V-belt drives are obtained for certain combinations of axial loading equipment at the driving and driven pulley. It is shown that some combinations give stable automatic drives while other combinations give unstable drives.

NOMENCLATURE

A	= Center distance
C	= Nondimensional spring constant
F_1	= Belt tension in the slack side
F_2	= Belt tension in the tight side
$F_z\left(F_{0z} = \frac{F_z}{F_2 + F_1}\right)$	= Axial force (nondimensional)
k_t	= Constant for torque ramp
k_v	= Constant for velocity sensing regulator
L	= Length of the belt
M	= Torque
R	= Pitch radius
$U = \frac{\omega_{dg}}{\omega_{dn}} = \frac{R_{dn}}{R_{dg}}$	= Velocity ratio
α	= Contact angle
β	= Groove angle
$\lambda = \frac{F_2 - F_1}{F_2 + F_1}$	= Coefficient of traction
μ	= Coefficient of friction
ω	= Angular velocity

INDICES

dn	= Driven pulley
dg	= Driving pulley
s	= Spring
z	= Axial direction
0	= Nondimensional
1	= Slack side, velocity ratio for U = 1
2	= Tight side

REFERENCES

1. W. S. Worley, "Designing Adjustable-Speed V-Belt Drives for Farm Implements." SAE Transactions, Vol. 63 (1955), pp. 321-330.

2. L. R. Oliver and D. D. Henderson, "Torque Sensing Variable Speed V-Belt Drive." SAE Transactions, Vol. 81 (1972), paper 720708.

3. B. G. Gerbert, "Force and Slip Behaviour in V-Belt Drives." Acta Polytechnica Scandinavica, Mech. Eng. Series No. 67, Helsinki, 1972.

4. B. G. Gerbert, "Tensile Stress Distribution in the Cord of V-Belts." ASME paper 73-DE-D, 1973.

5. B. G. Gerbert, "Power Loss and Optimum Tensioning of V-Belt Drives." ASME paper 73-WA/DE-7, 1973.

6. B. G. Gerbert, "Pressure Distribution and Belt Deformation in V-Belt Drives." Trans. Machine Elements Div., Lund Tech. Univ., Lund, 1973. (Accepted for publication in Trans. ASME).

7. B. G. Gerbert, "Scheibenspreizkräfte in Breitkeilriemengetrieben." Antriebstechnik, Vol. 13, No. 1 (1974), pp. 17-21.

8. L. R. Oliver, K. G. Hornung, and H. N. Shapiro, "An Automatic V-Belt Transmission with an Assymetrical Belt." Jrl. of Engrg. for Ind., Vol. 95, No. 3 (1973), pp. 771-779.

Microprocessor Controlled Continuously Variable Rubber V-Belt Transmission for an Automobile**

K. Srinivasan, D. R. Houser and G. L. Kinzel
Department of Mechanical Engineering
The Ohio State Univ.

D. L. Mattern
Union Carbide Corp.

**Paper 820745 presented at the Passenger Car Meeting, Troy, Michigan, June, 1982.

ABSTRACT

The technical advancements in microprocessors and their incorporation in motor vehicles has led to the feasibility of using continuously variable transmissions in automobiles. This paper discusses the design, development and testing of a rubber V-belt continuously variable transmission for a production vehicle. Overall speed ratios in excess of 7.0 were achieved with a hydraulically actuated driver sheave and a spring and cam loaded driven sheave. The drive was controlled with an 8-bit microprocessor which operated in three different control modes. The test results show that the rubber V-belt is viable for automotive use and has potential for improved fuel economy while maintaining good vehicle drivability.

THE RISING COST OF PETROLEUM has stimulated efforts to reduce fuel consumption of automobiles because atuomobiles and light trucks account for nearly one-third of the total U.S. oil usage. The automotive companies have been responding by downsizing automobiles and engines, reducing vehicle weight, rolling resistance, and aerodynamic drag. More efficient engines and transmissions are also being designed. It has been projected that with current technology, the average fuel economy of the entire U.S. automobile fleet built in 1995 could be raised to more than 60mpg. With more advanced technology, this figure might be increased to 90 mph [1].*

One of the areas of improvement being investigated concerns the efficiency variation of the internal combustion engine for changes in speed and load. For any given power level, there is a unique engine operating point yielding maximum fuel economy. Therefore, maximum engine efficiency can be obtained by varying the transmission ratio such that, whatever the road speed and demanded power, the engine achieves maximum fuel economy. The present stepped ratio gear transmissions have a finite number of fixed ratios and do not have a wide enough ratio range to allow the engine to run at peak efficiency for a wide range of vehicle operating conditions. A properly designed continuously variable transmission (CVT) can operate at an infinite number of speed ratios over a relatively large ratio range and would enable more flexible engine and vehicle control strategies [2]. The control strategies can be optimized not only for fuel economy but for emission control [3], vehicle performance or any other suitable criterion.

Various prototype cars using CVT's have shown efficiency increases of 15% to 30% and higher when compared with manual and automatic transmissions. The CVT's tested thus far have included adjustable rubber V-belt drives [4,5], adjustable steel V-belt drives [6,7], toroidal traction drives [8,9] and hydrostatic transmissions using the power-split principle [10]. The increasing demand for smaller sized vehicles [1] has also improved the feasibility of CVT's using traction drives.

While many of the CVT's mentioned above have used all-mechanical control systems, it is clear that the use of a microprocessor for implementing the transmission control strategy [4,6,9] results in considerable versatility and flexibility. The application of microprocessors to engine control in production automobiles and recent advances in sensors for automotive use [11] have made this a viable approach.

This paper discusses the design, development and testing of a rubber V-belt continuously variable transmission for a Volvo 343 vehicle. The control scheme chosen for the transmission control is described as are the control hardware, software, and instrumentation. Experimental results from chassis dynamometer tests and road tests are also described. The overall performance of the transmission is evaluated and recommendations made for further work.

MECHANICAL DESIGN OF TRANSMISSION

A schematic drawing of the drive train of the Volvo 343 is shown in Figure 1. The CVT uses a single V-belt, hydraulically actuated

*Numbers in parentheses designate References at the end of the paper.

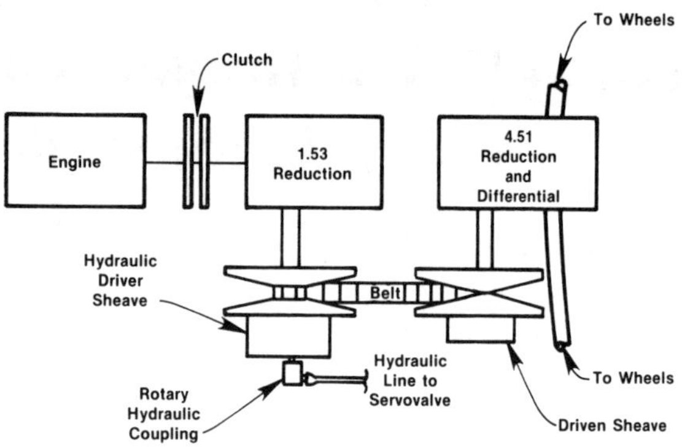

Fig. 1 - Schematic of the drive train

sheave. This drive replaces a two belt, mechanically-actuated drive used by Volvo, but uses the Volvo forward and rear gearbox assemblies. Figure 2 shows the assembled drive on a small test stand used for drive checkout.

Fig. 2 - V-Belt CVT assembled on a test stand

Both the driver and driven sheaves are machined from aluminum and have an outside diameter of 10.0 inches and a groove angle of 30 degrees. The minimum pitch diameter of both the driver and driven sheaves is 3.50 inches. For the V-belt used, which is nominally 0.8 inches thick, this gives a maximum theoretical speed range of about 7.75 to 1 and a minimum center distance of 10 inches. Tests on the automobile were conducted for speed ranges up to 7.5 to 1.0. The minimum required operating volume for the system is a box with dimensions 10x20x9.5 inches. In a production vehicle, the theoretical speed range would be reduced to improve belt life and reduce the overall package size. Based on discussions with the research engineers from one belt manufacturer [15], an overall ratio of 6 to 1 using two eight-inch sheaves could be achieved with reasonable belt life, and this size is within the size constraints imposed by compact automobiles. The sizes used for the research vehicle here were chosen to accomodate the volume available when the original Volvo sheaves were removed. Also, the large sheave sizes made possible experiments over relatively large velocity ratios.

The driver sheave is hydraulically actuated by four piston cylinders as shown in Figure 3. In this design, the cylinders move axially relative to the pistons which are stationary relative to the shaft. The pistons were machined from aluminum and are nominally 0.75 inches in diameter. The hydraulic cylinders were machined into the sheave block, and hydraulic fluid was supplied to the cylinders through a 1/8 inch hole drilled along the center line of the piston. Pressurizing the cylinder chambers causes the driver sheave to move inward, causing the transmission to shift up and the speed ratio to increase. The four cylinders are symmetrically located relative to the sheave center line and are supplied from the same manifold. A rotary hydraulic coupling is used to transmit the hydraulic fluid to the pistons. The pistons are enclosed in an outer protective sleeve of aluminum tubing. Torque from the moving half of the sheave is carried through a 1.5 inch splined shaft.

A schematic drawing of the hydraulic system actuating the driver sheave is shown in Figure 4. The hydraulic pressure is maintained by a pump/motor combination powered from a 12 volt battery. A pressure switch is adjusted to turn the motor on if the pressure drops below 450 psi and to turn the motor off if the pressure exceeds 600 psi.

The driven sheave has a torque ramp with a torsional coil spring and axial compression spring in parallel to apply an axial force to the belt. The torque ramp consists of a dual cam and follower arrangement that applies an axial force to the belt as a function of the load torque on the driven shaft. The ramp is a 40-degree helical cam machined from oil-hardening steel. Standard 3/4-inch cam followers were used. When the driver sheave is depressurized, it will open. At the same time, the spring force on the driven sheave causes it to close. The transmission therefore downshifts and the speed ratio decreases.

Two springs are used because it was found that one torsional spring could not give adequate axial force for rapid downshifting. This limitation was due to the volume in which the spring must fit and the stress limits for the spring material. On the other hand, a single larger axial spring alone was not used because the variable axial force provided by the torsional spring and torque ramp reduces the average force which would be required on the belt. Therefore, the total belt tension is regulated by the load rather than being maintained at the high level required to transmit the maximum load [12]. This greatly increases belt life.

In the vehicle tested, a 51-inch V-belt with a 30 degree wedge angle and a pitchline width of two inches was used. The belt was corrugated on both the top and bottom to improve heat transfer characteristics and to

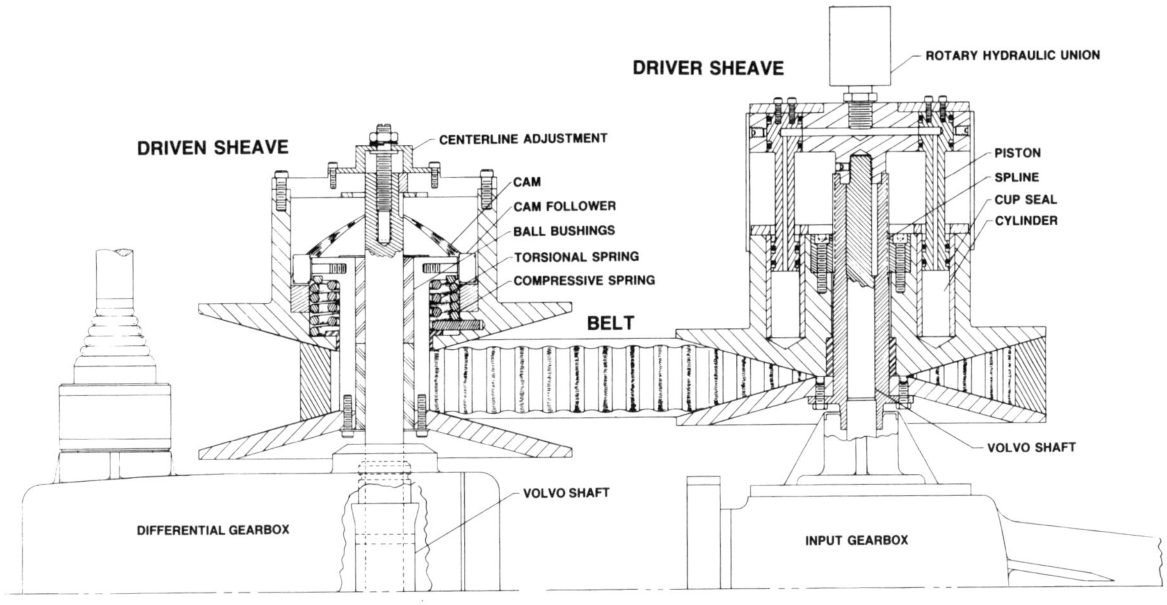

Fig. 3 - Driver and driven sheaves

reduce the minimum radius about which the belt could be run. All of the experimental belts were provided by the Dayco Corporation. A center distance of 15 inches was used. For most of the tests, the CVT speed ratio was set between 0.38 and 2.4 for an overall range of 6.3. However, greater ranges were achieved.

CONTROL STRATEGY

The design criterion for the controller was to achieve a fuel economy improvement over that for the production model of the Volvo 343 without degrading the performance of the vehicle. Since the CVT was a prototype, it was realized that there would be many parameters requiring adjustment in both the control and the hardware design. Therefore, the control scheme was kept simple. Simplicity was also important due to the lack of any accurate models of transmission behavior, particularly in the transient condition. In the interest of simplicity, it was decided to automatically control only the transmission ratio. The throttle is controlled manually by the driver.

To obtain an increase in fuel economy, the transmission ratio was adjusted such that the engine would operate in a region of low brake specific fuel consumption (BSFC). Figure 5 shows the engine map of a Volvo 343. The contours correspond to constant specific fuel consumption. The ideal engine operating schedule corresponding to maximum fuel economy is indicated on the figure by the dashed line. The constant torque line (85Nm or 63 ft-lb) on the engine map is a straight line approximation to the ideal engine operating line. It should also be noted that better approximations to the ideal engine operating line can be implemented on the microprocessor.

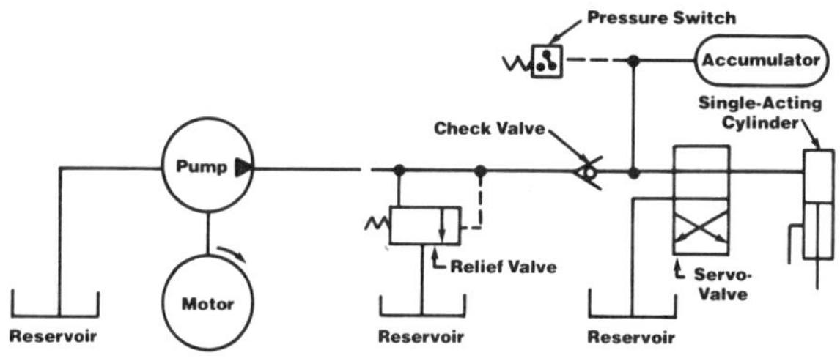

Fig. 4 - Schematic of hydraulic system

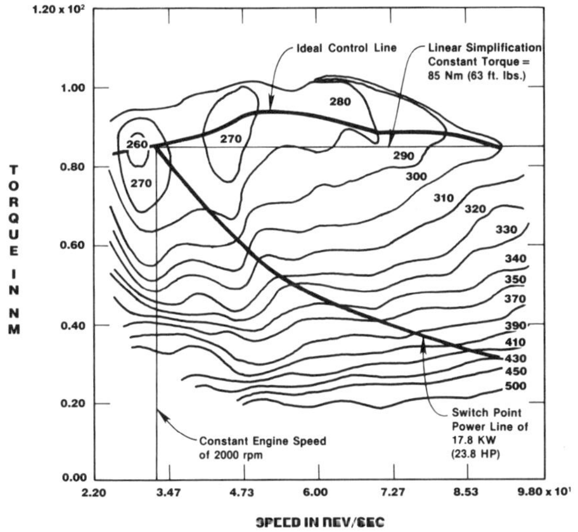

Fig. 5 - Engine map of Volvo 343 with control lines indicated (lines of constant specific fuel consumption in gr/KWh are shown)

A low speed cutoff was necessary to ensure smooth engine operation at all required power levels. The centrifugal clutch used in the Volvo 343 started to engage at an engine speed of about 1050 RPM and became fully engaged at engine speeds of about 2350 RPM. A clutch plate of different thickness was used and the low speed cutoff was set at 2000 RPM and in subsequent operation has run successfully at 1800 RPM. The low speed cutoff is also shown in Figure 5.

The controller therefore implemented two control modes: one at constant engine speed (speed control) and one at constant engine torque (torque control). Under speed control, the transmission would shift up for engine speeds above 2000 RPM, which increases the load on the engine and brings the engine speed back towards 2000 RPM. For engine speeds less than 2000 RPM, the transmission would downshift, bringing the engine speed back towards 2000 RPM. Under torque control, the transmission downshifts for engine torques higher than 85 Nm (63ft-lb). For an engine torque lower than 85 Nm (63ft-lb), the transmission shifts up causing the engine torque to increase.

The switch point power line shown in Figure 5 was used to switch smoothly between the torque and speed control modes. If the computed engine power is greater than 17.8 KW (24.2 HP), torque control is implemented. If it is lower than 17.8 KW (24.2 HP), speed control is implemented. In the absence of the switch point power line, large transients would occur in the hydraulic pressure whenever mode switching occurred. This was especially true if the engine speed increased over 2000 RPM at low levels of engine torque.

A third control mode, referred to as the speed ratio control mode, was added to the speed and torque control modes to improve vehicle performance. In the absence of this control mode, during deceleration, the transmission would downshift too rapidly causing the V-belt to go slack and the driven sheave to oscillate. Lower controller gains could be used in the other control modes to better control the deceleration, but there exists a tradeoff between good down shifting during deceleration and good steady state operation. The speed ratio control mode was added to avoid such a tradeoff by making the speed ratio during deceleration a function of the vehicle road speed as shown in Figure 6. Complete retraction of the throttle actuated this mode. The speed ratio control ensures good performance and is not related to fuel economy considerations.

The three control modes employed an error driven feedback control system as shown in Figure 7. The controlled variables were different in the three control modes. However, in all cases, the controller output was amplified by a current amplifier and actuated the spool of a servovalve controlling hydraulic flow to the driver sheave (Figure 7).

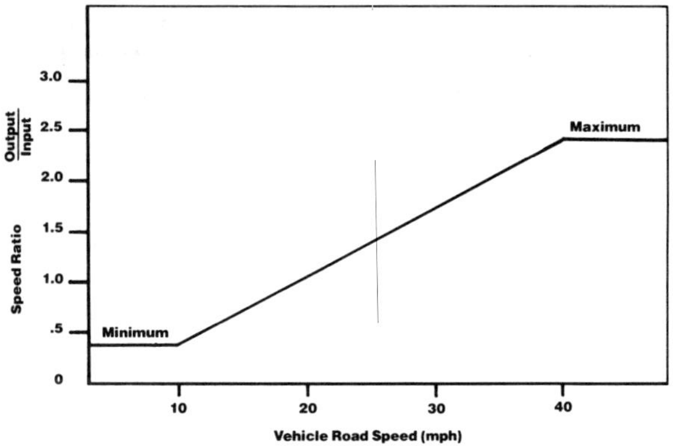

Fig. 6 - Deceleration mode shift schedule

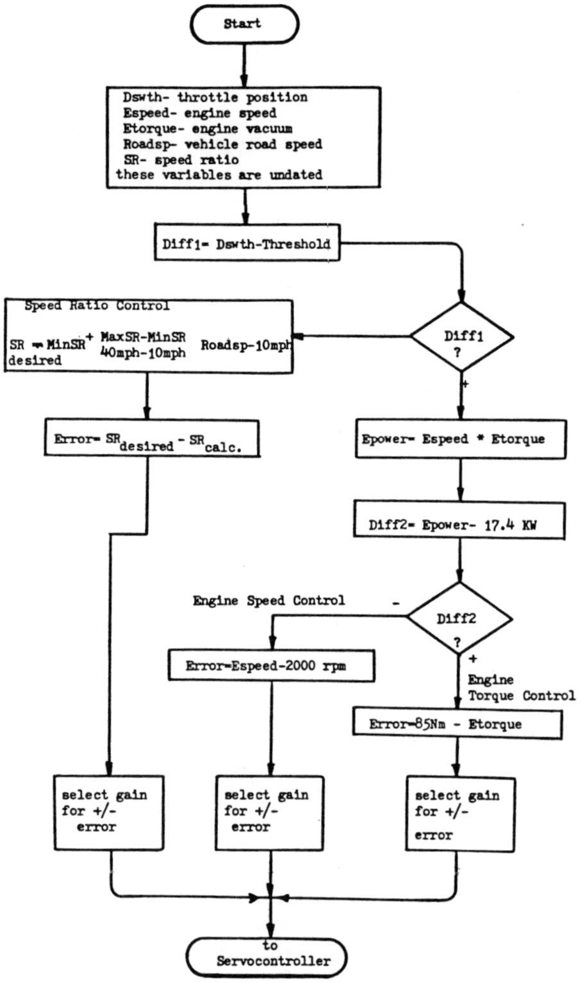

Fig. 7 - Description of CVT control scheme

Proportional control algorithms were used in all of the control modes. In the absence of a better understanding of the transient behavior of adjustable speed rubber V-belt drives, it was believed that more sophisticated control algorithms need be attempted only if the simpler algorithms proved deficient. The controller gains were different in the different modes and were obtained by trial and error. For all of the control modes, the controller gains were chosen higher for positive errors than for negative errors. The error, in all cases, was defined such that a negative error indicated a need for the transmission to downshift and a positive error a need for shifting up (Figure 7). Because the driver sheave piston is single acting, it often opens more rapidly than desired. The lower controller gain corresponding to negative errors lowers the rate of decrease of sheave pressure and hence slows down the outward motion of the driver sheave. This in turn prevents rapid downshifting of the transmission in the different control modes and improves overall control.

The advantages of using a microprocessor to implement the logic and computation required by the control scheme shown in Figure 7 are obvious. It should also be noted that the following variables need to be measured continuously: engine and rear wheel speeds, engine torque, and throttle position.

CONTROL HARDWARE, SOFTWARE AND INSTRUMENTATION

Board-level bus-compatible microcomputer components were selected for implementing the controller hardware. The functional modularity of the boards helps in hardware development because individualized cards make it easier to debug hardware problems and simplify external connections. The resulting hardware configuration is modular, but is not minimal. It was believed, however, that for a prototype application such as the one being considered, such an approach was well suited. Once the application requirements are defined better from experience gained from the prototype, the computer hardware design can be optimized.

Considering the anticipated speed of response of the mechanical system to be controlled, it was felt that an 8-bit microprocessor would have adequate speed for the real-time computations. The major bus systems supporting 8-bit microprocessors were Intel's Multibus, the S-100 bus and the Pro-Log/Mostek STD bus [13]. The STD bus was chosen because it was widely supported by a number of manufacturers. Additional reasons for the selection of the STD bus were compactness of board size (6.5 inch x 4.5 inch), low cost, and industrial quality components and construction.

The STD bus standardizes the physical and electrical aspects of the microprocessor card system. The bus is reserved for internal communication only. All other connections were made via suitable connectors on the I/O interface card edge. The Z-80 microprocessor, was selected for use in the STD bus system. The CVT control system hardware is summarized in Figure 8.

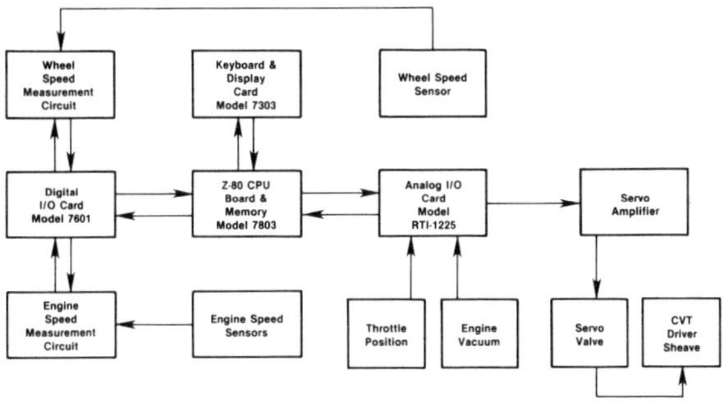

Fig. 8 - CVT control system hardware

The Z-80 processor card has a Z-80 microprocessor, associated devices for system control, adequate Random Access Memory (RAM), and Erasable Programmable Read Only Memory (EPROM). The TTL I/O port card provides 8-bit gated input ports and 8-bit latched output ports. This card is used for digital I/O. The input ports are used to accept the two 13-bit outputs from the speed measurement circuits to be described later. The digital output ports are used to send control signals to the speed circuitry. The analog I/O card accepts analog inputs and has analog outputs. A 10-bit Analog-to-Digital Converter (ADC) is used. The ADC has a conversion time of 40 microseconds and is adequate for the application. The Digital-to-Analog Converters (DAC) have 8-bit resolution and a settling time of 50 microseconds to within $\pm$ 1/2 LSB for a 20 volt step. Two's complement notation was used for number representation.

The speed measurement circuits were specially designed to yield quick and accurate speed measurement over a wide range of speeds. Magnetic proximity pickups were used, in combination with toothed wheels, to generate pulse trains of frequency proportional to the engine speed and wheel speed respectively. Direct measurement of these frequencies, either by analog frequency-to-DC converters or by digital counters would have resulted in a tradeoff between response time of the speed measurement circuit and accuracy of the speed measurement. Thus, if the counting period is long enough for the count to be large enough at low speeds, the response time of the measurement system is too high. If the counting period is low enough for quick response, the count is inaccurate at low speed.

It was decided, therefore, to measure the half-period of the proximity probe output using the microprocessor clock as a time base (Figure 9). The response time of the measurement system increases as the speed decreases, but it is always approximately equal to the period of the probe output. The response time of this measurement system is thus considerably shorter than in the other speed measurement methods. The measured half-period is inversely proportional to the speed being measured. The speed measurement is, therefore, least accurate at high speeds. However, by designing circuits to have a 12-bit output, the accuracy at high speeds is high enough. For example, if the range of speeds being measured is 32:1 and the speed measurement circuit has a 12-bit output, the accuracy at the high speed end is about one percent, which is acceptable. The accuracy at lower speeds is better, of course, since the generated count increases. Figure 10 shows a schematic of the speed measurement circuit. An STD bus compatible printed circuit board version of the circuit was fabricated and used. Identical circuits were used for engine speed and rear wheel speed measurement.

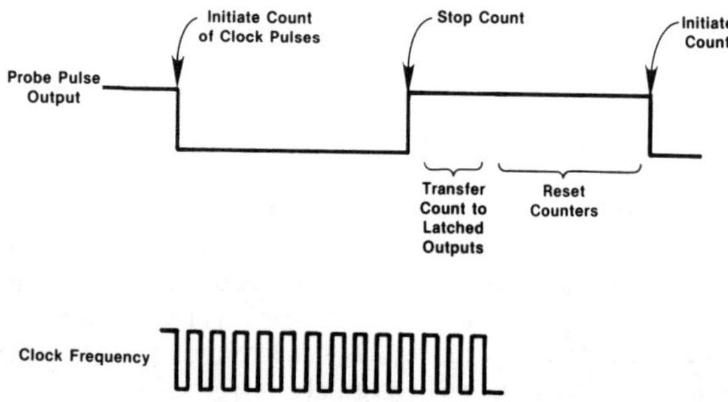

Fig. 9 - Principle of speed measurement circuit

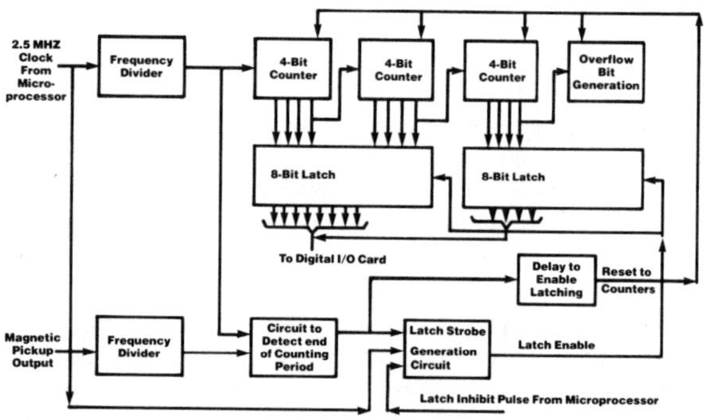

Fig. 10 - Schematic of speed measurement circuit

Since control computations were expected to result in a wide range of numbers, a floating point representation was chosen for internal number representation. A 16-bit format (8-bit mantissa, 8-bit exponent) was chosen primarily to ensure shorter computation times. The floating point numbers were also normalized following any arithmetic computation. A normalized representation merely requires the absolute value of the mantissa to have its leading bit to be non-zero.

The arithmetic routines used for addition, subtraction, multiplication, division, and normalization are modified versions of available routines for 32-bit floating point representation [14]. Worst case computation times for one iteration of the control program using 16-bit and 32-bit floating point representation are 7.2 msec and 63.2 msec respectively. The control program occupies 1085 8-bit bytes of EPROM and requires some additional RAM for scratch pad memory.

The magnetic proximity probes for measuring engine and wheel speeds were mounted at the flywheel and at the differential housing respectively. The speed measurement circuits and their interface to the microprocessor have already been described. The speed ratio is computed by the microprocessor whenever required.

Direct measurement of engine torque is rather difficult. Therefore, engine vacuum was used to determine engine torque since steady state dynamometer testing indicated a linear static relationship between the two variables (Figure 11). The engine vacuum was measured by a strain gage differential pressure transducer. The bridge output was amplified by an instrumentation amplifier, low pass filtered and input to the analog I/O interface card for the microprocessor. The throttle position was sensed by a linear potentiometer connected to the carburetor-gas pedal linkage and was used primarily to indicate if the throttle opening was more or less then a specified threshold value.

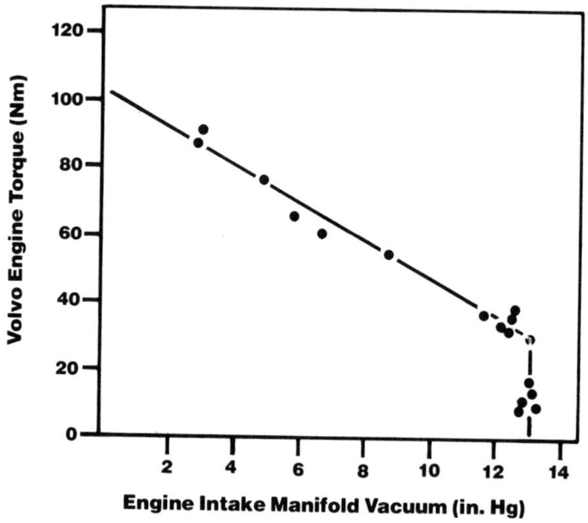

Fig. 11 - Calibration curve of engine torque versus vacuum

A two stage flow control servovalve was used for controlling hydraulic flow to the driver sheave. A servoamplifier controls valve spool position. The digital-to-analog converter output from the microprocessor is the input signal to the servoamplifier.

EXPERIMENTAL RESULTS

The original Volvo 343 transmission with mechanical controls was tested for fuel economy on a chassis dynamometer at conditions corresponding to road speeds of 15-60 mph and power levels of 5-30 HP. The results are shown in Figure 12 and compared with data for the microprocessor controlled CVT at similar test conditions. The data was taken early in the test program when the control system and sheave spring design had not been finalized. The manual mode indicates a manual speed ratio control mode where the desired speed ratio was set on a potentiometer to match the original Volvo ratios. Therefore, the manual mode does not necessarily approach optimum fuel economy conditions. The automatic mode indicates that the speed ratio was automatically selected by the microprocessor according to the control scheme in Figure 7. It appears from the limited data shown that the manually controlled single V-belt CVT has approximately the same efficiency as the original Volvo CVT. The automatically controlled single V-belt CVT seems to have slightly improved fuel economy over the original Volvo CVT for the data points shown. However, it should be noted that the torque control mode in Figure 5 is effective only for power levels above 17.8 KW (23.9 HP) where data was not taken. It is under torque control that maximum fuel economy can be obtained.

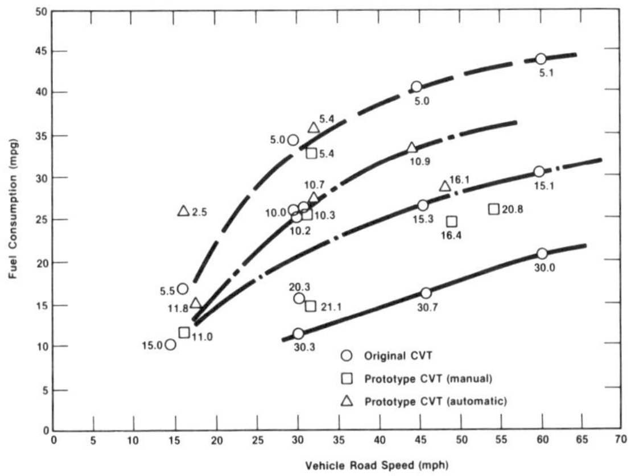

Fig. 12 - Fuel economy data from dynamometer tests (numbers beside points indicate horsepower)

During road tests, the following signals were recorded on an oscillograph: engine and rear wheel speeds, the engine vacuum, the calculated speed ratio, the controller output in the closed feedback loop and the driver sheave pressure. Analog signals proportional to the engine and wheel speeds were generated by frequency-to-analog voltage converters. The speed ratio computed by the microprocessor was converted to an analog voltage by a digital-to-analog converter. Driver sheave pressure was measured with a pressure transducer.

Figure 13 shows data for a moderate acceleration from zero to 30 mph followed by sharp braking to a stop. The speed ratio control mode is active through most of the sequence with engine speed being held constant at 2000 rpm as the speed ratio changes from 0.38 to 2.44.* Note that the sheave pressure varies considerably through the sequence.

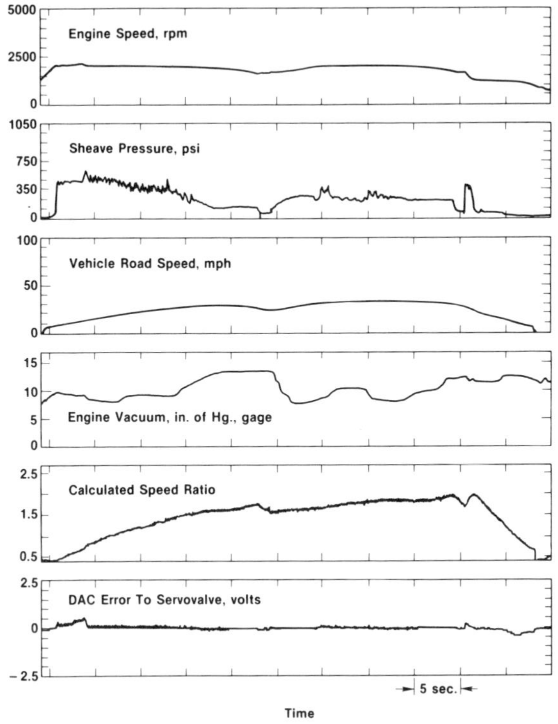

Fig. 13 - Engine speed control during moderate acceleration

During braking, the ratio closely follows road speed as is desired for ratio control.

Figure 14 shows the performance of the transmission when the car encounters a hill while cruising at 55 mph. At the beginning of the sequence, the transmission is in speed control. The large error indicates the ratio is at its maximum. Upon encountering the hill, the engine vacuum reaches the torque control level of 3.5 in. Hg at which point the speed ratio continually adjusts itself such that the vacuum (torque) holds constant. As the ratio decreases, the engine speed rises to a maximum of 3500 rpm.

Figure 15 shows the capability of the transmission to shift down rapidly when the throttle is suddenly depressed while traveling at 35 mph. The throttle's depression results in a sudden increase in torque (vacuum shifts from 9.7 in. Hg to 1.75 in. Hg and goes into torque control) which results in a sudden change of speed ratio from 2.0 to 0.8. The change of speed ratio results in an increase in engine speed from 2000 to 4500 rpm. The sharp changes in the error signal at times

*The overall transmission ratio from the engine to the rear drive shaft is computed by dividing the 6.9 combined reduction ratio of the Volvo input transmission and differential by the CVT speed ratio.

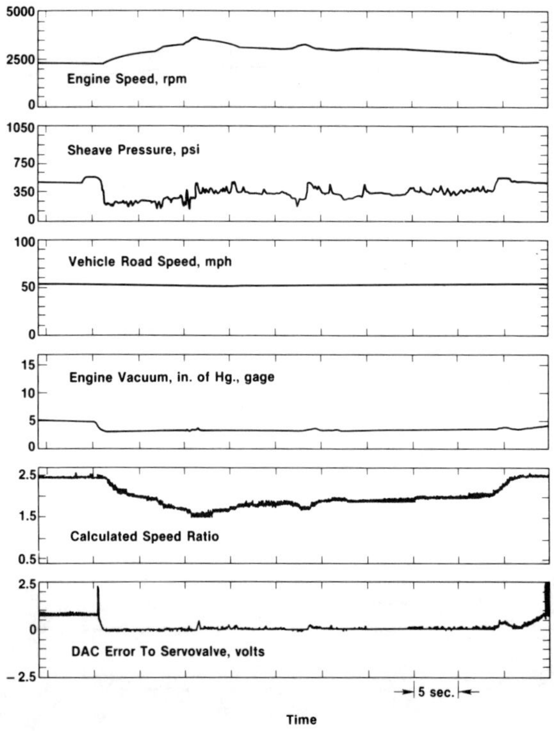

Fig. 14 - Torque control while climbing a hill at constant speed

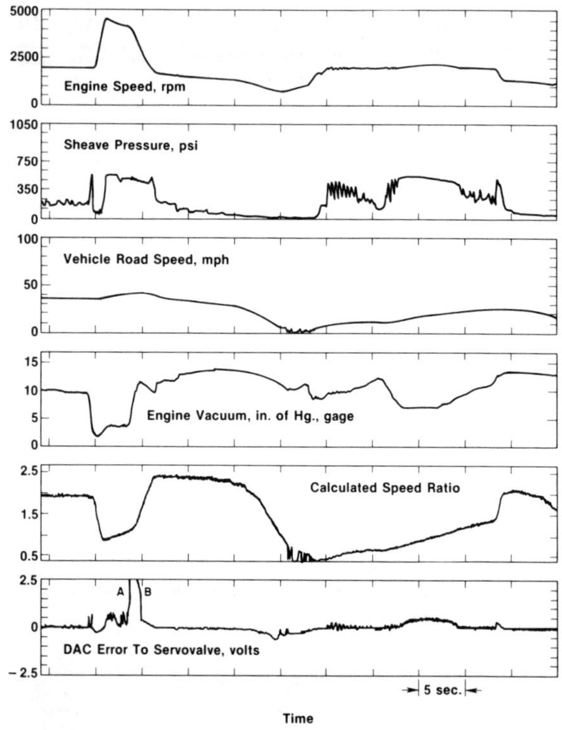

Fig. 15 - Sudden downshifting due to throttle depression followed by a moderate acceleration

A and B respectively indicate a switching from torque control to speed control and then a switching to ratio control as the driver releases the throttle. During this condition, it is possible to very rapidly reach the engine's maximum power output at any driving speed. This provides definite engine sizing advantages over discrete ratio automatic transmissions which can reach maximum power at only one road speed per ratio range.

SUMMARY

This paper has described the development and testing of a microprocessor controlled rubber V-belt CVT drive on a Volvo 343 vehicle. The belt has functioned successfully up to highway power levels of 60 HP. The belt manufacturer projects that a belt similar to that used in these tests could be successfully used on a 50 HP vehicle [15]. The rubber belt provides a low-cost, unlubricated transmission. The transmission shifts very smoothly with no noticable jerking and provides the ability to get maximum engine power at any road speed. The microprocessor controller works very well and could be further improved with more advanced control algorithms.

In order to further develop this concept, it is believed that a better understanding of the belt and sheave transient shifting dynamics is needed. It is recommended that both sheaves should be hydraulically actuated in order to better control belt tensions and that piston areas be increased so that lower hydraulic pressures could be used.

In conclusion, the authors feel that a rubber V-belt transmission provides a viable alternative to conventional automatic transmissions and steel V-belt and traction CVT's.

ACKNOWLEDGEMENT

The authors would like to thank Dayco Corporation for their financial and technical support of the research described in this paper.

REFERENCES

1.C.L.Gray Jr. and F. Von Hippel, "The Fuel Economy of Light Vehicles," Scientific American, Vol.244, No.5, May 1981.

2.B.C.Christenson, A.A.Frank, and N.H.Beachley, "The Fuel Saving Potential of Cars with Continuously Variable Transmission and an Optimal Control Algorithm," ASME Paper No.75 - WA/AUT-20, Nov.1975.

3.R.R.Racke, L.E.Unnewehr, and R.J.Freedman, "Optimization of a Continuously Variable Transmission with Emission Constraints," SAE Paper No.810107, Feb.1981.

4.D.Scott, "Belt-Drive Automatic Preprogrammed for Economy," Automotive Engineering, Aug.1978.

5.H.H.J.Ludoph, "Stepless Automatic Transmission for Cars," Paper presented at the 10th International Automobile Technical Congress, Japan, 1964.

6.P.Baudoin, "Continuously Variable Transmissions for Cars with High Ratio Coverage," SAE Paper No.790041, Feb.1979.

7.D.Scott, "Belt-Drive CVT for '82 Model Year," Automotive Engineering, Feb.1981.

8.J.H.Kraus, "Traction Drive Shows Automotive Promise," Machine Design, Oct.1973.

9.P.W.R.Stubbs, "The Development of a Perbury Traction Transmission for a Motor Car Application," ASME Paper No.80-C2/DET-59, Aug.1980.

10.A.A.Frank and N.H.Beachley, "Evaluation of the Flywheel Drive Concept for Passenger Vehicles," SAE Paper No.790049, Feb.1979.

11.F.A.Russo and B.E.Walker, "Automotive Sensors-Design/Production," SAE Paper No.790141, Feb.1979.

12.L.R.Oliver, K.G.Hornung, J.E.Swenson, and H.N.Shapiro, "Design Equations for a Speed and Torque Controlled Variable Ratio V-Belt Transmission," SAE Paper No.730003, Feb.1973.

13.M.J.McGowan, "From the S-100 to CAMAC: The Diversity of Digital Buses," Control Engineering, April 1979.

14.N.Wadsworth, "Z-80 Software Gourmet Guide & Cookbook," Scelbi Publications, 1979.

15.L.R.Oliver, Dayco Corporation, Private Communication, July 1981.

A Rubber Belt CVT for Front-Wheel-Drive Cars**

Richard F. Stieg and Wm. Spencer Worley
The Gates Corp.
Denver, CO

**Paper 820746 presented at the Passenger Car Meeting, Troy, Michigan, June, 1982.

ABSTRACT

A wide ratio, rubber V-belt driven, continuously variable transmission was designed, built, and operated in a 1.7 liter engine, front-wheel-drive car. A new V-belt under development with transverse stiffening elements is used in the demonstration CVT. Engine operation is controlled to a specified operating line by the CVT mechanical control system. The analytically determined belt side forces are varied with the torque load to maximize belt life.

Dynamometer tests showed a 17.7% improvement in mechanical efficiency over a 3 speed automatic transmission. Vehicle fuel economy tests showed a 15.5% improvement for the EPA driving cycle. Driveability is improved by the smoothness of operation and the absence of discrete ratio changes. The design method described is adaptable to different types of sheave control systems.

A DEMONSTRATION PROGRAM was undertaken to build a rubber V-belt continuously variable transmission (CVT) for a transverse-engine, front-wheel-drive car. The goals of the program were: (1) to show that a V-belt CVT designed for adequate life will fit in this type of car, (2) to demonstrate that the CVT has significantly better fuel economy than existing automatic transmissions while maintaining performance, (3) to show that a self-actuated mechanical control system is a valid method for controlling a CVT. A rubber V-belt CVT has been designed, built, and tested in a 1979 Plymouth Horizon to demonstrate the feasibility of such a transmission.

CVT ADVANTAGES

Recently many papers have been presented showing that wide-ratio range automatic transmissions have better fuel economy than the more common 3.5-to-4.0 ratio range (or variation) transmissions. (1-4)* The improvement is a result of being able to operate the engine at more economical operating conditions. It can be achieved with either a CVT or a discrete ratio transmission. The CVT has the advantage of stepless ratio changes and eliminates shift harshness. Studies show a discrete-ratio automatic transmission would require four or five ratios with lock-up in all ratios to approach the fuel economy of a CVT assuming comparable efficiencies for both transmissions. (1, 2)

A CVT can be controlled to shift at a constant engine speed for a fixed throttle position (Fig. 1). Each throttle and speed response pair identifies a specific engine power output; thus, the locus of operating points for the full travel of the throttle is a line, referred to as the operating line. For best fuel economy, the operating line is a monotonically increasing line with respect to power through the minimum brake specific fuel consumption (BSFC) contours on an engine power and speed map (Fig. 2). A comparable analysis with a discrete ratio transmission would show the transmission operating on constant throttle lines on the BSFC map, and the envelope of these lines for all throttle positions would be a band that becomes wider as the throttle opening increases. Therefore, a discrete-ratio transmission cannot operate close to the optimum line at all times.

A rubber V-belt CVT has a potential mechanical efficiency advantage over a discrete-ratio automatic transmission owing to its lower parasitic losses. A discrete-ratio automatic

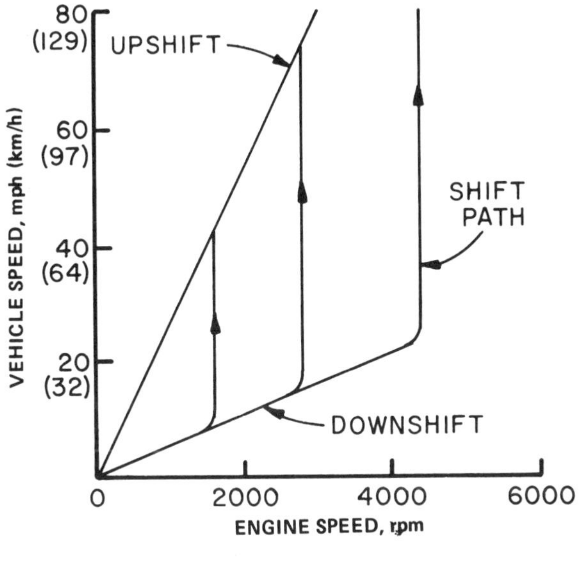

Fig. 1 - Vehicle speed vs. engine speed shift paths for a CVT designed for a constant engine speed during fixed-throttle upshifts

*Numbers in parentheses designate References at end of paper.

transmission requires a hydraulic pump to actuate the clutches and brakes, to pressurize the torque converter, and to engage the converter lock-up clutch. The V-belt CVT can be designed to operate without any hydraulic pump for control functions. It is believed that the improvement in fuel economy obtained by this demonstration CVT is partially due to the elimination of the pump.

Another area of comparison between the V-belt CVT and current automatic transmissions is the manufacturing cost. A cost study made by a consultant familiar with automotive manufacturing practices of a proposed CVT based upon this demonstration CVT shows a significant advantage over a three-speed automatic transmission. The cost advantage is due to design simplicity and fewer parts in the CVT.

THE HIGH CAPACITY V-BELT

The V-belt used in this CVT is the Power Trac™ belt currently under development by The Gates Corporation. It is a rubber-based belt that runs dry on steel sheaves. The belt is constructed of trapezoidal blocks fastened to a flat-belt tension member. The blocks are molded of rubber and fabric and have internal reinforcing members to resist the transverse or axial forces. The flat-belt tension member is made of aramid cords and rubberized fabric. This V-belt has substantially more power capacity than a conventional rubber variable speed belt owing to its high transverse stiffness. The reason for the improvement in capacity is that a conventional variable speed V-belt is constructed of a layer of tensile cords surrounded by rubber and fabric with a trapezoidal cross section. At high loads, the conventional belt will buckle transversely, resulting in slip and uneven cord loading.

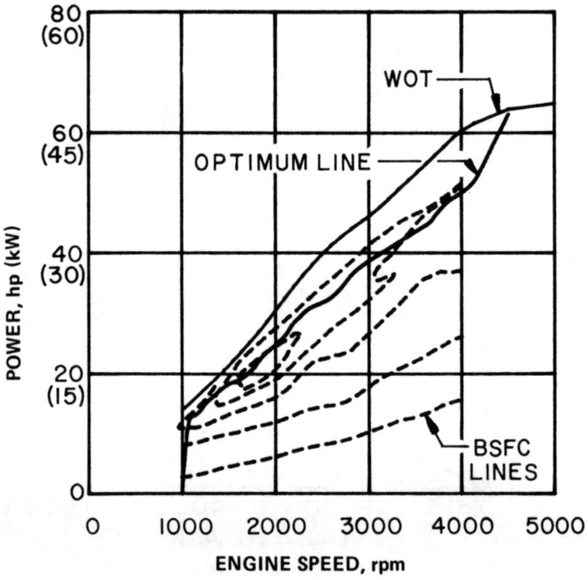

Fig. 2 - Engine power vs. speed showing the optimum operating line

Belt life is highly dependent upon the load cycle since the textile tension member is subject to cyclical forces: tight and slack side tension, bending loads, and centrifugal loads. The life of the tension member is a function of the cumulative fatigue due to the load conditions, so it is important that the belt be tensioned just enough to prevent slip at each torque load condition. For maximum belt life, then, the sheave actuators must tension the belt as a function of the torque load and the belt pitch diameter. In this CVT with a fixed center distance between sheaves, the belt is tensioned by applying the necessary side force to each of the variable pitch sheaves.

RATIO RANGE SELECTION

The basic design parameters for a CVT vehicle are the minimum and maximum N/V ratios (engine rpm/vehicle mph), which affect fuel economy, performance, and belt life. The conflicting requirements of fuel economy, performance, and belt life must be compromised to reach acceptable levels for all three. Note that a CVT does not have the problem of ratio placement within the overall range as does a discrete ratio transmission. The minimum N/V affects fuel economy, and for the vehicle used in this demonstration, a minimum N/V of 31 would be required to obtain the maximum level road fuel economy (Fig. 3). However, for the EPA urban driving cycle, the improvement is negligible

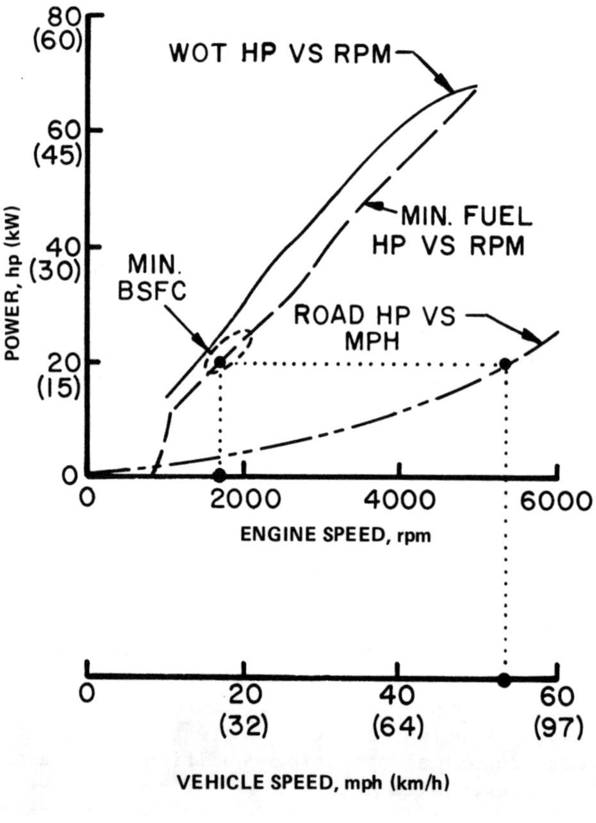

Fig. 3 - Power vs. engine speed and power vs. vehicle speed to show the minimum N/V for lowest level road fuel consumption

below 32 N/V and small below 36 N/V (Fig. 4). This is due to the limited time and low power expended during conditions where an N/V of 32 or less could be used.

Maximum N/V affects acceleration and gradeability. The 0-to-60 mph acceleration time is reduced as maximum N/V increases, but for maximum N/V ratios from 180 to 200, there is not a large difference (Fig. 5). Start-up gradeability is almost directly proportional to the maximum N/V. The relationship between N/V and percent grade is:

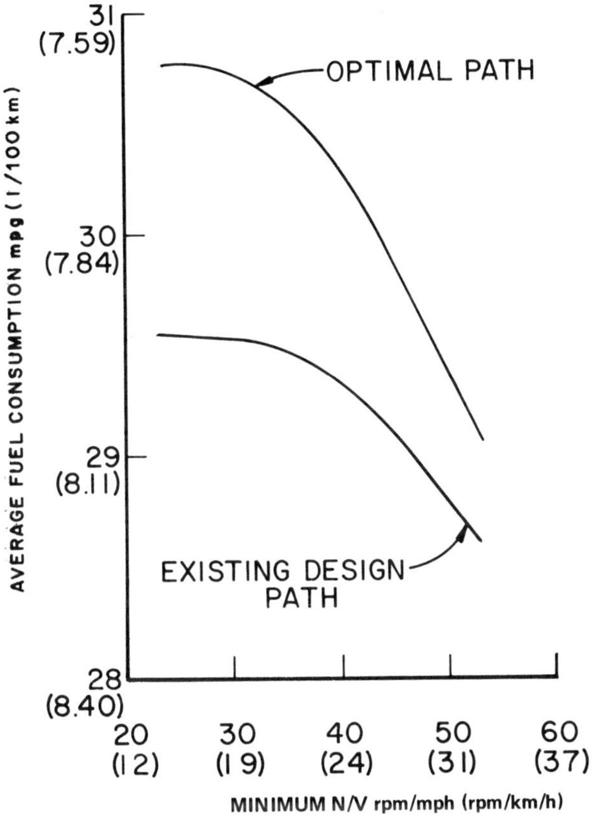

Fig. 4 - Predicted effect of minimum N/V on fuel economy for EPA urban cycle with maximum NV from 180 to 200

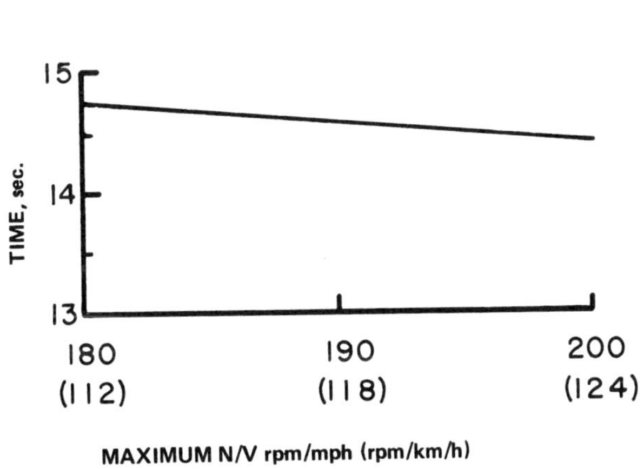

Fig. 5 - Predicted 0 to 60 mph time vs. maximum N/V

$$N/V = \frac{44}{\pi} \cdot \frac{W}{Q} \cdot \frac{100}{E} \cdot \sin\left(\tan^{-1} \frac{G}{100}\right) \quad (1)$$

For the Horizon:

W = 3000 lb. - gross vehicle weight
Q = 75 lb-ft. - WOT engine torque at 1500 rpm
G = 27.5% - maximum grade at sea level
E = 80% - assumed driveline efficiency at stall

The maximum N/V should be at least:

$$N/V = \frac{44}{\pi} \cdot \frac{3000}{75} \cdot \frac{100}{80} \cdot \sin\left(\tan^{-1} \frac{27.5}{100}\right) = 185.7$$

Belt life is normally reduced as the pitch diameters decrease because of the increase in tension and bending stresses. For a given maximum pitch diameter, an increase in the speed variation (ratio range) tends to decrease belt life because it reduces the minimum pitch diameter. In the Horizon, available space limited the maximum pitch diameter to 9.8 inches. The maximum N/V was set at 186 to satisfy the 27.5% grade requirement. The minimum N/V was set at 37.5, which is close to the maximum fuel economy for the design operating line (Fig. 4). The variation of 4.96 (186 ÷ 37.5) gives a calculated belt life of 70,000 miles at the 10% failure level (Fig. 6).

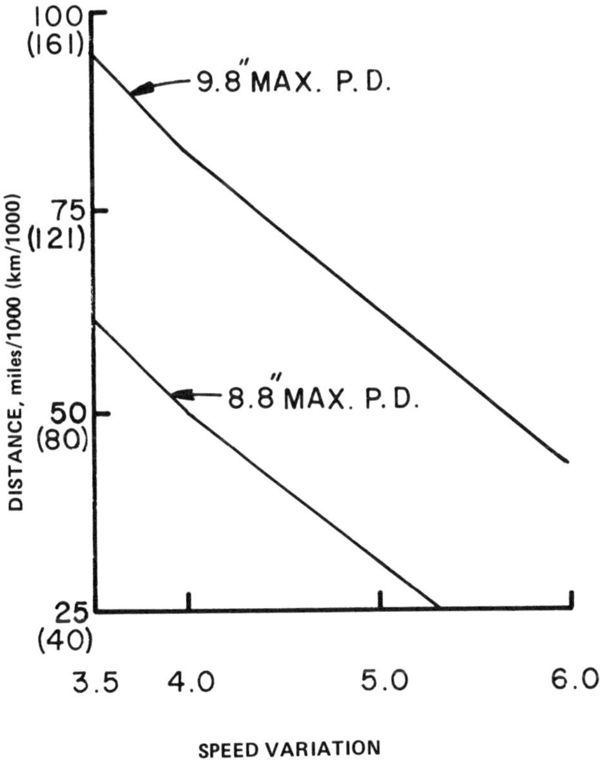

Fig. 6 - Calculated belt life (10% failures) vs. CVT variation for a maximum N/V of 186 rpm per mph

The data for the fuel economy vs. minimum N/V plot (Fig. 4) were obtained from a computer simulation that calculates the instantaneous power needed for each condition of a driving cycle. It then calculates the fuel required at that power level based upon a particular operating line for the engine (described later in the shift path section). This fuel consumption simulation is used to compare different operating lines rather than to predict actual vehicle mpg. The data for acceleration time vs. maximum N/V are computed by a numerical integration of the instantaneous acceleration of the vehicle. The belt life vs. variation data are obtained from a computer simulation that computes the fatigue load for each condition of a specified driving cycle. It weighs each fatigue load on a time basis and calculates belt life from the experimentally determined fatigue data of the tensile cord. The fatigue life is sensitive to the minimum pitch diameter, so it is important to choose a maximum diameter along with the N/V range that will result in an acceptable belt life.

The required fixed ratio in addition to the V-belt ratio needed to obtain the desired overall N/V is given by:

$$R = \frac{60 \cdot (N/V)}{U \cdot T} \qquad (2)$$

U = 2.227 - max. dR/dN ratio
T = 896 - tire revolutions per mile
N/V = 186 - max. rpm/mph

$$R = \frac{60 \cdot 186}{2.227 \cdot 896} = 5.593$$

In this demonstration a final drive ratio of 3.76:1 is used because of parts availability so a 1.49:1 (5.593 ÷ 3.76) additional reduction was necessary. This was placed between the engine and V-belt to allow an existing final drive assembly to be used without change. Even without constraints on the final drive ratio, some speed reduction between the engine and the V-belt is necessary to limit maximum belt speeds in this application.

OPERATING LINE

As described earlier, the envelope of the operating points of a CVT is a line on the BSFC map if shifting is done at a constant engine speed. Before the sheave control system can be designed, an operating line must be chosen. The optimum operating line is a monotonically increasing line on the engine BSFC-horsepower map (a single power value for each speed and a single speed value for each power) which results in the minimum BSFC at each power. This line could require a lower minimum N/V (thus larger sheaves) than space permits (Fig. 3). Most control systems require that the operating line is single-valued on the engine torque map; i.e., there is just one torque value for each speed and one speed value for each torque (Fig. 7).

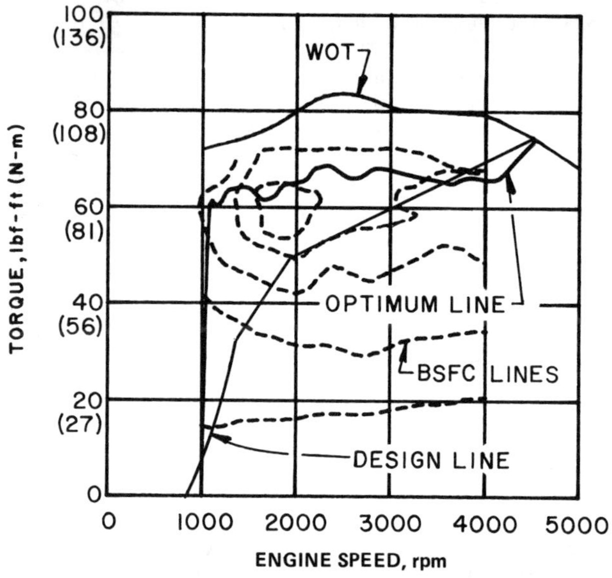

Fig. 7 - Optimum operating line on an engine torque plot

This requirement plus the need to terminate at or near the peak power point defines an operating line spaced away from the optimum line.

After the operating line is established, each engine power level is related to an engine speed. For a specific operating condition, the vehicle speed, engine power, and the engine speed are determined; thus the transmission ratio and N/V can be determined. If the computed N/V is outside the transmission N/V range, then the transmission is operating at its ratio limit, and the engine speed is computed from the vehicle speed and the limiting N/V value. The fuel consumption can be computed for either case, within the N/V range or at the N/V limit, since these points can be located on the BSFC map. A fuel economy simulation was used to compare different operating lines. The "existing" design" line (Figs. 4, 7) used in the demonstration vehicle results in 1.0 mpg (3.4%) lower mileage than the optimum over the EPA urban cycle with a 37.5 minimum N/V.

AXIAL FORCE DETERMINATION

After the ratio range and pitch diameter range have been selected, the sheave diameter and center distance must be checked with the available space envelope of the vehicle. When they are acceptable, the axial forces required to sustain the peak torque loads are calculated. The axial force calculations represent the numerical integration of closed-form solutions to the V-belt force and slip equations given by Gerbert. (5, 6) The minimum axial force that will support the applied torque for the full range of pitch diameters is calculated for both the driveR and

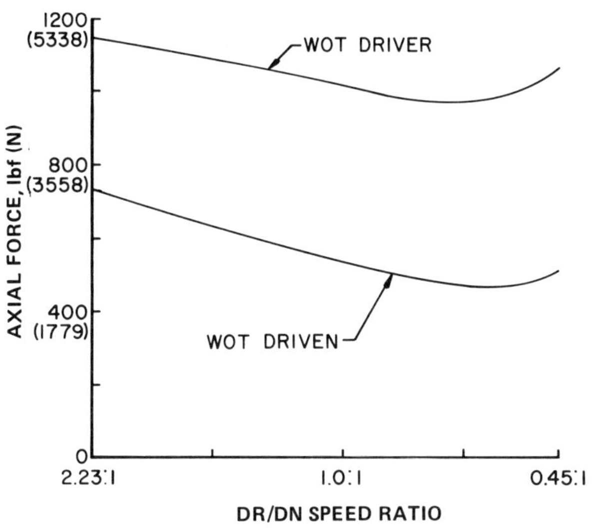

Fig. 8 - Desired axial force of driveR and driveN sheaves at WOT

driveN sheaves (Fig. 8). When these axial forces are applied to the sheaves and the specified torques are applied to the shafts, the belt system is balanced and stable and will not tend to upshift or downshift. If, however, the output torque is large enough to accelerate the vehicle, then both sheaves will tend to increase in speed but if the driveR force is a function of speed, its axial force increases while the driveN axial force remains the same, so the transmission upshifts (for a constant throttle setting). In response to a change in throttle (and engine torque), an axial force unbalance is created, and when the sheave axial friction is overcome, the transmission shifts up or down until balance is achieved at a new speed ratio. In addition to improving belt life, another reason for designing for minimum belt tension is that it minimizes the axial force unbalance needed to initiate shifting, resulting in rapid shift response.

SHEAVE CONTROL

There are two design approaches to sheave control: One is an open-loop force-balance system for both the driveR and driveN actuators; the second approach is to use a closed-loop position or ratio-control actuator on one sheave and an open-loop force actuator on the other. Either approach can work. The driveN open-loop force-balance actuator supplies an axial force slightly in excess of that needed to prevent slipping at WOT and proportionally less at part throttle. The open-loop force-balance driveR actuator supplies an axial force that is a function of engine speed to balance the driveN axial force that is a function of driveN torque. Since driveR torque is related to driveN torque by the transmission ratio (sheave position), the driveR axial force is a direct function of engine speed and an indirect function of engine torque. This control system holds the engine to a specified engine power and speed operating line unless the CVT is at its upper or lower ratio limit.

If a closed loop driveR controller were used with this torque responsive driveN actuator, it would position the driveR sheave so that engine speed is a direct function of throttle position and manifold vacuum and, therefore, an indirect function of engine torque. The open-loop controller has no parasitic losses because it uses a centrifugally actuated force vs. speed mechanism on the driveR sheave. A closed-loop controller would have the advantage of flexibility in programming the operating line and the ability to control "kickdown" and downgrade ratio control conditions but would require a full-time external power source for a hydraulic, pneumatic, or electromagnetic actuator with its attendant parasitic loss. The open-loop control system was chosen to obtain maximum efficiency and minimum complexity.

The driveR and driveN sheave controllers must provide side or axial forces to satisfy three basic requirements:

1) The belt should not slip at any load.
2) For the longest belt life, the axial force should be as low as possible for each load.
3) The CVT speed ratio should be controlled to obtain the desired engine operating line.

Requirements 1) and 2) for slip and life are met by supplying axial forces equal to those described in Fig. 8 for each torque load. The driveN sheave actuator is a mechanical torque actuator plus a diaphragm spring. The torque actuator mechanism is equivalent to a "torque ramp" with a varying cam angle.

A diaphragm spring provides a restoring force to downshift the transmission as the vehicle stops. The spring force is nearly constant, ranging from 90 to 140 pounds over the shift range. This is sufficient to downshift the CVT as the vehicle stops. If a conventional positive rate coil spring were used, it would not be possible to obtain the desired axial force curve (Fig. 8) over the full torque range because a positive rate spring supplies its maximum force at upshift (driveN sheave opened) where the required total force is at or near a minimum. A positive rate spring plus a variable angle cam could supply the desired axial force at maximum torque but would supply excessive axial force at part throttle, and the result would be reduced belt life.

The driveN torque actuator provides the difference between the required WOT axial force (Fig. 8) and the diaphragm spring force. The axial force-to-torque factor for the driveN actuator is the ratio of the required axial force to the torque on the movable driveN sheave half (one half the driveN torque). This factor varies over the shift range.

For example:

$$K = \frac{(F - S) \cdot 2}{Q \cdot R \cdot U} \qquad (3)$$

$$K = \frac{(785 - 50) \cdot 2}{(74 \cdot 1.49 \cdot 2.227)}$$

$= 5.98$ lbs/lbf-ft (20.04 N/N-m)

where:

F = 785 lbf - WOT driveN axial force at U
S = 50 lbf - driveN spring force at U
Q = 74 lbf-ft - engine torque at WOT
R = 1.49 - engine-to-driveR speed ratio
U = 2.227 - driveR-to-driveN speed ratio

Fig. 9 shows the desired force-to-torque factor and the actual mechanism factor over its full range. The mechanism provides an axial force no less than the desired force at any point to prevent slip. The sinusoidal shape of the desired curve is due to the influence of the diaphragm spring. The total axial force on the driveN sheave is therefore proportional to torque load and it varies from 90 pounds at closed throttle to 780 pounds at WOT.

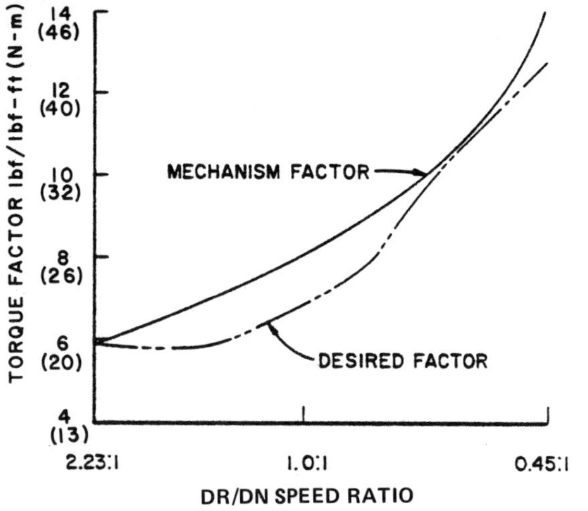

Fig. 9 - DriveN actuator torque factor, design and actual mechanism

The driveR actuator is a centrifugal weight actuator that is somewhat different from typical centrifugal actuators. As described earlier, the driveR axial force must match both the axial force vs. torque plot of Fig. 7 and the requirements of the engine operating line plot of Fig. 8. From Fig. 8 it is seen that the intersection of the operating line and the WOT line occurs at 4600 rpm. Therefore, since the shifting occurs at a constant engine speed, the WOT line of Fig. 8 can also be identified as a 4600-rpm condition. The same can be done for each fractional throttle (or percent WOT torque) condition. A cross plot of this information to show driveR axial force vs. engine speed at fixed sheave position can be made to help characterize the driveR actuator (Fig. 10). A normal centrifugal actuator gives an axial force proportional to the square of the speed, a parabola, which is appreciably different from Fig. 10. A modified centrifugal mechanism was devised, which closely approximates the desired curve.

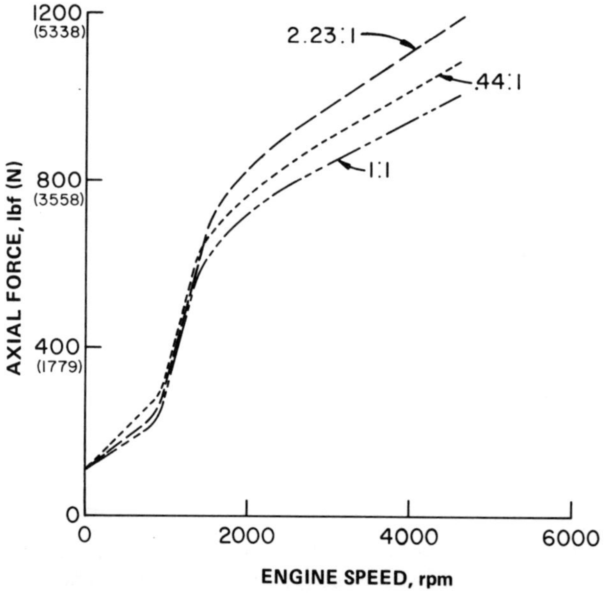

Fig. 10 - DriveR axial force vs. engine speed for three transmission ratios

TRANSAXLE DESIGN

The transaxle, Fig. 11, consists of a centrifugal dry clutch, a 1.49:1 speed reduction using a Gates Poly Chain® synchronous belt, a forward-neutral-reverse gearbox, the driveR sheave and actuator, the V-belt, the driveN sheave and actuator, and the final drive and differential gearing. The basic specifications for the unit are given in Table 1. Since this CVT is a demonstration unit, existing components were used wherever possible.

Table 1 - CVT Vehicle Specifications

Pitch diameter range	4.4 in. to 9.8 in. (112 to 249 mm)
Sheave o.d.	10.64 in. (270 mm)
Center distance	10.87 in. (276 mm)
V-belt speed ratio	2.227:1 to 0.449:1
Variation	4.96
Vehicle N/V	186 to 37.5 rpm/mph (116 to 23.3 rpm/km)
Engine to driveR ratio	1.49:1
Final drive ratio	3.76:1
Wheel size	896 rev/mile (557 rev/km)
Clutch	7 in. dry centrifugal (178 mm)
Engine torque at WOT shift	74 lbf-ft. @ 4600 rpm (100 N-m)
DriveR torque at WOT shift	107 lbf-ft. @ 3042 rpm (145 N-m)
Vehicle weight	2250 lb. (1023 kg)

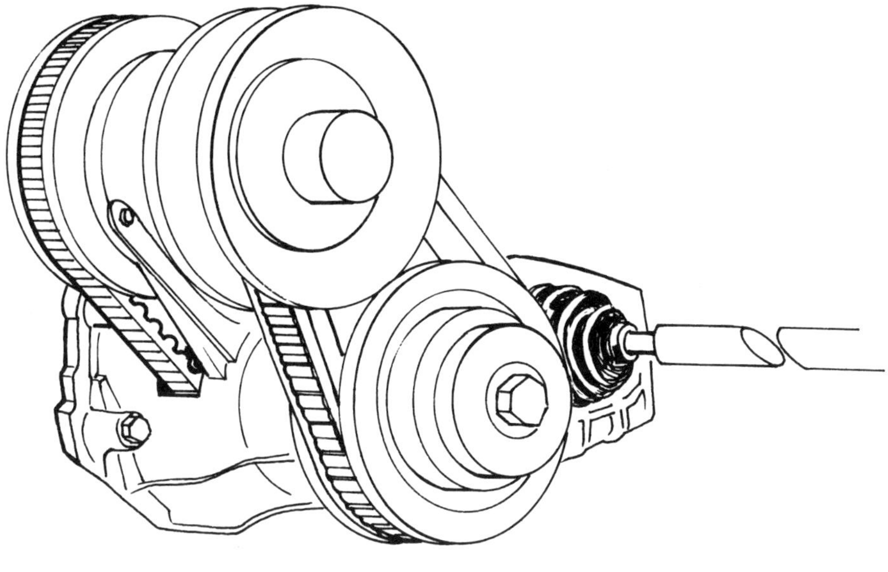

Fig. 11 - CVT transaxle arrangement; driveR sheave above and driveN sheave below

A centrifugal clutch was chosen because it is self-actuated and has no parasitic losses. It is placed between the engine and the fixed speed reduction. During the design stage, there was concern that clutch placement upstream from the V-belt would inhibit downshifting when coming to rest because the V-belt stops turning when the car stops. In practice the closing force of the spring on the driveN sheave downshifts the transmission during the short interval between 10 mph and stopping.

The forward-neutral-reverse gearbox is a planetary unit engaged through synchronizers from a manual transmission. It does require complete clutch disengagement to allow shifting into forward or reverse.

The vehicle did not need any major modifications to accept the CVT (Fig. 12). To install the transaxle it was necessary to relocate the battery and the electric radiator fan (Fig. 13). The engine was also moved one inch to the right; however, the need for this change could be eliminated by some repackaging of the CVT. A pan was added underneath the V-belt to shield the belt from direct road splash (Fig. 14).

CVT PERFORMANCE

A transmission dynamometer was instrumented to measure the mechanical efficiency of the CVT (Fig. 15). Input and output power was measured by speed and torque sensors on the input and output shafts. The CVT was powered by an internal combustion engine similar in power to the one in the car. The aerodynamic drag, rolling friction, and mass of the car were simulated by a water brake, friction brake, and inertia wheel. To evaluate the effects of shifting and wide variations in power on efficiency, a simulated EPA driving cycle was used. The CVT was "driven" through the EPA cycle, and the speed and torque were

Fig. 12 - Horizon CVT vehicle

Fig. 13 - Top view of CVT installation

recorded on tape each second. The tape was processed by a computer to accumulate the total mechanical work in and out. The same test was run on an automatic three-speed open-torque converter transmission. The test includes idle time during which there is some input and no output power. The improvement shown by the CVT in Table 2 can be attributed to the high belt efficiency and the absence of parasitic losses.

Table 2 - Transmission Mechanical Efficiency, EPA Urban Cycle

CVT	85.0%
3-speed automatic	72.2%
CVT % improvement	17.7%

Fig. 14 - Bottom view of CVT installation with splash shield removed

Fuel efficiency of the vehicle was compared between the CVT and a three-speed automatic transmission by running the EPA emission tests before and after the CVT installation into the vehicle. Both tests were conducted by the EPA-certified Automotive Testing Laboratory in Denver; the results are summarized in Table 3.

Table 3 - Vehicle Fuel Economy, MPG, EPA Cycle

	Urban	Highway	Combined*
CVT	26.80	36.11	30.32
3-speed automatic	23.42	30.79	26.25
CVT % improvement	14.43%	17.28%	15.51%

*Combined on distance basis, 55% urban, 45% highway.

The CVT was designed for a wider N/V range (lower minimum N/V) than the three-speed automatic, but for the EPA test the minimum N/V of the CVT was mechanically limited to 50. Since the minimum N/V with the three-speed automatic was 52, the engine was running over nearly the same speed range for both tests. This means that the improvement shown by the CVT in this test is mainly due to its better mechanical efficiency and an improved engine operating region. Some additional improvement would be realized with the CVT operating over its full N/V range.

The 10-to-60 mph WOT acceleration is about equal to that of a three-speed automatic transmission. WOT acceleration from rest is equivalent to that of a manual transmission with a 2000-rpm clutch-engagement speed. A careful match of the clutch torque curve to the engine torque curve is necessary for good engagement quality at all throttle positions.

Downshift response at WOT is a function of the sheave unbalance force and the sheave axial friction. The unbalance force is proportional to the torque differential between the operating and the WOT line (Fig. 7), which generally decreases as engine speed increases. WOT downshifts are good throughout the operating range. Increasing the unbalance force would increase the amount and rate of ratio change during a WOT downshift.

Most drivers have described the driveability of this CVT as very good. The engine sound is different because of its constant speed during acceleration. Belt noise was not considered objectionable. Smoothness of operation is particularly noticeable going up a winding road because of the absence of frequent shifting. A means of limiting the CVT ratio for engine braking may be added, although long, ten percent downgrades have not caused any difficulty. This CVT-equipped vehicle has been driven over 9500 miles in a wide range of temperatures and weather conditions without adverse effects.

Fig. 15 - CVT on dynamometer for operational and efficiency tests

SUMMARY

The design and successful installation of this rubber V-belt CVT into a Horizon was done without major modifications to the car. It shows more than a 15% improvement over an automatic transmission in both mechanical efficiency and vehicle fuel economy while maintaining equivalent acceleration performance and providing good driveability. The mechanical self-actuated sheave controllers successfully fulfill the basic control system requirements, contribute to high efficiency by minimizing parasitic losses, and maximize belt life by tensioning the belt in proportion to torque load. This project also demonstrates a design method to specify the sheave controller axial force requirements to match a desired engine operating line. The method would be usable for the design of various types of variable-speed V-belt drives. A rubber V-belt CVT has the potential of combining good fuel economy, smooth operation, and a competitive cost in an automatic transmission for front-wheel-drive cars.

REFERENCES

1. H. E. Chana, W. L. Fedewa, and J. E. Mahoney, "An Analytical Study of Transmission Modifications as Related to Vehicle Performance and Economy." Paper 770418 presented at the SAE International Congress, Detroit, February 1977.

2. R. H. Thring, "Engine Transmission Matching." Paper 810446 presented at the SAE International Congress, Detroit, February 1981.

3. R. R. Radtke, L. E. Unnewehr, and R. J. Freedman, "Optimization of a Continuously Variable Transmission with Emission Constraints." Paper 810107 presented at the SAE International Congress, Detroit, February 1981.

4. B. C. Christenson, N. H. Beachley, and A. A. Frank, "The Fuel Saving Potential of Cars with Continuously Variable Transmissions and an Optimal Control Algorithm." Paper 75-WA/Aut-20, presented at the ASME Winter Meeting, Houston, November 1975.

5. B. G. Gerbert, "Force and Slip Behaviour in V-Belt Drives." Mech. Eng. Series No. 67, Acta Polytechnica Scandinavica, Helsinki, 1972.

6. B. G. Gerbert, "Adjustable Speed V-Belt Drives--Mechanical Properties and Design." Paper 740747 presented at the SAE National Farm Machinery Meeting, Milwaukee, September 1974.

Development of a Flat-Belt, Continuously-Variable Transmission*

Emerson L. Kumm and Theodore C. Kraver
Kumm Industries
Phoenix, AZ

**Paper 850488 presented at the International Congress and Exposition, Detroit, Michigan, February, 1985.

ABSTRACT

A unique, flat belt, continuously variable transmission with hydraulic actuators is presented. The CVT was designed and tested under the NASA/DOE electric vehicle program. Performance of the 75 KW, 4.3:1 speed ratio range, 46 KG CVT over the 600 to 6000 input rpm range tested at efficiencies of over 90%. Design modifications can increase the speed ratio range to greater than 6:1. Development issues include sliding surface wear, belt durablity, actuation control systems, temperature effects and scaling. Potential applications include automotive, truck, off-road, aerospace, and industrial.

HISTORY

THE FLAT BELT is inherently a very efficient device for transmitting power. Losses as low as 1% can be expected from a well designed two pulley system. The challenge has been to design a variable diameter pulley system that would maintain adequate but not excessive belt tension at all times. An intensive technology study and design effort in the late 1970's resulted in a patented device that promised to provide a solution, Figure 1.

In the late 1970's the Department of Energy (DOE) initiated the Electric and Hybrid Vehicle Program. Project management was delegated to NASA. Over 20 proposals were submitted in the design competition for a variable speed assembly (VSA). Emerson L. Kumm, Mechanical Engineer, was one of the four winners. After successful completion of the VSA design (DEN 3-114) in 1980, Mr. Kumm was the only contractor awarded support for a follow-on development phase. Under NASA supervision, the prototype transmission and test facility was built. The feasibility testing was completed in the spring of 1984.

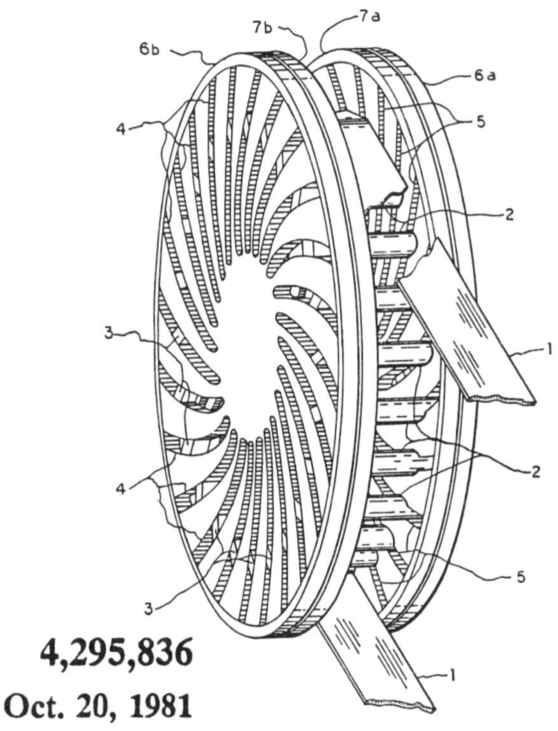

Kumm Flat Belt Pulley

Figure 1

Gasoline prices have now stablized at much lower prices than forecast in the late 1970's. The electric and hybrid vehicle market for the CVT is no longer a near term consideration. However, this CVT technology holds promise for a number of other vehicular and industrial applications.

HOW THE KUMM FLAT BELT CVT OPERATES

The flat belt runs on segmented, barrel-like surfaces generated with 24 belt drive elements on each pulley, Figures 1, 2, and 3. These belt drive elements are parallel to the pulley axis of rotation. At the smallest radius, these elements are close together, forming a nearly solid surface. As the belt drive elements move out to a radius that is 2 to 3 times longer than the minimum radius, gaps are created between the belt drive elements.

Two guideway discs form each side of a pulley. These discs have logarithmic spiral guideway slots which support the ends of the belt drive elements. The set of guideways in one disc have a clockwise curvature. The second disc's guideways have a counter-clockwise curvature. The cavities generated by the guideways' intersections provide precise positioning of the belt drive elements. The two inner discs are locked to move together and the two outer discs are also locked to each other.

As the pulley rotates at high speed, the two sets of pulley discs make small rotations with respect to each other. This relative disc movement causes the syncronized radial movement of the belt drive elements.

BELT DRIVE ELEMENTS

Figure 2

GUIDEWAY DISCS (One side)

Figure 3

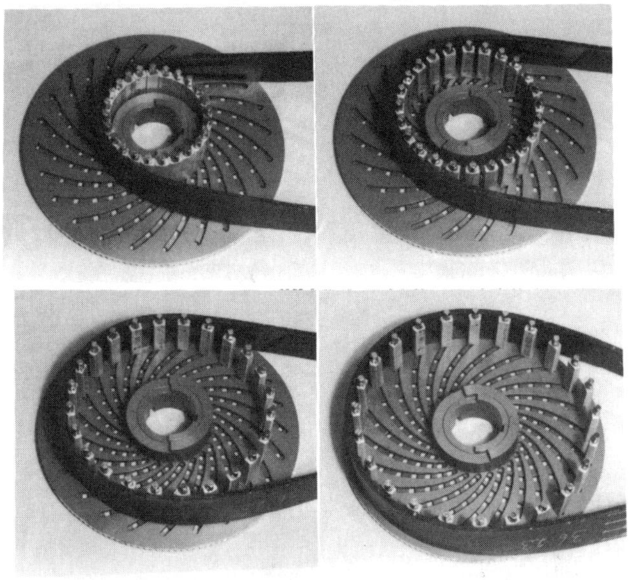

RADIAL POSITIONING OF BELT BY DRIVE ELEMENTS IN GUIDEWAY DISCS
(One set of discs are removed)

Figure 4

The pulley hydraulic actuator rotates the inner set of discs with respect to the outer set of discs. The two rotary actuators, one on each pulley, generate the required belt tension at the desired speed ratio.

The hydraulic actuator is built into the pulley system and rotates with the pulley shaft, Figure 5. The pulley shaft, with either two or three actuator vanes (or shaft struts), is connected to the outer guideway discs. The rotary actuator case (or housing) is attached to the inner guideway discs. Hydraulic control oil, at pressures typically of .3 to 1.7 MPa (50 to 250 psig), is supplied to the cavities on either side of the actuator vanes. The control oil enters through separate passages inside the rotating shaft.

The vanes are positioned with respect to the actuator case by the differential hydraulic pressures in the pulley system's two actuators. These pressures control the belt tension to prevent belt slippage. An increase or decrease in the differential hydraulic pressure in either rotor actuator changes the speed ratio to another equilibrium position.

Centrifugal forces on the belt drive elements will increase belt tension, and become significant at high pulley speeds. The differential hydraulic pressure in the rotary actuator is used to counteract the centrifugal force to avoid excessive loads on the belt.

ROTARY HYDRAULIC ACTUATOR

Figure 5

Figure 6 gives a graphic illustration of how the output speed is varied for a constant input speed.

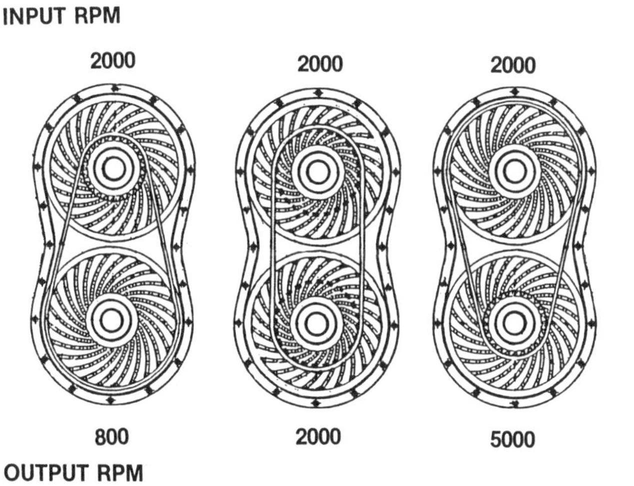

SPEED RATIO CHANGE

Figure 6

MAJOR COMPONENTS FOR VSA PULLEY SYSTEM

The guideway discs were NC machined from 7075-T6 aluminum. They were then treated with a Teflon impregnated anodized process (Nituff). This low friction, 0.05 mm (.002 in) thick surface has a hardness of approximately 60 Rockwell C.

The pulley shaft in the rotary actuator rotates 81 degrees with respect to the actuator case. A 4.3:1 speed ratio range for the VSA design results. Buna N rubber strip seals were used to seal the shaft struts to the case internal surfaces. Internal seal bypass flows were approximately 2.0 liters per minute at 1.4 MPa (200 psid) in the VSA design. Current designs eliminate nearly all of this bypass leakage flow.

The belt drive elements should be light to reduce their centrifugal loads. They must also support high radial loads. The VSA design used a ZK 60A-T5 magnesium extrusion from Dow Chemical as the housing, Figure 7. A high strength 7075-T6 aluminum rod was inserted to carry the loads to the guideway slots. The slide surfaces were machined at 84 degrees to fit within the guideway slot. The final VSA successfully tested design used 1.59 mm (.063 inch) roll pins and epoxy cement between the housing and rod to prevent twisting due to torque. The aluminum rod was plated with 0.025 mm (.001 in) electroless nickel to give abrasion resistance.

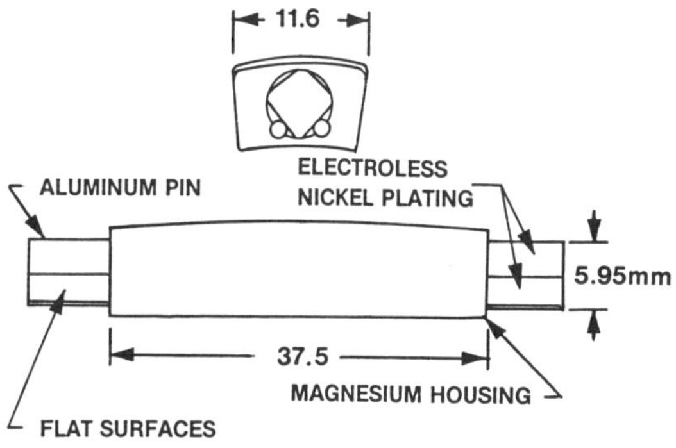

BELT DRIVE ELEMENT

Figure 7

Two belt manufacturers (Gates Rubber and Dayco) provided belts. The final tests were made with Gates belts modified by our surface coatings, Figures 8 and 9. The 34.9 mm (1.375 in) wide and 4.76 mm (.188 in) thick belts have Kevlar (Dupont) cords for strength. Breaking load is of the order of 33,000 N (7500 lbf) for one side. This is 6 to 7 times the maximum design load. The Kevlar cords operate at an infinite cycle stress level.

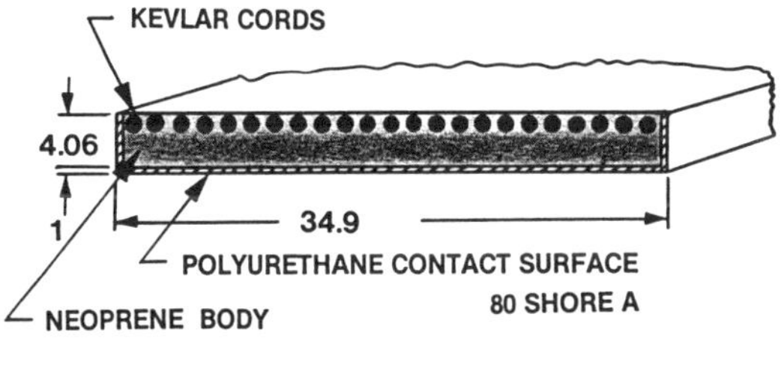

ELASTROMERIC FLAT BELT CROSSECTION

Figure 8

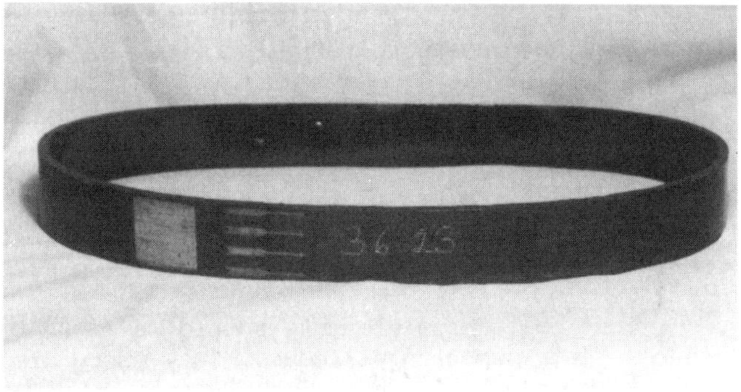

CVT FLAT BELT

Figure 9

A high belt contact friction coefficient is critical for effective design and operation. Neoprene rubber has a maximum friction coefficient of about 0.45 against a magnesium surface, whereas the polyurethane belts had only 0.20 to 0.25. A new rubber material (ML-6 of Meridian Laboratories) was tried as a coating and gave friction coefficients of 1 to 1.2 at slip. Unfortunately, the bonding of the ML-6 to the neoprene was inadequate, resulting in separation from the base belt material, Figure 10. With proper surface treatment, a thin coat (1mm) of low hardness polyurethane rubber, Flexane 80 (Devcon Corporation), has adhered well to the neoprene. It has excellent low wear characteristics and gives a high slip friction coefficient. The friction coefficent at slip is of the order of 0.7 to 0.8 on the magnesium surface drive elements.

ADHESION PROBLEMS IN FLAT BELT COATINGS

Figure 10

NASA VSA DESIGN CRITERIA

The input to the electric vehicle VSA is a clutched flywheel having a maximum speed of 28,000 RPM and an electric drive motor. The VSA used a Kumm flat belt CVT designed with a 4.3:1 speed ratio range, Figure 11. The gearing provided a two speed mode with synchronized shift operation from high to low range. The planetary gearing arrangement is used in low range to provide a geared neutral and a limited reverse speed.

The design provided an infinitely variable transmission with -470 to +5000 RPM output. Maximum design power is 75KW (100 HP) with 450 N-m (330 ft-lb) maximum output torque. Design efficency was calculated to be 93% to 95% at the design point of 16KW, 3000 RPM output, 21,000 rpm input. Design life was 2600 hours (10% failure). Weight dry was 44.5 KG (98.0 LB), without flywheel, flywheel housing, and drive motor.

DOE/NASA VARIABLE SPEED ASSEMBLEY FOR ELECTRIC VEHICLE APPLICATION

Figure 11

NASA VSA DESIGN ANALYSIS

Design analyses were made for stress, vibration, and performance. Windage and lubication related losses as a function of rotational speed, and torque related losses were calculated for the seals, bearings, gears, and the pulley systems. The calculated efficiency across the entire transmission varied from 95.2 to 93.2 percent as a function of the the flywheel input speed, 14,000 to 28,000 RPM.

The critical speeds and torsional vibration modes were examined for each shaft assembly. All were significantly higher than the operating speeds of the VSA.

The bearings, gears, and shafts were designed for 2600 hours of normal duty cycle, using industry standard design practices.

The belt-pulley system requires special analysis. The maximum stress points in the guideway discs are the rim and the thin web section at the lowest radius of the quideway slot. The rim was designed to have sufficient radial width to contain the mass of the disc and the belt drive elements through hoop stress. The overspeed design condition is 16,000 rpm with failure calculated to occur at 24,000 rpm.

The shear stress in the thin web at the lowest radius of the guideway slot depends on the pulley operating torque at pulley radius ratios giving maximum pulley speeds. A design safety factor of 5 was used for the maximum torque load.

The belt drive elements are subject to the belt loads, their centrifugal load, and the end loads from the disc slots, Figure 12. The maximum bending stress of the pin is less than 83 Mpa (12000 psi) for the extreme torque condition. The maximum load is 21 MPa (3100 psi) for the end of the pin on the guideway slot.

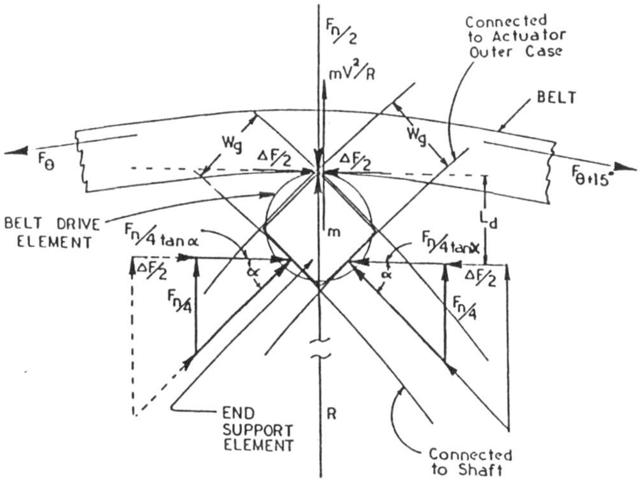

BELT DRIVE ELEMENTS LOADS

Figure 12

VSA TEST FACILITY

The test stand is driven by a 50 horsepower DC motor having a maximum speed of 2800 RPM, Figures 13 and 14. Eaton Lebow torque/speed sensors provide performance data. The test stand was originally built with a reverse operating VSA to load the test VSA with recirculating power in a back-to-back drive arrangement, Figures 15 to 16. Major dynamic and control problems were created by attempting to power the experimental transmission with a duplicate experimental transmission. The reverse operation VSA was removed. The final performance tests were conducted with a water cooled hydraulic caliper disc brake at lower powers.

VARIABLE SPEED ASSEMBLY TEST STAND

Figure 13

VSA IN TEST STAND

Figure 15

VSA IN TEST STAND

Figure 16

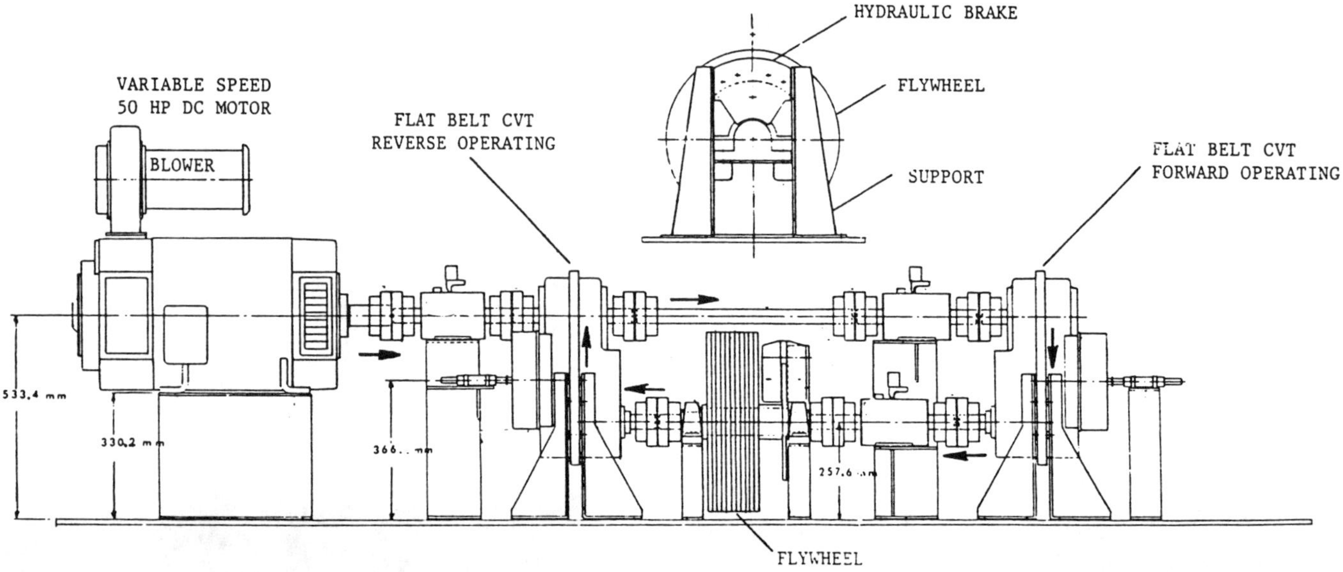

VSA TEST STAND

Figure 14

The hydraulic control rig supplies pressurized oil for lubrication, sychronizer shifting, and the hydraulic brake, Figure 17. Stepper motors adjust the three oil pressure requlators that operate the hydraulic actuators in the pulley system.

HYDRAULIC CONTROL RIG

Figure 17

The electronic control rig is managed by the Hewett Packard 9826 computer controller. It controls both the test speed and pulley system pressures. It also records, reduces, and displays test data. Belt slip or creep can be measured by Hall effect pickups triggered by small magnets attached to the shaft (outer pulley disc) and structure attached to the rotary actuator housing.

ELECTRONIC CONTROL RIG

Figure 18

VSA TEST RESULTS

A large number of tests were run to develop workable belt and belt drive element configurations. A few high horsepower steady state points were obtained from the back-to-back recirculating power rig. It was not practical to conduct transmission mapping with that configuration. By removing the second VSA, the rig was limited to 50 horsepower input, at the maximum rpm of the DC motor. At lower DC motor speeds the available power was also less. The caliper disc brake limited tests to much lower steady state horsepower loadings. Water sprays were used to cool the disc brake.

Previous detailed analyses (NASA CR-159822) indicated that speed dependent losses could be separated from torque dependent losses. At a constant low input speed, the VSA was tested over a range of speed ratios while varying the torque over a wide range. The torque dependent losses are on Figure 19.

Also, using a constant low output torque over the speed ratio range, power losses were mapped as speed varied over a wide range. The speed dependent losses are on Figure 20.

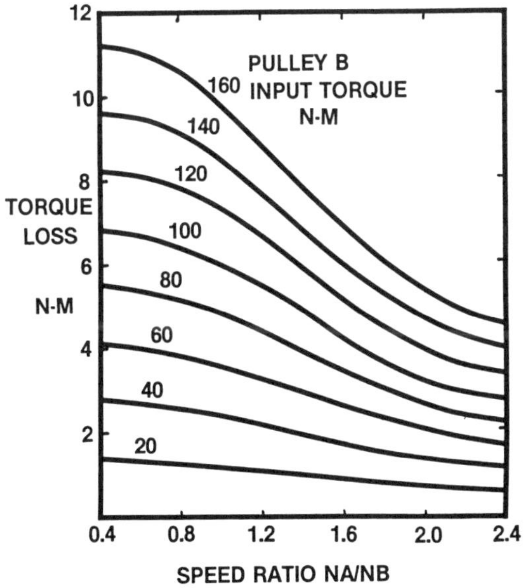

TORQUE RELATED INPUT TORQUE LOSS

Figure 19

These torque losses and speed losses were then combined to generate performance maps over a range of operating conditions, Figures 21, 22, and 23. Where the data has been extrapolated, a conservative approach was used to give the highest possible loss.

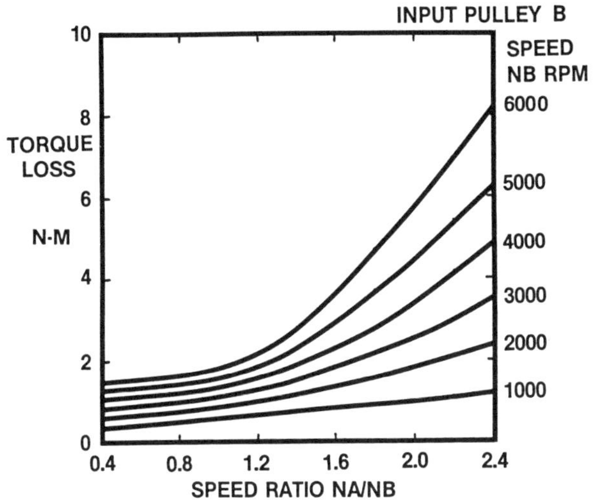

SPEED RELATED INPUT TORQUE LOSS

Figure 20

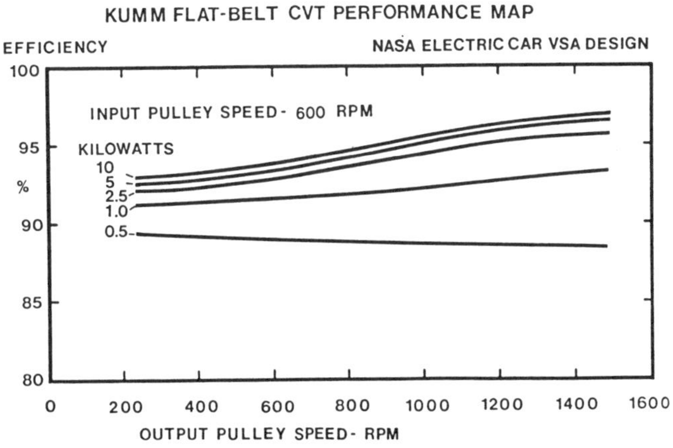

Figure 21

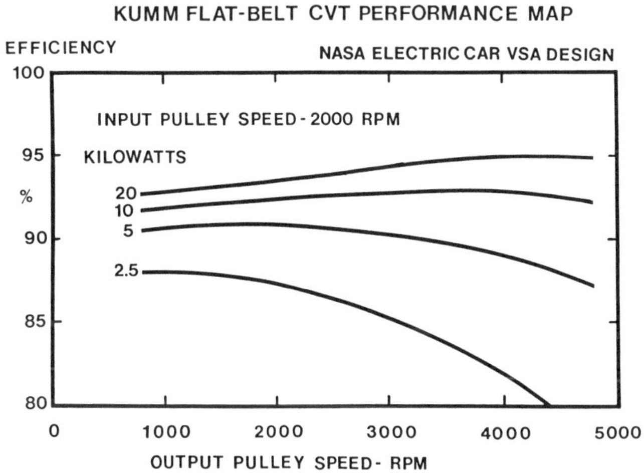

Figure 22

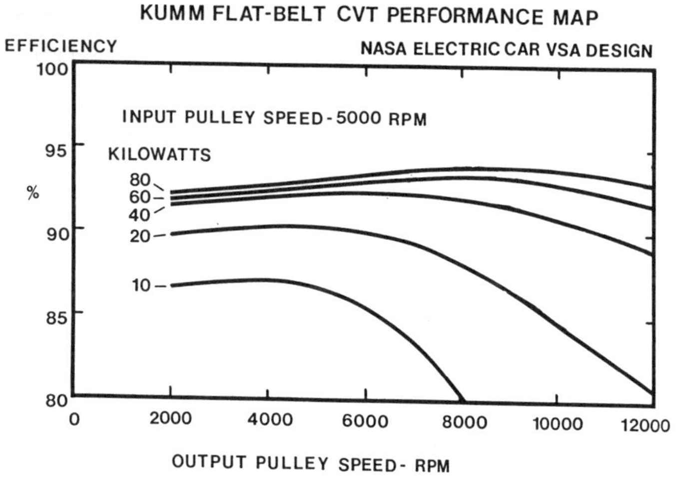

Figure 23

There is good correlation between measured experimental efficiencies and efficiencies computed using the torque and speed loss curves, Figure 24. This analysis used the higher power data points that had previously been obtained on the back-to-back test stand configuration.

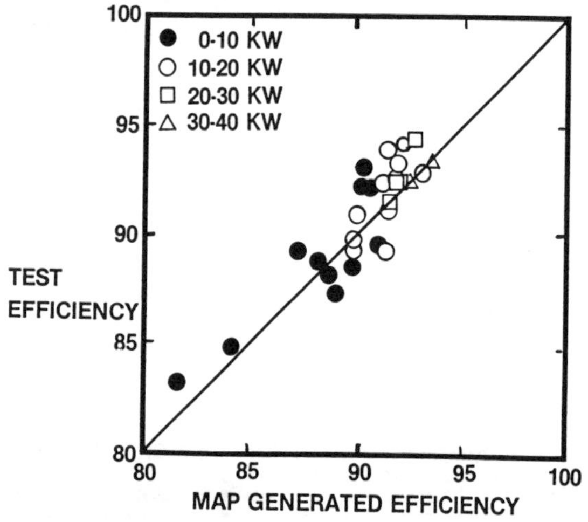

TEST DATA PERFORMANCE MAP VERIFICATION AT HIGHER POWER

Figure 24

MAJOR DEVELOPMENT ISSUES

There are a number of component and system issues that must be successfully addressed before the Kumm flat belt CVT technology will mature to the level for production design.

SLOT/PIN END WEAR-The dry sliding surface between the belt drive elements and the guideway slots is subject to wear. Under normal test operation, the hard coat Teflon (Nituff) and electroless nickel has performed with no noticable wear over 10's of hours. Using data on a number of low friction, high strength coating materials, analyses indicate that there are designs and materials that will give adequately low wear over the life of an automotive application.

A wear test rig was designed and constructed, Figure 25. Its operation permits determining the wear characteristics of various low friction dry sliding surface combinations. The loads and surface velocities simulate the operating conditions in the automotive application. A computer simulation of two different driving cycles for 100,000 miles (3,000 hours) was applied to the critical sliding surface. The sliding velocity times surface pressure integrated over time is 2 to 4 Newton/mm2*M/sec*hours for a 1200 KG, 1.6 liter vehicle.

DRY SURFACE WEAR TEST RIG

Figure 25

BELT DURABILITY-The automotive industry has had successful experience with I.C. engine rubber cog timing belts. They have demonstrated adequate life in an environment and with loads similar to that facing the Kumm flat belt. The number of chordial impacts and releases with the 24 belt drive elements during 100,000 miles of automotive operation is in the range of 6 billion. These are spread over the entire belt surface, compared to the specific load points in of the timing belt. Localized shear and compression fatigue of the rubber is minimized.

The rubber in the CVT flat belt is compressed locally while transferring torque through shear. This reduces the shear strength requirements of the rubber, due to the cog-like effect. The shear forces are spread over a large belt cross section. Also, minor changes in the belt thickness due to the compression set of the rubber over the operating life are not critical.

The maximum tensile loading of the Kevlar cords is less than that of the infinite life level for repeated stress. Nevertheless, endurance testing of belts with adequate fricition coeficient will be a major focus.

TEMPERATURE-Because the flat belt is highly efficient, it generates little heat. The localized increase in belt temperature is low. After testing, the belt is only slightly warm to the touch.

However, the transmission environmental temperature can range from sub zero to 300+ degrees F. The belts will warm quickly in the cold weather, but nearly all of the higher strength elastomers require operating temperatures below 200 degrees F.

Under normal automotive operation, the conduction of heat from the pulley cavity to the air that flows about the transmission should prevent excessive temperatures. The flat belt pulley system can be placed outside the transmission housing. The heat conducting cover plate would transfer heat to the environment. Heat soak back from the engine and transmission gear case after vehicle shut down can be minimized with simple insulation. If necessary, outside filtered air can be discharged into the belt region using a thermostatically controlled fan.

ACTUATION DYNAMICS-Our test rig was designed to generate steady state performance data. To test for actuation dynamics, the input speed must change rapidly under load. An IC engine power source and a programmable hydraulic energy absorption device will be required.

With the restricted hydraulic lines to the VSA pulley actuators and at constant input speed, the VSA actuation has been tested. A shift across the full speed ratio range under belt tension was made in approximately one second. This, in general, confirms our dynamic analysis of the loads and movements within the pulley system.

HYDRAULIC POWER LOSSES-The hydraulic power requirements are a function of desired actuation times, geometry, and the transmission horsepower. Compared to traction and other metal-to-metal CVT's with friction coefficients of 0.1, the flat belt elastomer to elastomer has a design friction coefficient of 0.6 with a safety factor to slip of 1.7. The higher friction coeficients result in a proportional reduction of the hydraulic actuator power required to operate the flat belt CVT. Current analyses indicate that a hydraulic pump power requirement of less that 0.6 KW (0.8 HP) will provide adequate response rates for a CVT designed for a 2.8 liter automotive engine.

CONTROL SYSTEMS-The electric vehicle VSA was designed to operate with an all hydraulic control system. An automotive application will employ a microprocessor based control system. Rotor speed inputs (2) and throttle position will call for variations in pulley radius ratios. The rate of change would depend on the throttle position. Based on algorithms or stored maps, the appropriate hydraulic pressures can then be delivered to the actuators. The pressures would vary the radii of the pulley belt drive elements while maintaining belt tension above the slip point. Pressures would be controlled so as not to over stress the system. The proper transient response should prevent a hesitation at the start of the vehicle acceleration and a delay during engine assisted deceleration of the vehicle.

CLUTCHING INTEGRATION-The VSA did not include a clutch. A high speed electric clutch was considered for engaging the flywheel. The geared neutral used in the design provided the stop and reverse capability.

For an automotive application, a launching system will be required. This could be a lock up torque converter or fluid coupling, an electric clutch, or a multiplate wet clutch. The response characteristics should be similar to the current automatic transmissions. Driver feel of the initial acceleration is critical to the salability of the vehicle.

NOISE-Chordial noise in the frequency range of 100 to 2000 Hertz may be generated by two effects. The belt drive elements strike the belt. Also, as the belt loads and unloads the elements, the elements move rapidly to the opposite side of the guideway. With a small guideway to element end clearance (.03 to .05 mm) the impact energy is small. Experience during the open VSA tests indicates that an automotive application will have low noise emission.

SCALING EFFECTS-The VSA was designed for 75 KW (100 HP). The critical scaling elements are the belt loading, the maximum stress in the belt drive elements, and the wear of the guideway-belt drive element sliding surface. If the input and output speeds, stress, and belt velocities are held constant, the transmitted power is nearly proportional to the linear dimension to the 5/2 power for automotive applications.

$$P \propto D^{5/2}$$

Using a baseline 2.8 liter flat belt CVT automotive design the following estimates can be derived, Table 1.

Maximum Power KW (HP)	Input Torque N-m (ft-lb)	Engine Size Liters	Pulley Diameter mm (in)	Belt Width mm (in)
47.7 (64)	113 (83)	1.6	183 (7.2)	22 (0.88)
83.5 (112)	197 (145)	2.8	229 (9.0)	28 (1.10)
128.3 (172)	370 (273)	4.3	272 (10.7)	33 (1.30)

Over All Package Parameters*

Width mm (in)	Length mm (in)	Height mm (in)	Weight KG (lb)
122 (4.8)	361 (14.2)	284 (11.2)	14 (31)
152 (6.0)	450 (17.7)	356 (14.0)	27 (60)
180 (7.1)	536 (21.1)	424 (16.7)	46 (101)

Table 1

* Weight of the pulley system, hydraulic actuators, shafting, outboard bearing support, coverplate, and shaft bearings only. Hydraulic pump, input or output gears and their housing, clutches, or other power transfer means to the wheel differential gearing are not included in the package.

APPLICATION DESIGN

There are a number of applications that could have significant production potential for the Kumm flat belt CVT. They include on road (auto, truck, motorcycle), off road (construction, farm, recreation), industrial, accessory drives, and marine/aerospace. The automotive application will be considered in the following design/analysis. Based on the NASA test results, the design was extended to a 2.8 liter, General Motors front wheeled drive "X" car transmission configuration, Figure 26. The design input torque is 145 ft-lb at 2100 rpm. Maximum power is 112 HP at 4800 RPM. A clutch or fluid coupling with lockup, and a reversing gear to the differential is assumed, but are not detailed. The over hung bearing flat belt CVT arrangement is shown in Figure 27.

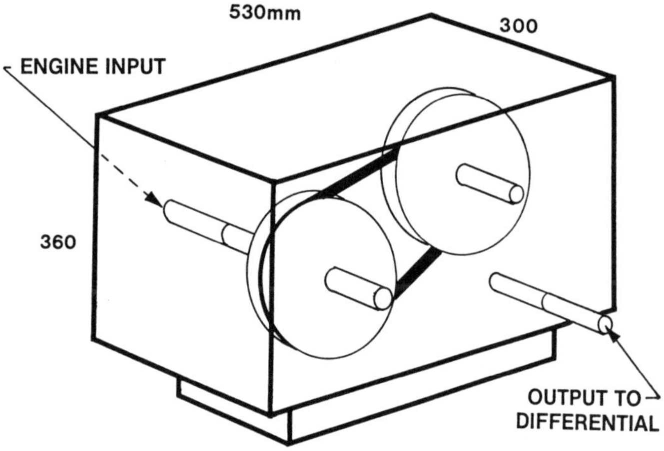

AUTOMOTIVE FRONT WHEEL

Figure 26

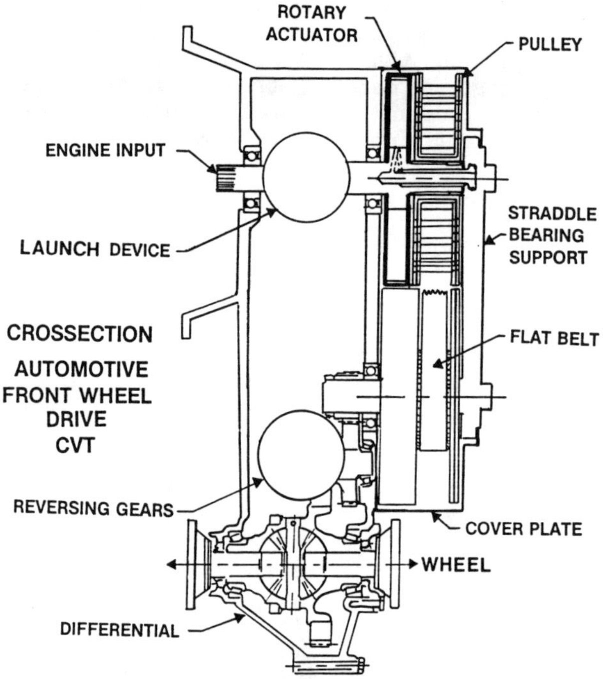

Figure 27

Performance of an automotive CVT with a single gear to the differential can be derived from the NASA VSA test data. A study revealed that the losses in the thirteen bearings of the VSA design are roughly equivalent to its four gear mesh losses, plus windage. The bearings and gear losses for automotive design would be expected to be about 44% of the VSA losses, Table 2.

	GEAR MESHES	BEARINGS (EQUIVALENT GEAR MESHES)	TOTAL
VSA DESIGN	4	4	8
AUTO DESIGN	1	2.5	3.5

Comparative losses = 100(3.5/8) = 44%

Table 2

The pulley and belt losses are estimated to be approximatly 1% of the transmitted power. The efficiency for the automotive design can be derived by removing the 1% pulley/belt loss from the VSA data, keeping 50% of the remaining VSA losses, and then adding back the 1% pulley/belt loss, Figures 28, 29, and 30.

AUTO% = 99 - (99 - VSA%)*.50 Efficiency

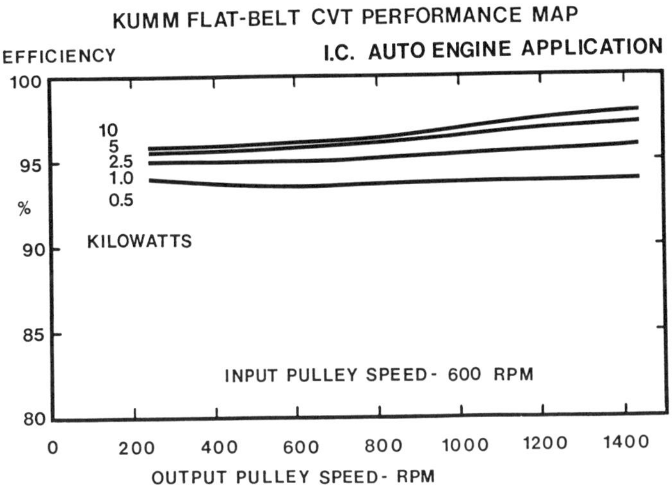

Figure 28

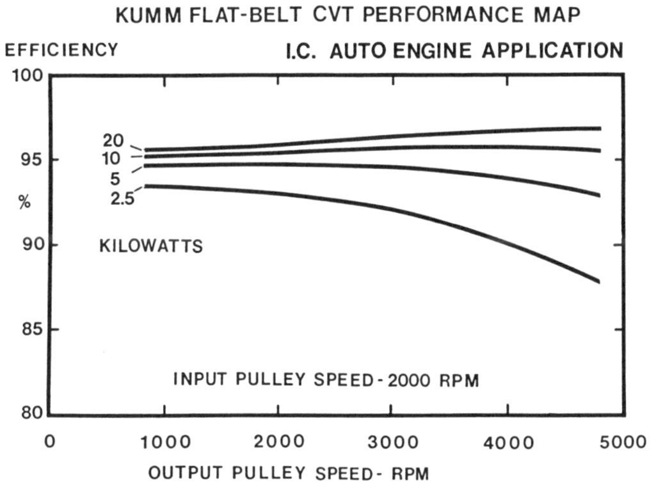

Figure 29

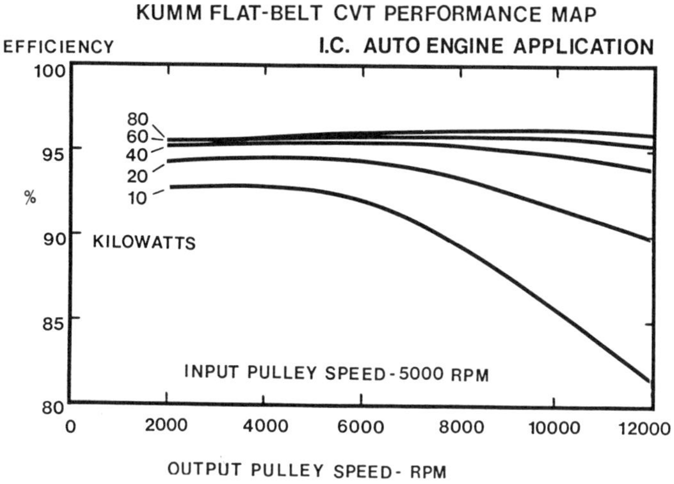

Figure 30

CONCLUSION

The attributes of the Kumm flat belt are high efficiency of the continuously variable component, low hydraulic pump losses, a large speed ratio range, ability to operate across a range of engine sizes, narrow width, easily changed belts, and relatively low weight and cost.

The Kumm flat belt CVT is infant technology, with less than 5 man years invested since conception. There are a number of technical areas that need to be addressed. We believe these engineering problems can be solved with current technology.

After many decades of industry effort to develop an automotive CVT, the Kumm flat belt CVT may be the innovation that provides a viable solution.

Rubber CVT Belts**

Serge Abrate
Dayco Technical Center
Springfield, MO

**Paper 850571 presented at the International Congress and Exposition, Detroit, Michigan, February, 1985.

ABSTRACT

The use of rubber CVT belts in automotive applications is presented. Basic drive design considerations are given so as to best take advantage of the belt capabilities in terms of capacity and durability. The influence of minimum pitch diameter and heat transfer conditions is demonstrated. Life testing of CVT belts is also discussed, and the need for better knowledge of the actual load spectrum is pointed out.

EXTENSIVE RESEARCH IN THE AREA of Continuously Variable Transmissions was motivated by the potential advantages of the system: the transmission will be automatic thus responding to a need and ensuring optimal functioning at all times, the stepless variation in speed ratio over a wide range will provide for better fuel economy by matching the engine and car characteristics in an optimal manner, and full power is available for acceleration over the speed range.

Similar features are thought to be achievable with a six speed automatic gear box (1)*, providing a wide ratio range with small steps between each ratio and thus achieving the same goals as a continuously variable transmission. In that case the influence of step changes when compared with a continuous variation is minimal.

* Numbers in parentheses designate references at end of paper.

Different CVT systems are being considered, a comprehensive review was presented in (2), the rubber belt being the first to have been in production on an automotive application with the DAF car in Holland starting in 1959 (3). The Van-Doorne metal belt which has been under development for more than 12 years is being considered by major car manufacturers (4). The PIV chain type and the Perbury traction drive are also being considered (5).

All systems obviously have their advantages and are currently being developed further. In this paper we will focus more specifically on the rubber belt in automotive applications. We first point out some of the advantages of rubber CVT belts followed by some basic considerations on the design of a CVT drive and how to best take advantage of the capability of the belt. Results on the capabilities of rubber CVT belts are then presented.

ADVANTAGES OF THE RUBBER BELT

The rubber belt has a long history of service. Variable speed drives for industrial, agricultural and leisure vehicle applications have been manufactured for years. On the DAF car, rubber belts proved to be reliable components. One drawback often mentioned was that the large size of the CVT system made it impractical for front wheel drive applications.

Progress in the development of new materialenabled to reduce the package size to meet the space limitations for engines up to 50 HP. Coefficients of friction between .25 and .40 allow to use axial forces much lower than those required for metal belts or PIV type chains with coefficients of frictions of the order of .1 due to the nature of the contact surfaces: metal-metal with lubrication.

Lower control forces allow for lighter mechanisms and will require less energy for actuation thus contributing to higher overall efficiency of the system especially at reduced power where efficiency is known to be lower.

Since a rubber belt requires no lubrication, the system is then simpler, less expensive, and more efficient.

DESIGN OF A CVT DRIVE

The matching of engine and car characteristics require that the total transmission ratio be low for best fuel efficiency at top speed and high at low speeds for adequate gradeability and acceleration. Practical applications suggest overall speed ratios of 5.0 to 6:1. Higher ratio presenting only minor fuel consumption advantages (1).

Selection of actual pulley sizes should account for the following constraints:

- the underdrive condition is most severe as the belt is bending over a small driver pitch diameter and high tensions are applied in order to transmit engine torque
- the maximum pitch diameter is limited by package size considerations and maximum allowable belt speed

One avenue to accommodate the two conflicting requirements of having a large driver P.D. in underdrive while keeping the package size at a minimum is to use a "non-square" drive. For a required overall ratio τ, the maximum underdrive ratio ρ_u and the maximum overdrive ratio ρ_0 are chosen:

$$\rho_u = \sqrt{\tau} / \theta$$

$$\rho_0 = 1/(\theta\sqrt{\tau})$$

so that:

$$\rho_u \cdot \rho_0 = \tau$$

where θ is a parameter larger than 1.0. Fig. 1 shows one practical example illustrating this point, the minimum driver P.D. is chosen to be 102 mm and the maximum and minimum pitch diameters, and the center-distance vary as the parameter θ sweeps the 1.0 to 1.2 range. The maximum pitch diameters and the minimum center distance required decreases as becomes larger thus allowing for smaller package size and smaller maximum belt speed for a given driver speed, while providing for a minimum driver P.D. that is sufficient to transmit the engine torque in the underdrive condition. The maximum value of θ should be limited so that the minimum driven P.D. does not become so small as to induce flex cracking of the belt in overdrive. For the particular example considered here, a proper choice would be $\theta = 1.15$ which will give the drive shown in Figure 2. For comparison a "square drive" ($\theta = 1.0$) will require a max. P.D. of 227 mm and a center distance of 243.7 mm.

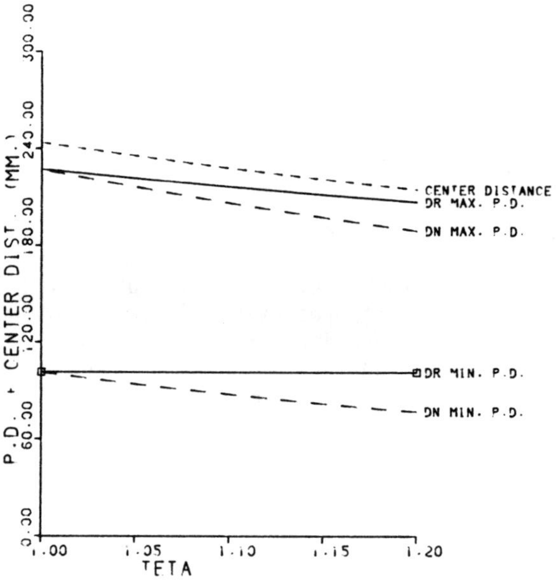

Figure 1: Evolution of Dimensions With Parameter θ

Belt size is basically governed by the choice of pitch diameters in the extreme conditions. The minimum belt width is determined so as to allow for the belt to travel from the max. underdrive position to max. overdrive (see Figure 3) and for a "non-square drive" ($\theta > 1.0$), both sheaves should be considered. Belt top width should be kept to a minimum as we shall see later. Belt thickness is to be chosen so as to provide sufficient support to the cord line in order to prevent dishing while at the same time allowing the belt to bend around small diameters without producing large strains and generating excessive heat.

In order to satisfy these conflicting requirements, two approaches can be used simultaneously:

- the material used can be tailored to provide high rigidity in the transverse direction to resist the axial forces and prevent dishing while in the longitudinal direction the modulus is much lower and therefore allows the belt to bend easily

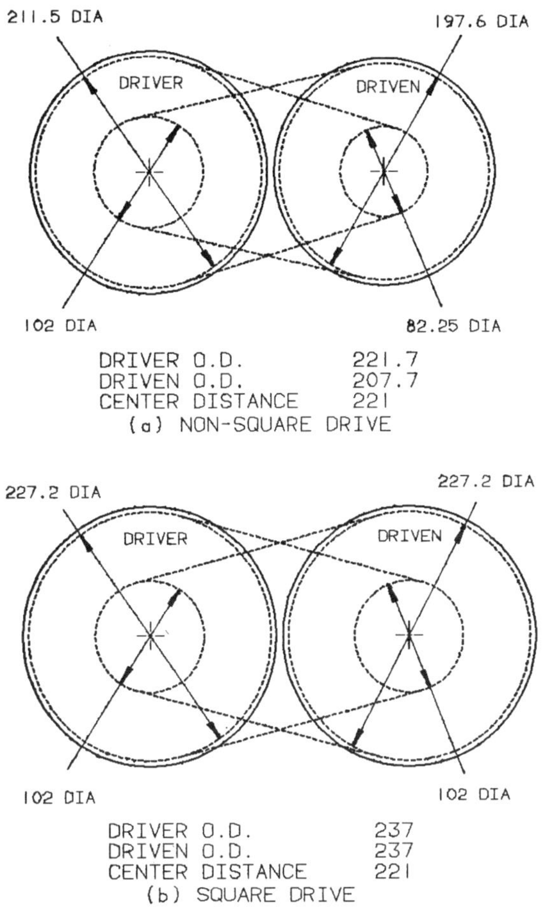

FIG. 2: EXAMPLE OF DRIVE DESIGN.
(ALL DIMENSIONS IN mm)

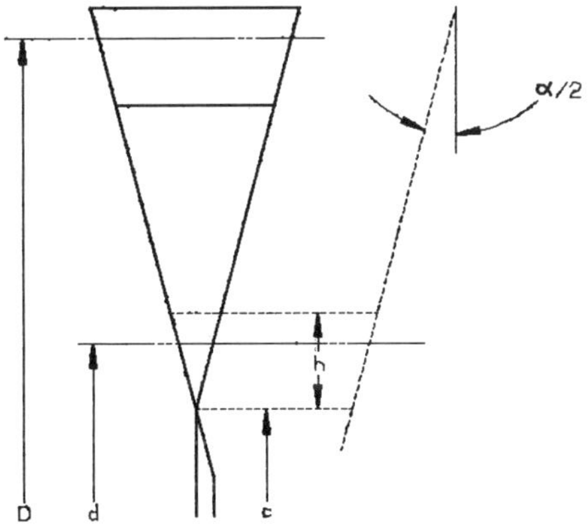

$TW = [(D-d)+2(h+c)] \tan \alpha/2$

FIG 3. DETERMINATION OF TW.

- the use of cogs in the belts so that the effective thickness of the belt is sufficient to prevent dishing while limiting the strain experienced by the material

Proper choice of the tooth profile provides the required additional support and does not create high stress concentrations in the bending thickness.

MATERIAL PROPERTIES - Requirements for CVT belts in terms of package size, power transmitted and expected durability are more stringent than on more classical variable speed belt drive. As mentioned before, a suitable material should have a high transverse modulus and a low longitudinal modulus. A typical CVT type material has an anisotropy ratio:

$$E_t/E_L = 6.78$$

Reinforced rubber being highly visco-elastic, energy is being dissipated as it is being deformed, this energy being eventually transformed into heat will in turn try to raise belt temperature and degrade material properties and possibly start a self-destructive process. Therefore hysteresis of the material has to be reduced.

The amount of reinforcement that can be obtained while keeping heat generation at acceptable levels is limited. Reference (5) shows that for short fiber- reinforced elastomers, heat generation is strongly dependent on the fiber volume fraction and fiber type. Therefore, there must be a limit after which more reinforcement improves the elastic modulus but because of the increased hysteresis makes the material unsuitable for dynamic application.

Table 1 shows a comparison of two materials, one being used in CVT belts while the other is used in snowmobile applications.

EVALUATION AND TESTING OF CVT BELTS

Evaluation of CVT belts should be conducted by considering the capacity of the belt to transmit torque, its efficiency over the range of operating conditions and the life expectancy.

TORQUE CAPACITY - The ability to carry maximum engine torque is particularly critical in the underdrive condition. Several factors are combined to make this condition the most difficult one as far as transmitting high torques is concerned: small arc of contact, high tensions, high temperature, and bending over small diameter. A study was conducted on a particular type of belt to determine torque capacity in the range of small diameters, in the under-drive condition. Here torque capacity is defined as the maximum torque transmissible, the belt having reached an equilibrium temperature.

Belt temperature being critical in the situation, several measures were taken during the tests in order to affect heat transfer. The pulleys could be equipped with fins and the entire drive could be completely enclosed. Results obtained with the various combinations are given in Fig. 4. Improving heat transfer is shown to be definitely helpful to belt operation and the capacity of the belt to transmit a load increases sharply with the driver pitch diameter. Another point to be considered is that while the equilibrium temperature in these tests is typically reached after one hour in steady state loading, the maximum torque maximum underdrive condition in the automotive

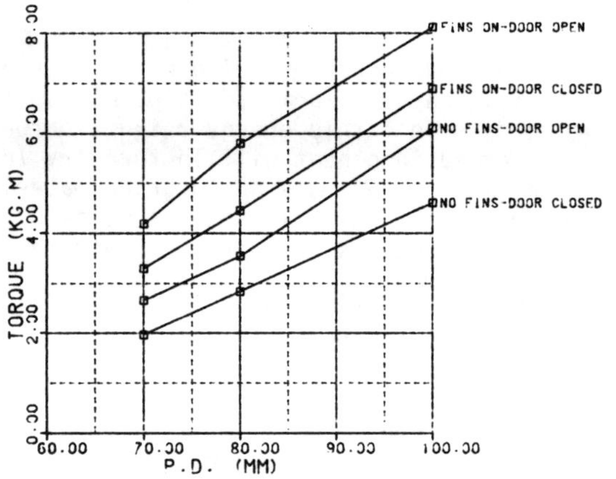

Figure 4: Torque vs. Min. P.D.

application is encountered for periods of much smaller duration. The belt is capable of transmitting higher torques for short amounts of time. For example, the belt referred to in Fig. 4, over a 102 mm diameter, can transmit a torque 18% higher for a 30 min. period. The top width of this belt was 42.8 mm an identical belt with a top width of 50 mm required a 110 mm diameter to transmit the same torque for 30 min., thus substantiating the claim that belt width should be kept to a minimum. Another study determined the capacity of the belt as a function of speed ratio. Figure 5 shows the variation of the belt maximum transmissible torque.

In conclusion the capacity of the belt while somewhat limited in underdrive, a condition in which the vehicle remains only for a very short time, is adequate for high torque engines.

BELT TENSIONING AND EFFICIENCY - Appropriate belt tension as a function of torque and speed ratio should be determined experimentally. Proper tensioning is determined by varying the centerline force over a wide range and selecting the value that will

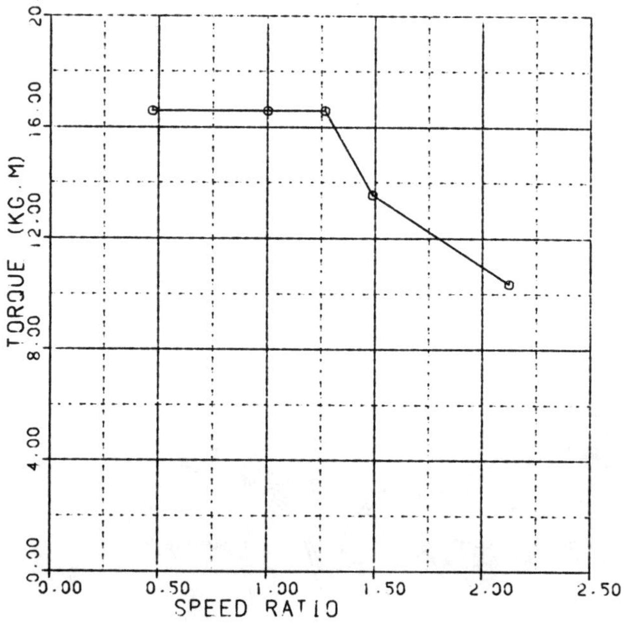

Figure 5: Torque Capacity

yield minimum belt temperature and slip. Starting tension value was determined from the following equation:

$$T_1/T_2 = \exp \mu\theta/\left(\sin\frac{\alpha}{2} + \cos\frac{\alpha}{2}\right)$$

where μ is the belt coefficient of friction, α the belt angle and θ the wrap angle. Experiments have shown that for a given speed ratio the optimum tension ratio is constant and that as the speed ratio changes, it follows the trend predicted by the above equation. Fig. 6 illustrates this point for a particular example. The axial forces required to provide a given level of tension can be determined using the equations given by Worley. Fig. 7 shows the axial forces required for the example discussed in Fig. 6.

The control system of the CVT transmission should ideally modulate the axial forces as shown in Fig. 7 therefore making sure that torque can be trasmitted efficiently and with the longest possible belt life.

Few references mention results from efficiency measurements (7, 8) but typically belt efficiencies are in the 92% - 98% range.

BELT LIFE - Belts in a CVT transmission are a critical element since they are stressed at all times as opposed to gears in a conventional transmission. The belt is subjected to a wide variety of operating conditions and its manufacturer should

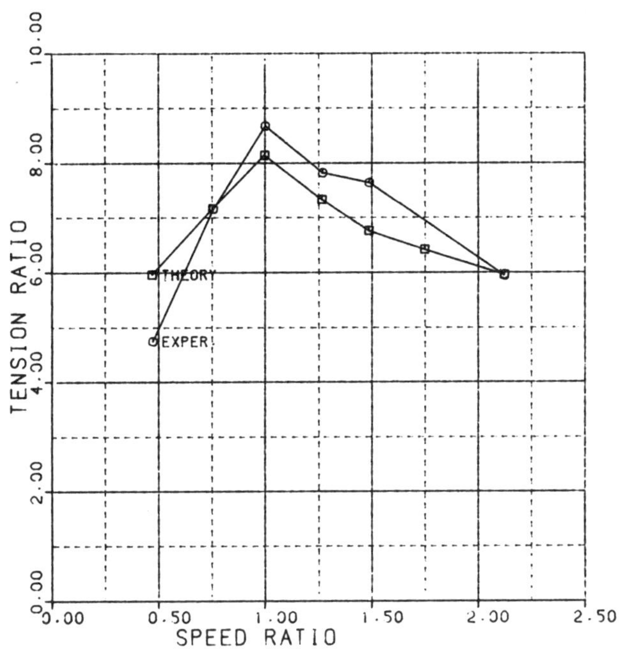

Figure 6: Tension Ratio vs. Speed Ratio

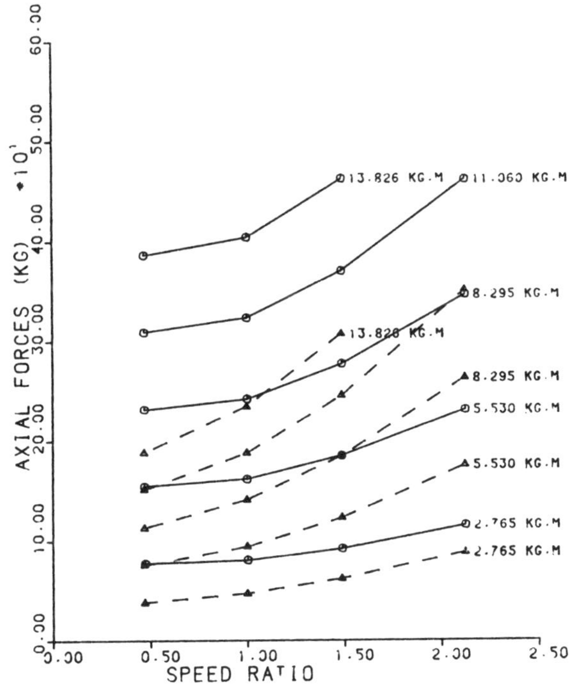

Figure 7: Axial Forces vs. Speed Ratio

ensure that its life will be adequate for car operation.

Test programs should be established so as to realistically approximate vehicle operations. To this end, a description of the expected operating conditions is needed possibly in the form of a table of joint probability of speed and torque as percentages of total time such as that given in Ref.(1). Fig. 8, 9 show experimental data from Ref. (1) on the distribution of vehicle speed and gearbox output torque as a function of time. This indicates that the combination of low vehicle speed with maximum output torque, clearly the hardest belt operating condition, occurs only for a small fraction of the time. Data available from past experience with vehicles equipped with conventional transmission is usually inadequate since they do not provide for a wide enough overall speed ratio and are not necessarily operated optimally all of the time. Practically the number of combinations of torque and speed ratio to be used have to be limited and a duty cycle is established so that a required number of hours is determined for each condition. Then two avenues are possible, the first one is to test several belts in each condition in order to determine belt life in each condition and then calculate whether a belt would pass the complete duty cycle based on a cumulative damage theory. The other approach being of subjecting each belt to the duty cycle in order to demonstrate that they satisfy the requirements.

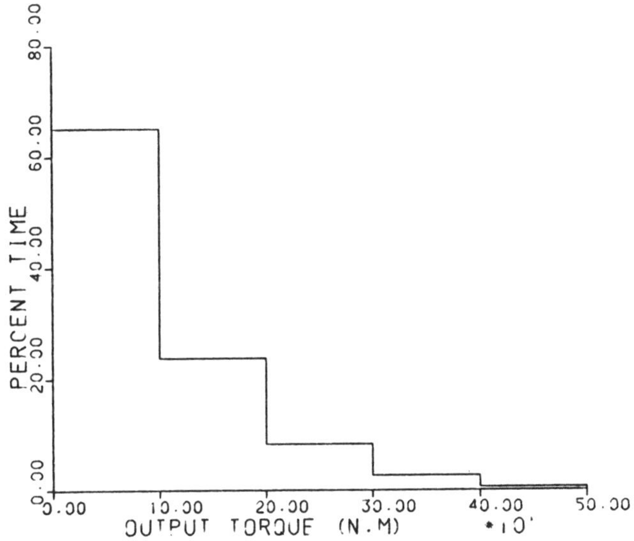

Figure 8: Torque Distribution

The first approach requires that an adequate cumulative damage theory is available. Another problem with this approach is that as the speed ratio decreases (tends to overdrive condition), belt life increases and so does the testing time. However this method allows for more flexibility since once data is collected, it can be used to predict belt life under different sets of operating conditions, and ways can be found to accelerate testing in order to reduce testing time.

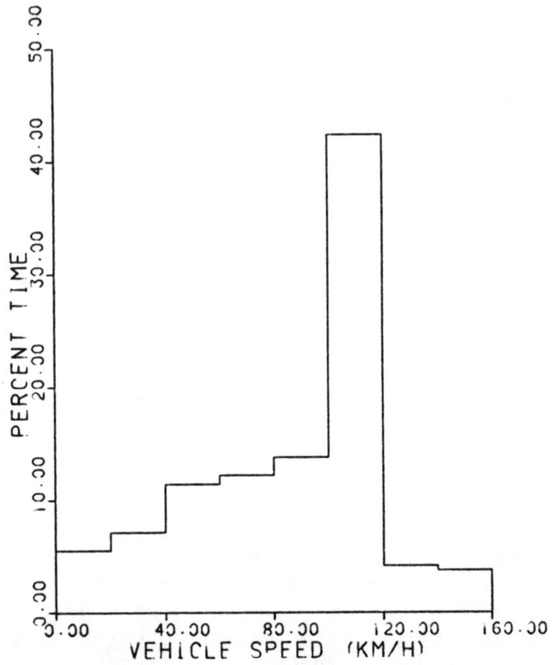

Figure 9: Speed Distribution

Subjecting a number of belts to a given duty cycle will do away with the need to develop a cumulative damage theory, and the related experimental data. However, the data collected on one particular case can hardly be used for a different one.

Tension levels being optimized for each condition, a suitable cumulative damage theory should account for variations in operating condition over the two dimensional domain spanned by the torque and speed ratio. Such a theory should also account for the variability in test results since the belt in a CVT transmission is an important element as far as the safety and dependability of the vehicle is concerned and therefore one should guarantee a minimum operating life. Another requirement is that the proposed theory should account for the order in the loading sequence. Experiments have shown that in two stage loadings, a low-high sequence is more favorable than a high-low. Research is being performed in this area and results will be presented at a later date.

CONCLUSION

In this paper we presented the capability of the rubber belt for automotive CVT applications. Some considerations on the basic drive design were presented in order to indicate how to best take advantage of the capacity of the belt. Belt life was discussed in reference to testing and elaboration of testing programs.
It is felt that rubber belts are suitable for automotive CVT applications for a certain class of vehicles.

REFERENCES

1. R. P. Jarvis: "Fuel Economy with Small Automatic Transmissions". Proceedings of the VECON '84 Conference, London, October 22-24, 1984, pp 227-234.

2. N. H. Beachley, A. A. Frank: "Principles and Definitions for Continuously Variable Transmission, with Emphasis on Automotive Applications", ASME Paper 80-C2/DET-95

3. J. P. Norbye: The Complete Handbook of Automatic Power Trains, Tab Books Inc., Blue Ridge Summit, PA, (1981).

4. Proceedings of the "VECON '84 - Fuel Efficient Power Trains and Vehicles" Conference, London, October 22-24, 1984.

5. C. J. Greenwood: "The Design, Construction and Operation of a Commercial Vehicle Continuously Variable Transmission", Driveline '84, I. Mech. E Conference Publications 1984-1, pp 89-102.

6. A. P. Foldi: "Reinforcement of Rubber Through Short Individual Filaments", Rubber Chemistry and Technology, Vol. 49, pp 379-389.

7. L. R. Oliver: "Demonstration Rubber Belt CVT for Tomorrow's Commuter Automobile", Proceedings of the National Conference on Power Transmission, Vol. 9, Illinois Institute of Technology, 1982.

8. D. J. Bents: "Axial Force and Efficiency Tests of Fixed Center Variable Speed Belt Drive", SAE Paper 810103.

TABLE 1
Comparison of CVT and Snowmobile Materials
CVT/Snowmobile Ratio

Properties	Longitudinal Direction	Transverse Direction
Tensile Strength	1.491	2.482
Elongation at Break	.478	1.0
10% Modulus	1.29	1.777

A "State of the Art" Chain Driven CVT Design*

Stanley Avramidis
PT Components, Inc.

*Paper 861354 presented at the Passenger Car Meeting and Exposition, Dearborn, Michigan, September, 1986.

ABSTRACT

The Continuously Variable Transmission (CVT) has emerged from being considered a novel automotive transmission concept to now being a fully developed design, ready to compete with current automatic transmissions. The CVT transmission is described in detail to demonstrate the advantages of its design as well as present the uniqueness of one such design, the PIV/PTC transmission. These features are then related to basic required driving functions.

IN RECENT YEARS increased interest has been generated in developing automotive continuously variable transmissions (CVT). This interest is primarily motivated by the prospect of fuel economy advantage of CVT's over current automatic transmissions. Greater fuel economy can be achieved by using the CVT's wide range of speed ratios and special controls to ensure the engine is operating at optimum speeds. An example of an actual fuel economy gain achieved with a CVT transmission is shown in Fig. 1. In addition, CVT's can provide quick and smooth response to a wide variety of driving conditions. Electronic controls provide an additional dimension to speed ranges and response times. Last, but not least, CVT's are simpler in design than current automatic transmissions and thus offer the potential for lower cost and improved durability.

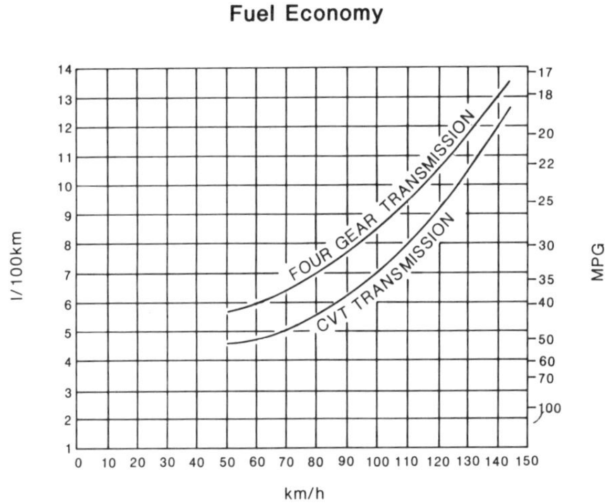

Fig. 1

CVT DEVELOPMENT

Although work on the CVT transmission has received increased attention recently, the design described herein has been used for many years. Starting from industrial applications of a continuously variable gear box, PIV/PTC developed automotive CVT transmissions dating back to the 1960's. Such transmissions have been successfully operating in vehicles for more than 50,000 km. In addition, numerous prototype automotive transmissions have been developed for many different automobile manufacturers.

CVT DESIGN

CVT designs can vary as greatly as current automatic transmission designs. However, a typical PIV/PTC arrangement will be described in Fig. 2. The engine connection may be in the form of a dry clutch, a fluid coupling to absorb some torsional shock, or a torque converter to absorb shock plus add some torque multiplication. Mode selection can be either similar to a manual transmission forward-neutral-reverse shift engagement, or operated hyddraulically through a planetary gear arrangement with a selector level as in an automatic transmission.

Front Wheel Drive CVT Transmission

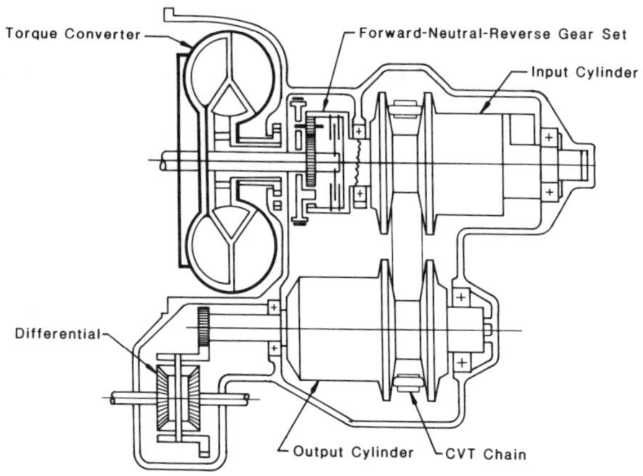

Fig. 2

Common in all designs is a set of fixed sheaves with opposing movable sheaves, one for the input and one for the output shafts. The power is then transmitted from input to output shaft by a flexible steel belt or a steel chain. The continuously variable speed ratio results from moving one set of sheaves together, thus forcing the steel chain radially outward while simultaneously allowing the other set of sheaves to be forced apart and the chain moving radially inward to a smaller radius. Thus, in going from underdrive to overdrive, the input sheaves would be forced together while the chain forces the output sheaves apart. Total ratio ranges can be as low as 2:1 or as high as 7:1. Even with the higher ratios, a final drive gear set is usually required.

Because the chain-sheave design is crucial to CVT operation, it will be described in some detail. Because chain/sheave contact is metal-to-metal, special precautions must be taken to eliminate slippage. Any slippage would reduce power transmission efficiency and greatly accelerate wear. In the normal CVT operating mode, a squeezing schedule must be developed to insure that the exact sheave clamping force is applied for a variety of speeds, torques and drive ratios. Less than the correct squeezing force would allow slippage, too much squeezing would lower efficiency. In addition, a special hydraulic-mechanical torque sensing device is installed to instantaneously increase clamping force to take care of torque spikes. Torque spikes occur under certain driving conditions, such as spinning driving wheels on ice then hitting dry pavement. The torque sensor also provides a reservoir of oil which can be pumped to the system for emergency ratio changing. During normal driving conditions, the torque sensor assures that the correct squeezing pressure is applied to the sheaves.

Special chain and sheave geometry has been developed to optimize stresses and frictional forces at the contact points. These designs permit special functions such as stand still down shift while maintaining power transmission efficiency well above 90% (Fig. 3).

Efficiency of CVT Chain & Sheaves Unit

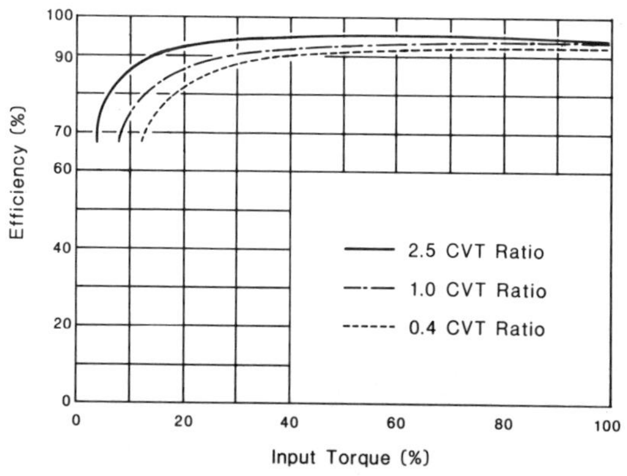

Fig. 3

Now concentrating on the individual parts (chain and sheaves) (Fig. 4) -- The heart of the PIV/PTC system is a steel chain as opposed to a belt. The chain engages smooth pulleys via the end of pins which serve a dual purpose. The first purpose permits the chain to articulate, i.e., it allows the chain segments or links that are joined by pins to rotate with respect to pivot centers (pins), thus cooperating with the smooth faced pulleys to transmit power. The second function of these pins is to contact the pulleys at their ends and because of the presence of hydraulic forces, a frictional connection between pins and pulleys is achieved. Most chains, belts and gears have an inherent capability to create undesirable noise at times, and create driving situations that are totally unacceptable unless properly addressed. The PIV/PTC chain used in this application is of the non-periodic variety, meaning that articulations and contact impact

Fig. 4

are controlled in a manner that results in white noise generation, as opposed to a fundamental frequency which could be very objectionable.

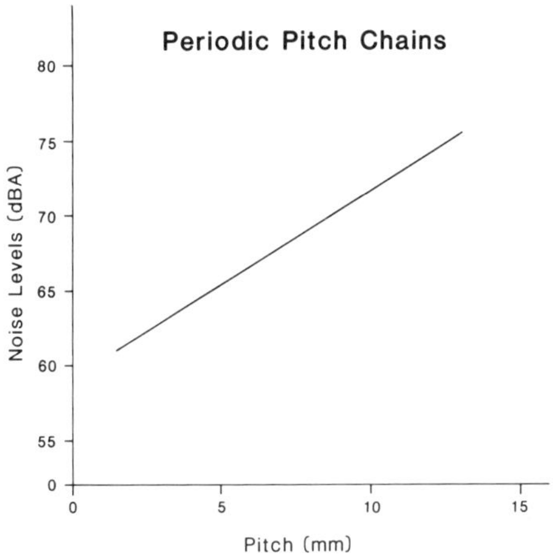

Fig. 5

Fig. 5 shows that chain noise is directly proportional to chain pitch. These test results indicated that as chain pitch increases, so does chain noise. Ideally, chain pitch would be limited to a few millimeters less than 5. However, in most applications chain pitch cannot be held within the required noise range because of strength and size requirements.

To mitigate chain noise PIV/PTC have developed a random pitch chain which breaks up the noise pattern. As shown in Fig. 6 the nosie "spikes" of the periodic chain assembled from links of the same pitch, i.e. the distance between pins is the same, have been considerably dampened by using a random pitch chain. Development is continuing in CVT chain noise reduction. However, in the final analysis each chain design must be tailored to an individual transmission design and unfortunately for each program, a satisfactory combination of long pitch and short pitch must be arrived at experimentally.

The final part of the CVT transmission to be discussed is the control system (Fig. 7). This device usually combines hydraulic, mechanical and electronic elements to modulate the transmission

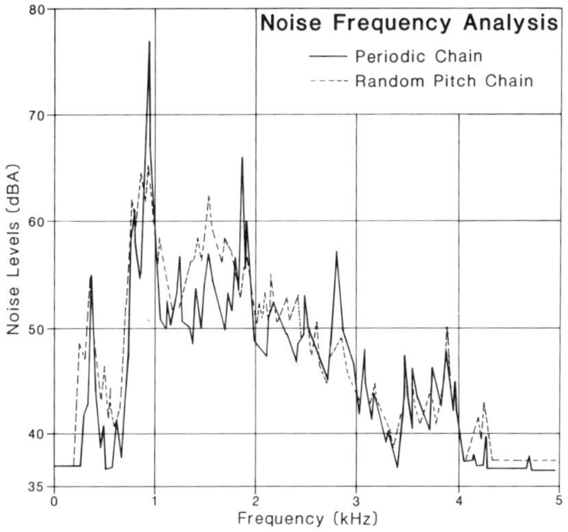

Fig. 6

operation for a wide range of driving conditions. Steady state freeway driving, down hill braking, passing acceleration and emergency stop and restarting must all be available from the control unit with optimum fuel economy under each condition. Because all of these features are difficult to achieve while simultaneously maintaining good "driveability," development activities in these areas focus on achieving desired results through electronics.

Speed Control System connected to the Accelerator Linkage of a Combustion Engine

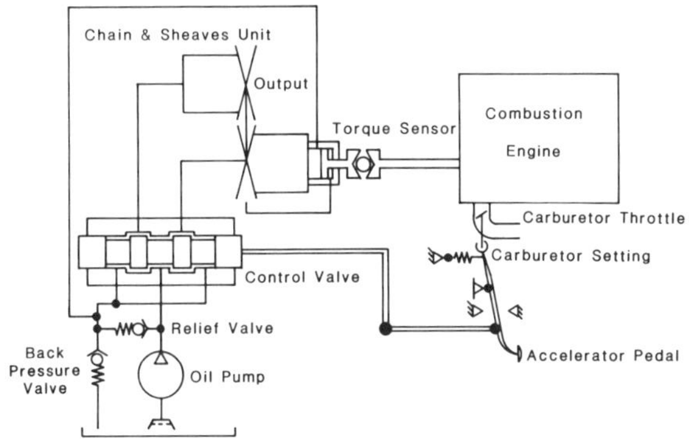

Fig. 7

Traction Drives

A High Power Density Traction Drive*

Yves Kemper
Vadetec Corp.

*Paper 790849 presented at the Off-Highway Vehicle Meeting and Exposition, Milwaukee, Wisconsin, September, 1979

A MECHANICAL CONTINUOUSLY VARIABLE transmission has been considered for many years to be the theoretically ideal transmission for self-propelled vehicles. Indeed, some of the early automotive transmissions were dry mechanical CVT's such as the right angle (disc and roller follower) drive. To overcome the life limitations of such drives, many mechanical CVT's designed to operate on an oil film were proposed and patented at the beginning of this century; but these so-called "traction drives" found commercial acceptance only in stationary applications. They have been widely used as machine tool and industrial drives, particularly in Europe and Japan, for nearly 50 years and have gained a reputation for long life and dependability.

In vehicular applications, however, the problem has been primarily one of power density. An industrial traction drive the size of a 100 HP automotive transmission is rated at only a few horsepower. In order to increase power density it was necessary to increase the stress levels in the contact areas. This required the use of costly alloys if surface fatigue failures (pitting) of the contacts were to be avoided. This factor, combined with the complex geometry of most of the proposed designs, tended to render traction drives less cost effective than other alternatives. In addition, the efficiency frequently proved to be less than was predicted because most designs involved multiple rollers which worked against each other if their positions could not be controlled precisely enough to be perfectly symmetrical for any given ratio. Therefore, in order to satisfy requirements for a CVT, vehicle designers have been obliged to resort to hydrokinetic, hydrostatic or electrical drives.

The energy crisis has renewed interest in mechanical CVT's because they offer the promise of better efficiency than non-mechanical drives, particularly at part load. Now that synthetic traction fluids with better coefficients of traction and standard bearing steels with improved fatigue life have become available, a completely new design approach could make the use of traction drives in vehicles practical for the first time.

NUTATING TRACTION DRIVE

The design is based on the kinematic principle of nutation, the wobbling motion of a body as it rotates about its axis. Hence the term, "Nutating Traction Drive" (NTD). It has the capacity to transmit higher torques than had heretofore been considered possible with a traction drive which explains why it can achieve the power densities required not only for automotive but even for off-highway vehicles.

As in all other traction drives, torque transmittal is by viscous shear of a thin film of oil between loaded rolling contact surfaces which act like large roller bearings. High coefficients of traction are possible due to

ABSTRACT

A new traction drive has been developed which has the power density required for use in both on- and off-highway vehicles. It avoids problems faced by other types of traction drives such as high contact stresses and roller positioning. The range extends from 0 to 1:1. Cost and efficiency are competitive with other automatic and manual transmissions.

the rheological behavior of the oil at high pressure. The viscosity of oils, and particularly traction fluids, increases greatly under pressure, becoming almost glassy under the stresses prevailing between the rolling contact surfaces. In the NTD, the rolling contact motion is provided by nutation of a roller member within a stationary reaction member.

The unit is made up of five basic components as shown in Fig. 1, which depicts a typical design arrangement:

- a transmission case or reaction member
- an input or driving member
- a nutating or rolling assembly
- two control rings or reaction members
- an output or driven member.

As viewed from the left end of this unit, the input torque to the driving member is in a clockwise direction.

The driving member casting has a cylindrical central section which has been hollowed out to accommodate the nutating assembly (see Fig. 2). The shaft of the nutating assembly, which is mounted on bearings in the input member, is skewed at an angle so that its axis intersects the axis of the input member at the center. As the input member rotates, the nutating shaft nutates, i.e. its axis describes two cones with their apexes at the point of intersection of its axis with the axis of the input member.

Mounted on the nutating shaft are two central collars, which are splined to the shaft, and two cones, which are free to both slide and rotate on the shaft. On one face of each central collar and on the base of each cone are two kidney-shaped grooves of semi-circular cross-section, which are sloped so that one end is deeper than the other to form a ramp angle as shown in Fig. 3. When mated the grooves form pockets, each containing a ball so that the cones are urged outward by these ball ramps when twisted on the nutating shaft. The ball ramps of one cone are indexed at 90° with respect to the other so as to provide a universal joint action. Between the two collars a belleville spring maintains a preload on the system.

The two cones are in contact respectively with two control rings through the openings in the sidewall of the input member. One of these points of contact is at the lower left of the left hand cone; the other is at the upper right of the right hand cone, as shown in Figs. 1 and 4.

The nutation of the cone and shaft assembly causes the cones to roll on the inside periphery of the control rings in an epicyclic or planetary motion, but in a direction opposite to input rotation.

The tangential force developed at each of the two contact points tends to twist the cones on the nutating shaft, thereby causing the entrapped balls to roll up the ramps and force the cones outward, which wedges them into the control rings. Thus, the contact pressure is a function of the torque transmitted and, by appropriate design of the ball ramp angle, is always maintained at the level necessary to provide the required traction.

The output planetary motion of the nutating shaft assembly can be viewed as orbiting about the main axis, at constant input speed and direction, while rotating about its own axis, or spinning, at speeds greater than input and in the opposite direction.

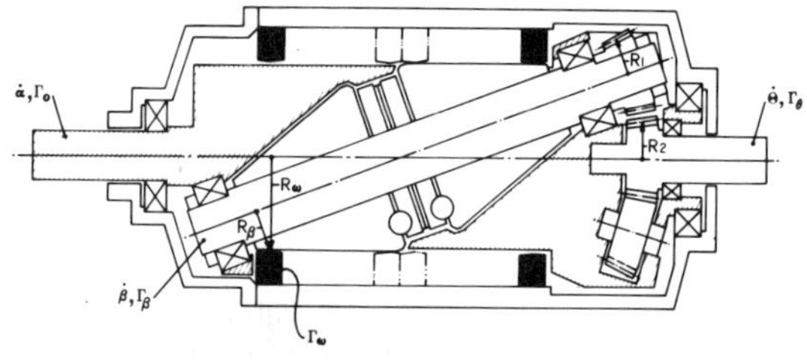

Fig. 1 - Schematic of the Nutating Traction Drive

Fig. 2 - The four major assemblies of the Nutating Traction Drive. Shown from front to rear are the nutating assembly, the input member, the control rings and the case

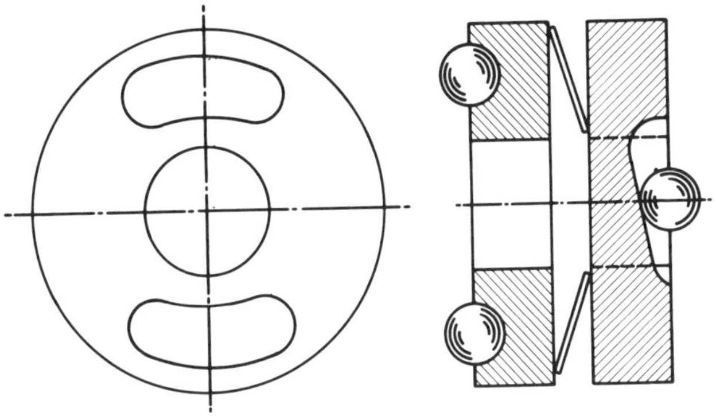

Fig. 3 - The ball-ramp system for loading the traction contacts

Fig. 4 - The nutating assembly and the control rings form two traction contacts

The rotational speed of the shaft about its own axis is a function of the geometric ratio represented by the perpendicular distance from the axis of the nutating shaft to the contact point, to the perpendicular distance from the axis of the input member to the contact point. The control rings are keyed to the transmission case, which is the reaction member, and can slide axially toward either the center or the ends of the case. Thus, by varying the position of the rings, the points of contact are shifted so that the rotational speed, and hence "speed ratio", is changed in accordance with the geometric ratio.

If the control rings are at the center of the case and the diameter of the base of the cones is, for example, close to the inside diameter of the control rings, the speed ratio is approximately zero (because the cones just wobble within the control rings without being able to roll).

If the control rings are at the ends of the case and the diameter of the small ends of the cones is, for example, equal to half the inside diameter of the control rings, the speed ratio is one. For this ratio of diameters, the nutating shaft is rotating at twice input speed, but in the opposite direction.

Since the output end of the nutating shaft orbits about the main axis of the transmission, it is necessary to bring its motion back into line with the main axis. This is accomplished through various output gear set arrangements, depending upon the specific application.

OUTPUT GEAR ARRANGEMENTS

Figs. 5 (a) and (b) show two possible output gear set arrangements. In arrangement (a), which is the one shown in Fig. 1, the drive pinion on the end of the nutating shaft drives the sun gear on the output shaft through an idler gear, which for clarity is shown in displaced section. The motion of the nutating

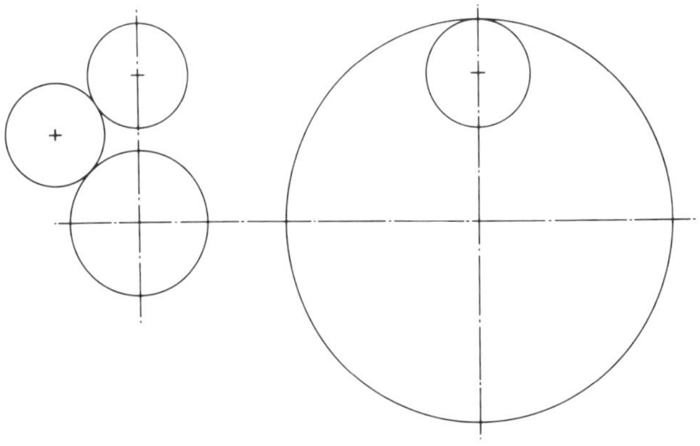

Fig. 5 - Two possible output gear set arrangements for the Nutating Traction Drive

shaft, in both orbiting about the output shaft (at direct input speed) and rotating about its own axis (at variable speed, opposite to and greater than input), determines the system output. Because of the idler gear, the relative effects of the rotational and orbital speeds are subtractive, thereby making the assembly a differential speed system. Maximum output speed, opposite to input rotation, occurs when the control rings are apart at the extreme end position; minimum speed occurs when the rings are together at the center.

The sun gear arrangement is versatile in that it provides either overdrive or underdrive outputs. By appropriate selection of the gear ratio a zero point can be reached within the range, so that both forward and reverse speeds are provided. In that case it is possible to pass smoothly from one direction to the other.

In arrangement (b), the drive pinion on the end of the nutating shaft meshes directly with an internal bevel ring gear mounted on the output shaft in place of the sun gear and idler arrangement shown in Fig. 1. The relative effects of the rotational and orbital speeds are again subtractive, but now the low and high ends of the ratio range are reversed.

Minimum output speed occurs when the control rings are apart, at the ends of their travel, because in this position the rotational speed of the nutating shaft is at its maximum and is subtracting the most from its orbital speed. Maximum output speed occurs when the control rings are close together because in this position the rotational speed is at its minimum. Thus, as the control rings move toward the center, output speed increases since the rotational speed of the nutating shaft is slowing down and subtracting less and less from its orbital speed.

The ring gear arrangement provides an underdrive output which, through appropriate selection of the gear ratio, can reach a zero point within the range so that both forward and reverse speeds are provided. This arrangement has the advantage of a high peak efficiency with a low contact stress level at maximum output speed, resulting in a very high life expectancy at the ratios where the NTD will operate most of the time. At maximum output speed, only about 40% of the power passes through the rolling contacts of the NTD.

ADVANTAGES OF THE NTD

The NTD has a number of advantages:

- The maximum Hertz stress in the rolling contact areas is on the order of 300,000 psi. This is due partly to the conformal surfaces (a convex surface rolling on a concave surface), which provide large contact areas, and partly to the fact that each contact area carries only half the load since the power passes through them in parallel instead of in series as in most other traction drives.
- The system is not sensitive to dimensional variations. Only roundness, concentricity and surface finish are important. The cones and rings can differ in dimensions without significantly impairing efficiency (see Fig. 6).
- The axial forces from the two contact areas cancel each other and, therefore, there are no heavy thrust loads. The radial loads are reduced by the couple about the midpoint of the machine provided by the gyroscopic effect which helps to apply the load required on the contact surfaces.
- The ball ramp method of loading the contact areas makes the system respond instantaneously to any torque variation. This feature, together with the small radius from the axis of the machine to the contact areas, accounts for the high power density.
- Manufacturing cost should be less than a fluid torque converter automatic transmission. There are fewer than 150 parts altogether, compared with over 1000 for an automatic. The primary ones are simple in shape, circular in cross-section and easily produced on standard machine tools.
- The peak efficiency is high. For the output gear set arrangement shown in Fig. 5(b) it is 94% at rated power.

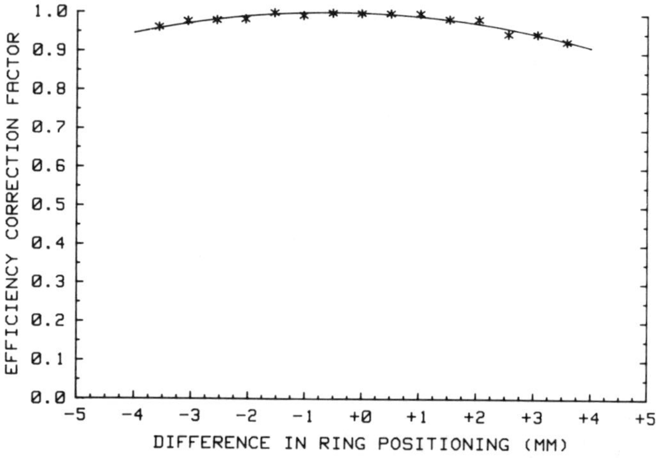

Fig. 6 - Sensitivity of the transmission efficiency to centering of the two rings

Fig. 7 - Bench testing of Vadetec NTD prototypes depends upon computer-controlled data acquisition and processing. (Photograph by Hal Anderson, Keyboard Magazine.)

CONCLUSION

The NTD is a practical, cost effective mechanical CVT with the necessary power density for use in vehicles. With a range extending from 0 to 1:1, it opens up many possibilities not only for new types of drive lines in vehicles powered by spark and compression ignition engines but also for power trains based on single shaft gas turbines, flywheel hybrids and electric hybrids for which the lack of such a transmission has been a major obstacle.

Nomenclature:

$\rho = \frac{R_\beta}{R_\omega}$ = cone/ring geometric ratio

$k = \frac{R_1}{R_2}$ = cone shaft terminal gear output ratio

Γ_o = input torque applied by power generating system

$\dot{\alpha}$ = speed of input member

Γ_θ = output torque resisted by the load (negative)

$\dot{\theta}$ = speed of output member

Γ_ω = torque reaction of the casing on the two control rings

Γ_β = torque applied by the cone shaft on its terminal gear

$\dot{\beta}$ = speed of cone shaft relative to main axis of drive

R_β = radius of cone

R_ω = radius of control ring

R_1 = pitch radius of cone shaft terminal gear

R_2 = pitch radius of output gear (sun gear if driven through idler, or ring gear)

APPENDIX

BASIC FORMULAS FOR CALCULATING THE VADETEC NTD - The following relationships exist for the ideal case where torque and speed losses are neglected.

Speed:

$$\dot{\theta} = \dot{\alpha}\left(1 - \frac{k}{\rho}\right); \quad \dot{\beta} = \dot{\alpha}\left(1 - \frac{1}{\rho}\right);$$

$$\text{Speed Ratio} = 1 - \frac{k}{\rho}$$

Torque:

$$\frac{\Gamma_\omega}{(-1)} = \frac{\Gamma_\beta}{(-\rho)} = \frac{\Gamma_\theta}{(\rho/k)} = \frac{\Gamma_o}{(1 - \rho/k)};$$

$$\text{Torque Ratio} = \frac{\rho}{\rho - k}$$

A New Approach to CVT Traction Drives in Automotive Applications*

Y. Kemper and J. L. Tevaarwerk
Vadetec Research Corp.

*Paper 861351 presented at the Passenger Car Meeting and Exposition, Dearborn, Michigan, September, 1986.

ABSTRACT.

This paper describes the operation and performance of the new Vadetec NTD-XA2 traction drive. The NTD-XA2 is based on the concept of a compound planetary where one stage is formed by variable ratio traction element and the other stage by a fixed ratio spur bevel gear set. This arrangement has the advantage that the power split between the traction element and the gear set can be controlled by the traction ratio.

The construction of the NTD-XA2 is such that it can be mounted directly on the engine crankshaft, replacing the conventional engine flywheel. Because the functions of flywheel, reverse, clutch and forward speed variation are all incorporated in one small package it allows for a great deal of freedom in the instalation configuration.

The manufacturing requirements for this drive fall well within current automotive practices. Automotive transmission fluid is used as the lubricating and traction fluid.

Computer simulation of the drive for a typical automotive application shows that the efficiency at full power is about 96% for most of the overdrive range, and remains high at part load.

1-0 INTRODUCTION.

The use of traction drives as the power transmission for propulsion of automotive vehicles has a long and checkered history. Some of the very early cars were equiped with one form or another of traction or friction drives. This was because of the inherent simplicity of operation of such drives and the simple manufacturing techniques that could be used.

The traction drives used at that time were based upon machine tool units. These often combined the functions of clutch, forward reverse and an infinitely variable speed ratio mechanism all in one unit. Friction materials such as rubber, leather and phenolic were used against cast iron for the main components. Manufacturing tolerances on these type of devices was very low an repair was simple. Soon however the increase in engine power outstripped the capabilityies of these simple traction drives. Also the increasing sophistication of gear manufaturing and design techniques made gearsmuch more reliable in operation than before, and so a rapid introduction of mechanically shifted fixed ratio transmissions followed. Since that time the traction drive has been trying to make a comeback but this has always been against a background of very well time tested standard and automatic transmissions.

While fixed ratio transmissions are solidly entrenched in the automotive field, there are severe inherent limitations in their use especially in the area of fuel consumption and perhaps more importantly in the emission control of automobiles. There are many variables that affect the above conditions and the transmission plays a very important role in these. Ideally the transmission for minimizing both fuel consumption and emmisions is one that would permit complete freedom between the engine and the car speed ratio and has a very high power efficiency. Engines operate most efficiently at a given power point and at a constant speed. Hence an ideal transmission must be infinitely variable and follow the constant power hyperbola as closely as possible with very low losses. Also part load efficiencies should be high because a good deal of the total fuel consumed is spent at part load conditions.

Studies have revealed that by using such CVT's (constantly variable transmissions) we could improve the fuel efficiency of todays vehicles by as much as 10 to 25%.

Further functions of the transmission are that it must allow for a smooth engagement and disengagement between the engine and vehicle at zero speed, must provide for a means of reversing the vehicle, must be silent in operation and have a smooth ratio change.

An ideal transmission therefore would integrate the above all in one functional unit. Further more the transmission will have a torque requirement that is only slightly above the wheel spin limit. This puts a cap on the maximum torque requirement for a practical transmission. Also because of maximum road speed restrictions the maximum output speed is limited. Figure 1 then shows the torque speed diagram of such an ideal transmission at a constant power setting with the maximum torque and speed limits indicated.

An essential requirement of the ideal transmission is that is constantly variable over the operating range. Several different approaches to develop CVT 's have been tried in the past such as traction drives, hydrostatic and belt drives. Even today each and every one of these is being developed and researched for possible automotive use. Also the current trend in the improvement of conventional automatic transmissions is to get closer and closer to the a CVT transmission by increasing the number of speeds and by the use of lock-up techniques to improve the efficiency. This however has increased the complexity and cost of these drives severly. Also these units do not directly integrate the functions of engagement and foreward reverse directly in one unit, but instead require separate components for these.

It is clear then that the search for an ideal mechanical power transmission for automotive use is continuing as strongly as ever and perhaps that the need for such a transmission has never been as high . Perhaps it is wise to outline the other principal requirements that such an ideal transmission must have in order to be a viable alternative to the current transmissions. These are;

- cost must be less than or the same as the current transmissions
- efficiency must be high over the entire operating range
- size and layout must suit current automotive design
- minimum weight
- high reliability and a predictable failure mode
- ease and smoothness of operation

The authors are currently working on the development of a transmission that satisfies these requirements and those for the ideal transmission very well sofar as can be ascertained at this time.

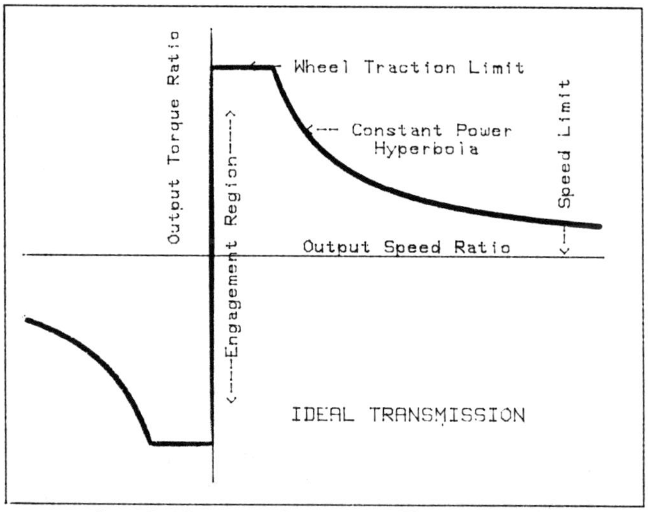

Fig. 1 The Speed Torque diagram for an Ideal transmission.

The transmission is based upon the same general design principals as are used in the NTD-CVT's as outlined by Elu et al (1), with the exception that a very much smaller total space is required for the same power transfer. The basic design strategies in these transmissions are the use of torque amplification during one mode of operation, and power split during the normal cruise range of operation. In this way good efficiency can be maintained throughout the speed range and the basic requirement of the ideal transmission are satisfied.

Furhermore there have been several advances in the traction drive technology, see Loewenthal(10) so that a new look at the use of traction drives for automotive applications is warrented.

2-0. DESCRIPTION OF THE TRANSMISSION.

The NTD-XA2 transmission as shown in figure 2 is essentially a compound planetary drive, where one set of the elements form a fixed ratio planetary, and the other set of elements form an infinitely variable ratio planetary. The use of the compound planetary techniques is common to all NTD's and give it key features that make it an extremely flexible device for use as a variable speed transmission. By using this compound planetary technique it is possible to combine several of the functions in one small package. Other forms of currently proposed variable speed transmissions do not share these advantages.

The NTD-XA2 consits of five main assemblies. These are:

---"Alpha body or Input member " This assembly rotates around the a axis at engine speed at all times and is the main support for the cone assembly. The a axis is concentric with the engine crankshaft and the output shaft. The assembly is equivalent to the carrier in a compound planetary.

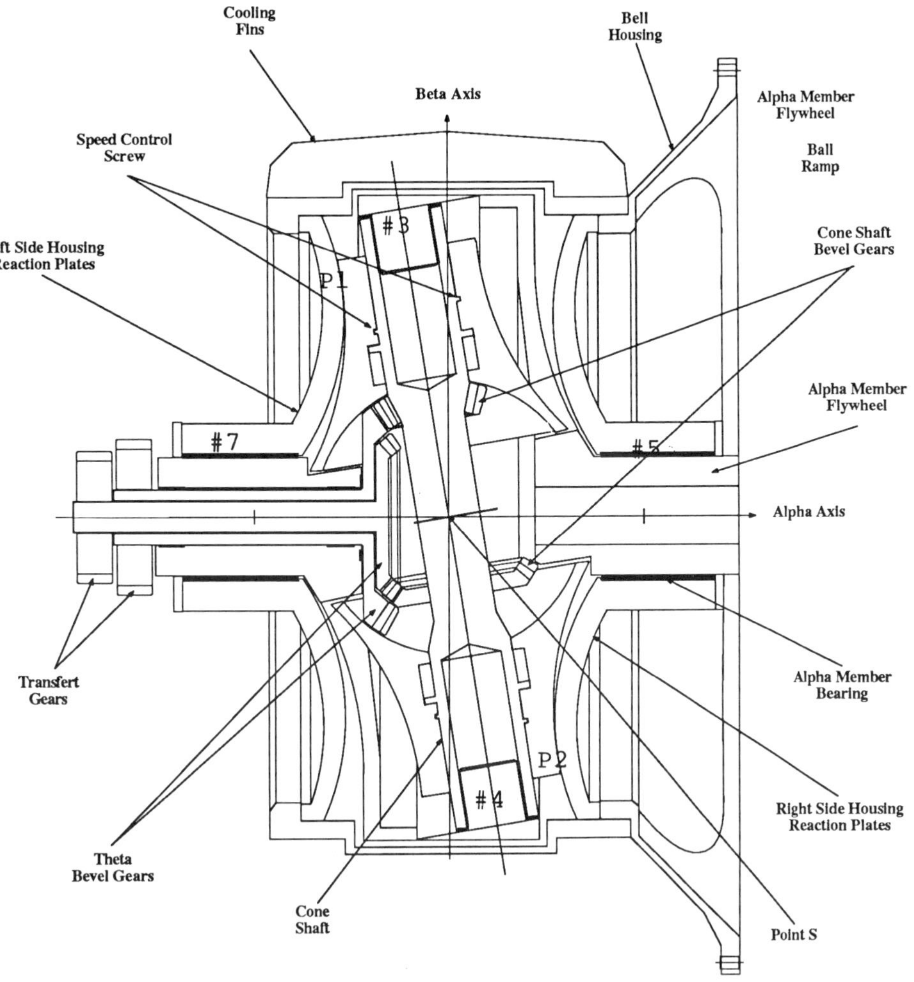

Fig. 2 Cross section of the VADETEC NTD-XA2 transmission showing the various details of the construction.

---"Beta body or Cone assembly" is supported by the Alpha body through two journal bearings #3 and #4, and has two conical rolling surfaces. This body is concentric with the b axis around which it rotates. It consists of two Cones, a Cone shaft and one or two bevel gears keyed to the Cone shaft. The whole assembly is equivalent to one of the planet assemblies of a compound planetary. The cones slide up and down the shaft but are keyed such that they both operate at the same speed as the shaft. The motion of the cones is controlled by the control mechanism. The cones are equivalent to one set of planet gears in a compound planetary. The gears on the coneshaft form part of the other planet set of a compound planetary.

---"Omega body or Plates", is part of the housing, and consists of two circular plates, each having a rolling surface machined on their inner side. These Plates are equivalent to a ring gear in a compound planetary.

---"Theta body or Output gear", consists of an output gear and its shaft. It is equivalent to a sun gear in a compound planetary. Depending on the particular variant this Output gear is in mesh with one of the two bevel gears that forms the two stage bevel gear reduction on the Idler gear set or directly with the bevel gears on the coneshaft.

---"Synchro shift mechanism", consists of a gear selection mechanism that allows for a random choice of engagement for the two bevel gear sets on the output shaft. The bevel gear ratios, the external gear ratios and the variable geometric ratio are selected such that at a given operating point the two output shafts rotate at the same speed.

This then allows for a "Synchro shift" between the two modes of operation.

The transmission is generally symmetrical about a point S. The rolling surfaces of the two Cones and the two Plates are respectively in contact with each other in two variable contact points P1 and P2, symmetrically located about the point S.

A schematic representation of the overall transmission is shown in figure 3. From this figure it is clear that we have two sets of rotating components that make up the overall transmission speed ratio. These are the two traction contacts between the cone and the plate which form the variable geometric ratio, and the fixed gear ratio, formed by the various gears in the drive.

The geometric ratio of the cone - plate combination may be changed in a number of different ways such as by shifting the plates, tilting the cone shaft, and by stroking the cones on the coneshaft. The particular variant that we are examining here is known as the XA2 transmission. For the XA2 transmission only the cones and plates are allowed to move in and out.

3-0 ANALYSIS OF THE NTD-XA2.

The analysis of the traction drive involves calculating the various speeds, torques and powers for the various components, and the ratios of these torques, speeds and power. Before we proceed with the real drive however, we will examine an no-loss form of the drive. Under a no-loss drive we assume a transmission that has no losses at all. While such an analysis has limited value in the comparison with real transmissions, it helps in understanding the origins of some of the unique features of the NTD-XA2.

3-1 ANALYSIS OF A NO-LOSS NTD.

The derivation of the basic equations for the no-loss drive is presented in Kemper et al (7), and the results are reported here. For an understanding of the no-loss and real drive properties we need the following three relations;

(1) $\dot{\Delta} / \dot{\alpha} = (1 - \text{Kint}\ \rho)\text{Kext}$ - speed ratio or the ratio of output to input speed.

(2) $\Gamma\alpha / \Gamma\Delta = - (1 - \text{Kint}\ \rho)\text{Kext}$ - Basic torque ratio of input to output torque.

(3) $Pc = - \text{Kint}\ \rho / (1 - \text{Kint}\ \rho)$ -Power split for the drive or the fraction of Input power transmitted through the contacts.

Where : $\rho = - Rg / Rz$ is defined as the infinitely variable geometrical ratio.

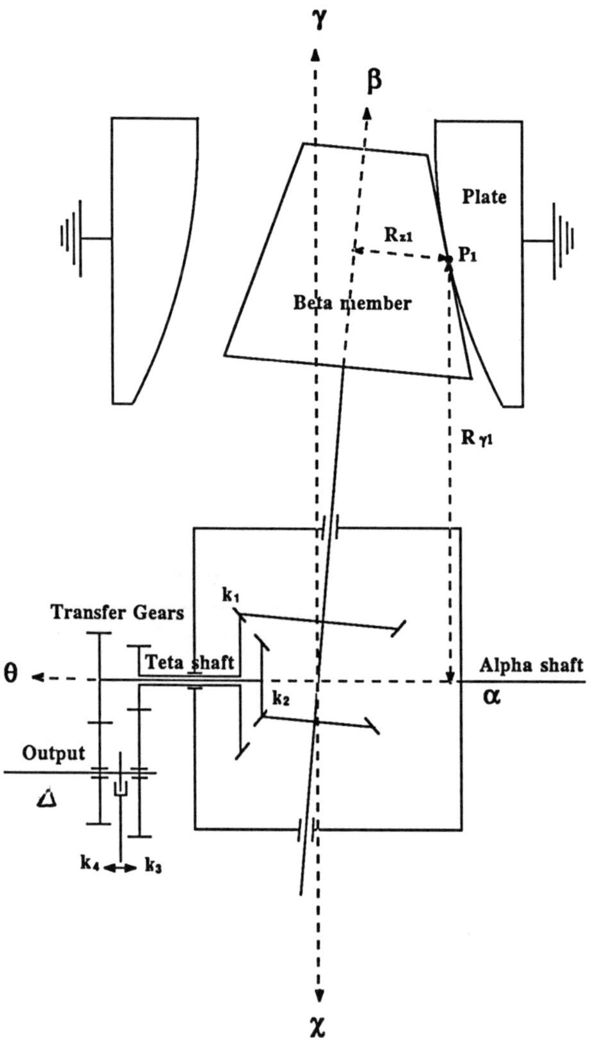

Fig. 3 Schematic representation of the VADETEC NTD-XA2.

$\text{Kint} = \dot{\theta}^* / \dot{\beta}^*$ is the fixed gear ratio for the drive in the rotating reference frame.

$\text{Kext} = \dot{\Delta}^* / \dot{\theta}^*$ is the fixed gear ratio for the reduction set in the fixed reference frame

$\Gamma\alpha$ = torque on the input or α member

$\Gamma\Delta$ = torque on the output or Δ member.

Furthermore it is instructive to look at the amount of spin on the traction contacts. When conditions of high spin exist the losses will be high in the contact and consequently the efficiency will suffer. The influence of the spin losses may however be minimized if we keep the amount of power through the contacts low at conditions of high spin losses. Loewenthal (8) suggests the following expression to calculate this spin,

(4) $\text{Spin} = \omega s / \dot{\alpha}$: where ωs = angular spin velocicty on the contact.

Unfortunately this parameter alone

does not reflect the actual spin losses in the drive because it lacks the contact geometry and the roller speed. An improved representation would be to divide the contact spin ωs by the rolling speed U as is required for the traction calculations. For most variable traction drives with spin the total contact losses are directly proportional to this grouping, see for example Tevaarwerk (4). For the ideal drive the contact rolling speed is given by;

(5) $U = \dot{\alpha} Rg$

Hence the definition of spin that we will use here is now;

(6) $Spin = \omega s \,/\, U$

This relationship now depends upon the size of the drive but has the contact geometry missing in it. It gives however a good indication of the total contact losses in this particular drive.

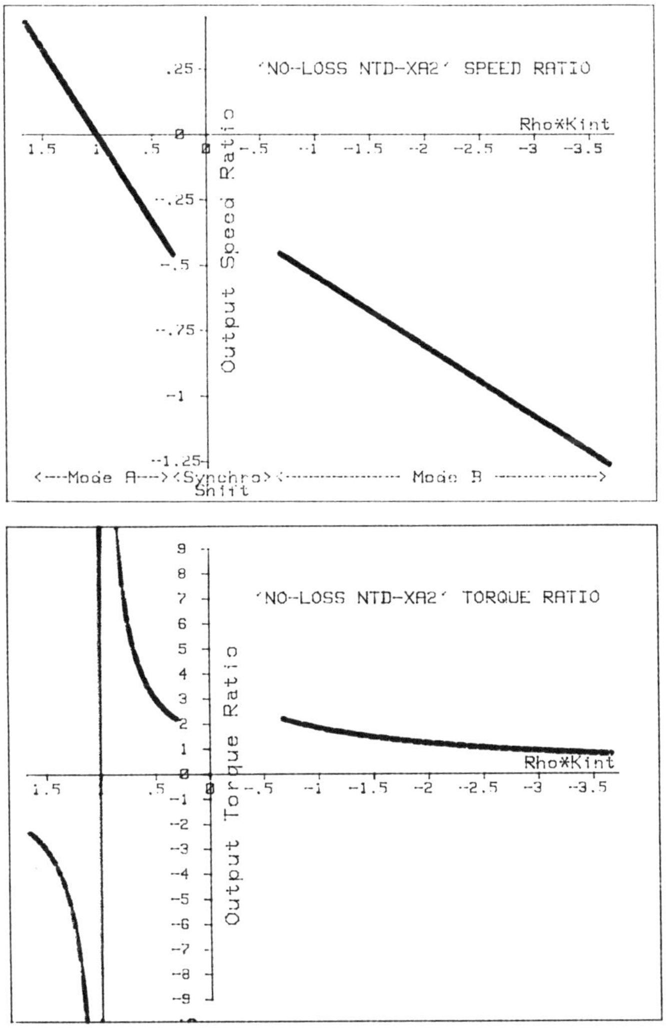

Fig. 4-1 and 4-2 Showing the Speed ratio and Torque ratio for the 'no-loss' form of the NTD-XA2.

The various parameters developed sofar may now be evaluated for the no-loss form of the NTD-XA2 transmission.

The speed ratio is shown in figure 4-1 as a function of Kint ρ. For the overall speed ratio as given by equation (1) it shows that the transmission has a Reverse and Underdrive region (or Mode A region), and an Overdrive region (or Mode B region). Overdrive or Underdrive refers to the ratio of output speed $\dot{\theta}$ to the input speed $\dot{\alpha}$. At the position of lowest Kint ρ on either side of this region of operartion we note that the output speed is the same. At this point it is possible to perform a synchro shift between Mode A and Mode B. Increasing ρ on the Mode A side will then reduce the speed through zero and into reverse. Increasing ρ on the Mode B side will increase the speed progressively.

In Mode A operation the drive has a very high torque multiplication as shown in figure 4-2. This is very desirable because only low engine power levels are required for vehicle manoeuvring and drive away conditions. At the zero output speed the torque multiplication for the no-loss drive is in fact infinite. In the real drive this multiplication is limited because a certain amount of slip is needed to generate any torque at all in the drive. Because no torque is generated at the ideal zero output speed point it is possible to operate this drive in the Geared neutral position where the drive is fully engaged but no forward or reverse motion results. It is also possible to operate in a true neutral fashion by bringing the contact load to zero. This breaks the torque loop in the drive.

Figure 4-3 shows the power split in the drive and figure 4-4 shows spin as per equation (6). The latter is for a typical geometry that we have determined for the cone and plate combination as being the best in terms of the range and spin.

Significant is the fact that power recirculation occurs in the Mode A operation. The highest recirculation occurs near the zero output speed. If at this point there are high losses in any of the internal bearings or traction contacts then the overall efficiency will be low. However because the torque multiplication is high in this region, the amount of input power required is low for the normal start-off conditions. The absolute power loss is therefore expected to be low in this region. Also the drive does not operate very long at this point so that the total power loss is not as important as in the Mode B operational range. Towards the end of Mode A the efficiency of the traction contacts drops rapidly because of the higher spin conditions. The power split is such however that the effect of this loss of contact efficiency is very low on the overall efficiency of the drive.

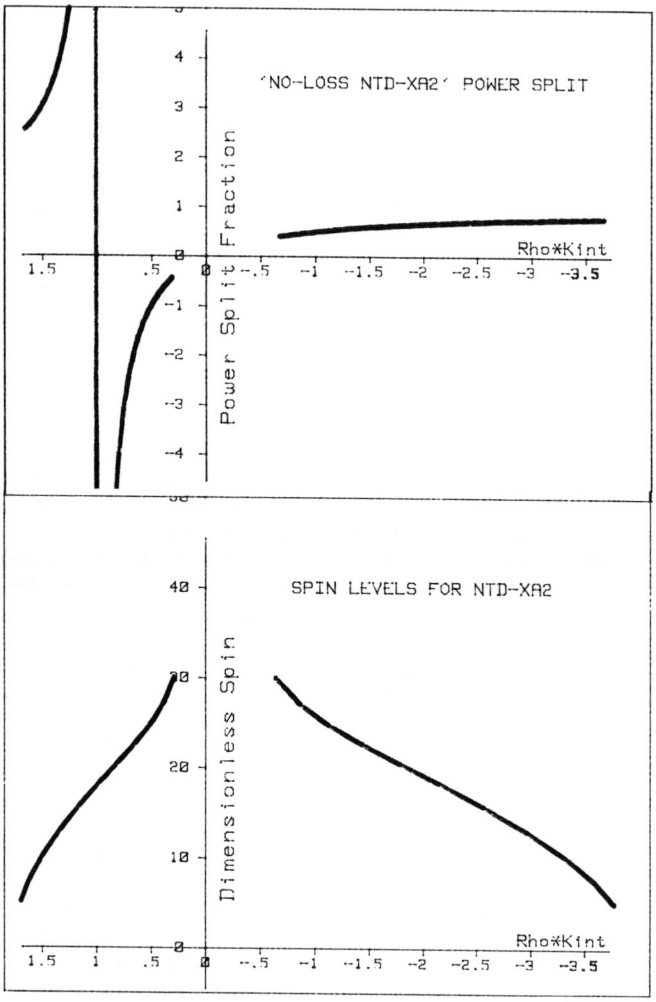

Fig. 4-3 and 4-4. Power split and Spin levels for the NTD-XA2.

In Mode B the same effect of the power split helps in keeping the overall efficiency high in the lower K_{int} ρ region. When a significant amount of power starts to be transmitted through the contacts, the contact spin conditions are such that a good efficiency is possible. This results in a good overall efficiency for the drive. In fact in all of Mode B the overall efficiency curve should be quite high and flat. These observations are experimentally confirmed, see Kemper and Elfes (6), by the extensive tests performed on various previous NTDs operating on this same principle.

3-2 ANALYSIS OF THE REAL NTD-XB2.

In the analysis of the no-loss drive we neglected all the sources of losses. In the real drive these losses consist of torque losses, speed losses and the product of these two namely the power losses. These losses are caused by the traction contacts (slip and torque losses), the bearings, gears, and due to the cone and alpha member rotation. The latter four sources all result in a torque loss. Further losses external to the mechanical components of the drive are the power losses associated with the hydraulic control pumps. Each of these losses has to be taken into account for the prediction of the overall performance. These various losses are load and speed dependent and so it is essential that the loads and speeds are known at every point or node in the drive.

For the analysis of all the loads, speeds and torques within the NTD-XA2, a computer simulation program "VERT" was developed. Many of the features and techniques used in this program are based upon experience gathered from performance evaluations of the other NTD's. It integrates all the elements necessary to evaluate the overall performance of the whole transmission, including the losses in the contact, the gears, the bearings and windage losses of various components. Currently we do use the approximation that the gear losses are introduced in the "VERT" program as a constant percentage of the load, with the friction losses based upon Buckinghams(2) techniques. Also the bearings are assumed to have a constant friction coefficient.

The traction calculations are based upon the thermal prediction techniques of Tevaarwerk(5) with full regard being given to all the losses in the contact. Windage losses due to the rotation of the cone and the Alpha member are based upon the techniques by Loewenthal (3).

4-0 RESULTS FOR A TYPICAL APPLICATION.

In order to proceed any further with the drive we need to specify a typical application. This will allow us then to determine the performance in terms of speed ratio range, efficiency, the size and estimated weight. The sizing of the drive components such as the bearings and gears is then also possible. Performance predictions are based on the real drive analysis concept as explained.

The following engine specifications were used in the drive simulation:

Engine speed	1500	3500	5000	rpm
Peak torque	200	224	200	N.m

The lubricating and traction fluid used in the simulation is a conventional automatic transmission fluid called ATF220. The traction characteristics of this fluid have been analysed for the range of pressures, temperatures and speeds encountered in the NTD-XA2. For the simulation it is assumed that the fluid inlet temperature is 100 °C.

4-1 DIMENSIONS.

The rotating Alpha member has an overall diameter of 276 mm, and a length of 258 mm. With these dimensions it is smaller than any of the currently used or proposed transmissions. It is interesting to note that Cuypers(9) states that traction drives have the smallest torque capacity of all CVT's. However a direct size comparison of the NTD-XA2 with a comparable capacity metal V belt transmission reveals quite the contrary. The NTD-XA2 is only about half the size of the current automotive metal belt drives. Cuypers ofcourse did not consider using compound planetary traction drives as we are proposing here.

The total rotating inertia of the transmission is estimated to be .111 kg.m². The rotating inertia of the Cone assembly around its own axis and the main axis of the transmission was vcalculated as .0590 kg.m².

The total inertia compares very favourably with flywheel inertias currently used on engines of the above specifications. Hence the concept of utilizing the NTD-XA2 as a flywheel is quite feasible from this point of view.

4-2 PERFORMANCE AND EFFICIENCY.

The torque speed performance of the NTD-AX2 is shown in figure 5. This particular curve is for full engine torque and at 1500 rpm input speed. It is clear from this figure that this particular transmission gets very close to the stated ideal transmission as discussed in the introduction.

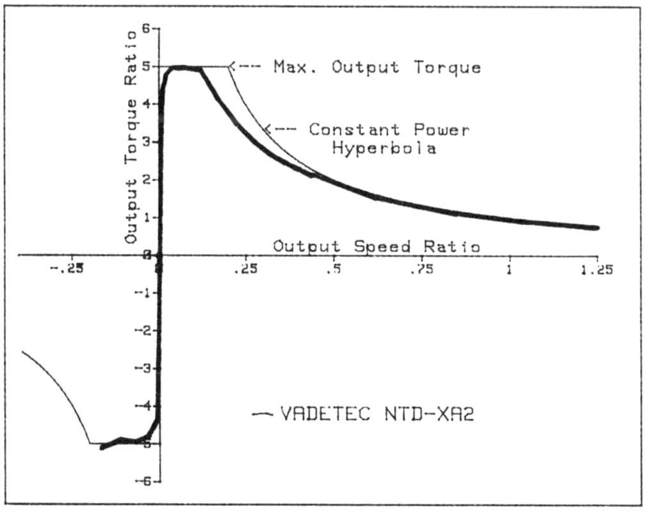

Fig 5. Torque Speed diagram for the NTD-XA2.

The efficiency for this operational curve is shown in figure 6. Note that the calculated efficiency is about 95% and that this is maintained for most of the Mode B operation. The reason for this lies in the power split arrangement of this drive and the fact that the losses diminish when more power is going through the contacts. This behaviour has been experimentally observed on the earlier NTD's operating in this regime. It is particularly advantages because most of the time spent during the operation of a car is in this Mode.

Reference to the Gear I,II,III,IV,V,R are to the ratios of an equivalent five speed manual transmission.

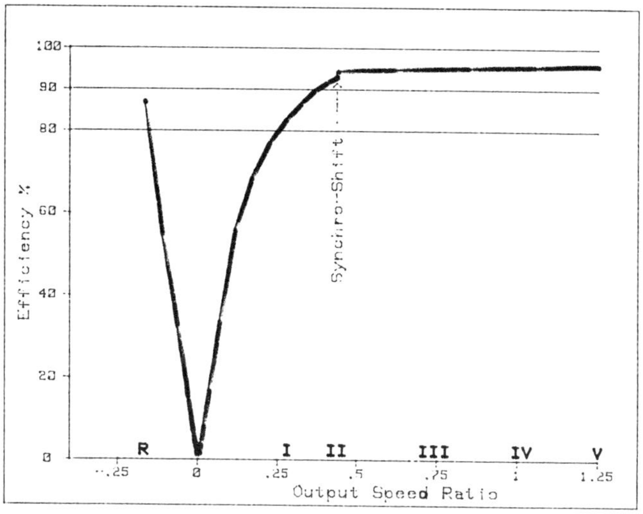

Fig. 6. Full load power efficiency for the NTD-XA2 at 1500 rpm input speed.

Part load efficiency at 1500 rpm input speed is shown in figure 7. It shows that the efficiency remains high even at reduced torque levels. The reason for this is that the losses in the traction contacts tend to decrease when the load is reduced.This offsets the increased component due to the fixed losses such as windage and bearings.

4-3 TRACTION CONTACTS.

The stress levels in the contacts are in general around 2 GPa. This contact stress level is well below that for the NTD-D transmission. No contact fatigue failures have ever been encountered in any operational NTD-D, including prototypes run through the 10,000 hour endurance tests. The stress levels in the NTD-D transmissions are in the order of 3.0 GPa. It is therefore expected that no fatigue failures will occur with the NTD-XA2 transmission.

A unique feature of the transmission is the low amount of heat dissipated in the traction contacts as a percent of the input power for the most extensively used ratio range. The heat generated within each traction contact at maximum input power is for Gear I, 0.96 kW or .9% of the input power resulting in a flash temperature of 136 °C. For Gear II this is 0.221 kW or .2% of the input power with a flash temperature of 125 °C and in Gear V it is 0.667 kW or .6% of the input power with a flash temperature of 117 °C. These values are all for full power input and are well below levels experienced in the NTD-D. No undue flash temperature distress has ever been observed in the NTD-D.

4-4 GEAR SET.

For the gears the root bending and contact stresses were analyzed under the worst case conditions. It was found that calculated L10 safety factors are within the normal automotive levels.

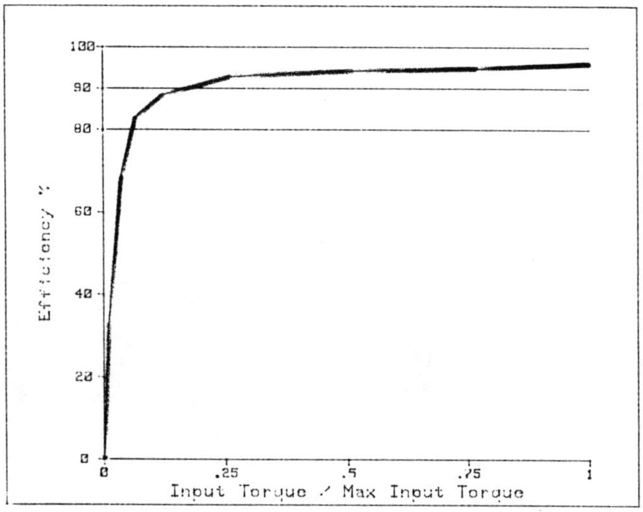

Fig 7. Part load efficiency of the NTD-XA2 at 1500 rpm input speed and at an output speed ratio of 1.25 .

4-5 BEARINGS.

All the bearings in the transmission are of the hydrodynamic type. For this particular application they have all been sized and analyzed according to the worst load/speed conditions. The traction fluid is used as the lubricant and the inlet temperature is assumed to be 100 °C. Adiabatic solutions were used with a clearance ratio of 1000. All bearings were found to have adequate load capacity at acceptable values of filmthickness.

4-6 WEIGHT AND COST.

The estimated weight of the NTD-XA2 used for this simulation is 39.5 kg. This is 62 % of the weight of a comparable four speed lock-up transmission (64 kg). The weight of the cones and the cone shaft (6 kg) are calculated directly by the drive simulation program. The weight of the plates (16.3 kg) is calculated by approximating them as two flat plates. The Alpha member is estimated to be 9.25 kg. The housing and all remaining parts are estimated to be 20% of the total (7.9 kg). The Plates themselves are part of the housing, and therefore most of the housing consists of a ring around the two Plates. This ring has three local reinforcement ribs where the ramps are situated, and together with the bell housing, is bolted to the back of the engine.

A detailed calculation of the manufacturing cost of this transmission was out of the scope of this study.

REFERENCES

1 "Performance of a Nutating Traction drive.", P. Elu and Y. Kemper. ASME 80-C2/DET-63,1980.

2 "Analytical Mechanics of Gears", by E. BUCKINGHAM, Dover Publications, 1963.

3 "Design of Spur Gears for Improved Efficiency", N. Anderson and S. Loewenthal., Journal of Mechanical Design, Vol. 104, Oct. 1982, pp. 767-774.

4 "Traction Drive Performance Prediction for the Johnson and Tevaarwerk Traction Model.", J.L. Tevaarwerk, NASA TP-1530, Nasa Lewis Research Centre, Cleveland, Ohio, 1979.

5 "Thermal Traction Contact Performance Evaluation under fully Flooded and Starved conditions.", J.L. Tevaarwerk, NASA CR-168173, Nasa Lewis Research Centre, Cleveland, Ohio, 1984.

6 "A continuously variable Traction Drive for Heavy-Duty Agricultural and Industrial Applications", by Y.Kemper and L. Elfes. SAE Publication 810948, 1981

7 "The NTD-XA2 : Description and dimensioning for Automotive Applications.", Y. Kemper, J.L. Tevaarwerk, and H. Pouliot, 1986. Report is available upon request from VADETEC Research Corp. P.O. Box 217, Bloomfield Hills, MI 48013.

8 "Spin Analysis of Concentrated Traction Contacts", S.H. Loewenthal. NASA TM 83713, Nasa Lewis Research Centre, Cleveland, Ohio, 1984.

9 "Numerical Evaluation of the Torque Capacities of Mechanical CVT Drives", M.H. Cuypers., ASME Journal of Mechanical Design Paper 80-C2/Det-120, 1980.

10 "Advances in Traction Drive Technology", S.H. Loewenthal, D.Rohn and N.E. Anderson. SAE Paper 831304, 1983.

Automotive Traction Drive CVTs—
An Overview*

Lubomyr O. Hewko
Power Systems Research Dept.
General Motors Research Laboratories

*Paper 861355 presented at the Passenger Car Meeting and Exposition, Dearborn, Michigan, September, 1986.

ABSTRACT

This paper is based on a prepared discussion which was presented at the CVT Panel (Session #1PC26) during the 1985 SAE International Congress and Exposition. The paper reviews the background of this rather intriguing class of variable-ratio mechanical drives and briefly summarizes the past worldwide research which led to the development of several automotive traction drive CVTs. The "toric" CVT, which was a subject of extensive research and development effort at General Motors and elsewhere, is used as an example for illustrating typical traction drive characteristics. Further, discussion is presented of the desirable attributes of traction drives, as well as of the various technical concerns arising from the power transfer through traction forces of a rolling contact under conditions unique to automotive transmission applications. The current state-of-the-art traction drive technology and several new developments in the area of automotive traction drive CVTs, reported in recent literature, are also discussed.

TRACTION DRIVES are neither new nor rare. A steel wheel propelling a locomotive along a rail or a tire driving a car on the pavement are probably the earliest and most frequently seen examples of traction drives. These are, however, fixed-ratio drives and can only become continuously variable-ratio transmissions, or CVTs, if the radius at which the tractive force acts can be varied. A good example of a simple traction drive CVT is a rubber roller driving, by traction, a steel disc. By moving the roller further from or closer to the center of the disc, the drive ratio can be changed in a continuous manner. One of the early automobiles (i.e., the 1906 Carter car powered by a 12-hp engine) was in fact produced for a number of years with exactly such a CVT. Some low-power yard equipment successfully utilizes disc/wheel CVTs even today.

Traction drives are one of the first CVT concepts ever invented. The high level of creativity exhibited by engineers at the dawn of the industrial revolution is reflected in the numerous, often most ingenious, kinematic arrangements of rollers, discs, cones, and other bodies of revolution, ratio control schemes, and materials which have been used in order to produce more perfect traction drive CVTs. Figure 1, taken from Reference 1, is but a small sample illustrating the various concepts employed. However, literally dozens of additional configurations exist. Some of these drives can be traced through patents to the second half of the 19th century. Even today, many of them are being manufactured by a number of companies and successfully marketed for industrial use. Generally, these industrial drives are heavy, bulky, limited in power capacity, and utilize ratio control mechanisms which are slow-responding and not readily suited for automotive use.

Although industrial traction drive CVTs can claim a modest success, it is not so with automotive applications. The traction drive configuration which almost exclusively was experimented with for automotive applications is the "toric" drive, represented schematically in Figure 1 as drive "12." The name comes from the toroidal cavity machined between two discs which are facing each other. Three equally spaced rollers (only one is shown in Figure 1) transmit the torque from the input disc to the output disc in parallel with each other. This increases the power-to-weight ratio of this type of drive above that of other configurations and into a practical range of automotive transmissions. The concentric and compact geometry of the toric drive, when coupled with a long list of intrinsic desirable attributes of traction drives (to be discussed later in this paper), have provided further incentive to use this particular configuration for automotive transmissions.

The early 1930s saw a flurry of activities on automotive CVTs. The first toric CVT was

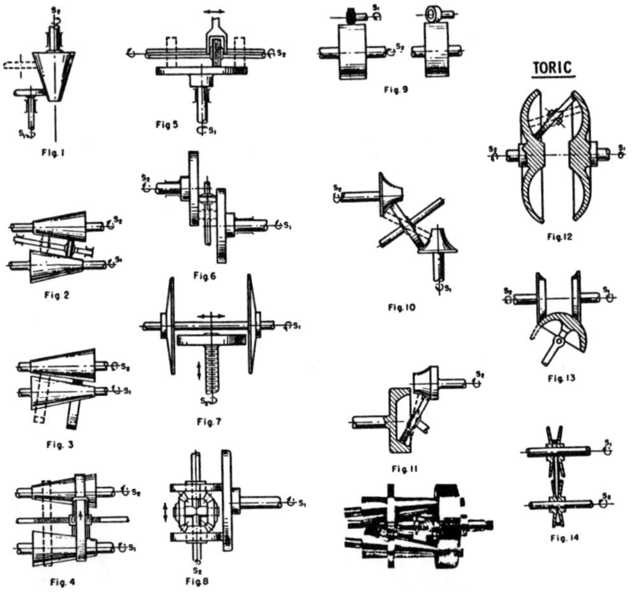

Figure 1 - Examples of kinematic arrangements for traction drive CVTs (taken from Reference 1).

manufactured in the United States by Hayes and was used for several years circa 1934 in England in a production Austin automobile [2,3].* The high cost, poor reliability, and inadequate torque capacity resulted in dropping the Hayes CVT from production a few years thereafter. There were also several other very serious attempts to develop traction drive CVTs to the production stage. For example, the Austin work was paralleled by an extensive research and development effort at the General Motors Research Laboratories (GMR) between the years 1926 and 1935 [4]. First, this effort produced an industrial CVT for use in tool machinery (lathes, boring mills, etc.). A whole line of low-power units was marketed by GM under the tradename of "Transitorq." The drive concept used in the "Transitorq" was then extended by GMR to automotive transmission applications. Jointly with the Buick Motor Division, a rear-wheel-drive transaxle CVT was developed for planned introduction into the 1936 Buick sedans. Some 30 prototypes of these transmissions were built and extensively tested in vehicles. However, concerns for ratio control, durability, weight, producibility, and higher cost precluded release of the toric CVT for production. Instead, the decision was made to produce a more conventional, step-type automatic transmission which had been undergoing concurrent development at the Detroit Transmission Division in Livonia, a precursor of the Hydra-matic Division [4].

There were also other attempts to adapt traction drives to automobiles [3]; however, broad use of traction drive CVTs as automotive transmissions even today has not yet materialized.

We will review several more recent traction drive CVT developments later in this paper; but first, in order to fully understand the potential that these drives offer for use as automotive CVTs, one must understand in greater detail how these drives operate; what are their desirable attributes; what are their limitations; and finally, what are the unresolved technical issues that so far have precluded their wide-spread automotive use.

TRACTION DRIVE DEFINITION

It will be helpful in subsequent discussion to first define a generic traction drive as "a power transmission device which utilizes hardened, metallic, rolling bodies for transmittal of power through an elastohydrodynamic fluid film." Although this definition applies to both the variable-ratio and fixed-ratio traction drives, this discussion will center primarily on the variable-ratio drives -- or CVTs -- since they are the principal candidates for use as automotive transmissions. It should be noted that variable-sheave-drive CVTs, such as the Van Dooren metal belt or PIV metal chain, are excluded from discussion by this definition. In these drives, the sheave contact is "static" rather than "rolling;" thus, it does not exhibit significant elastohydrodynamic fluid film phenomena. For the same reason, drives which operate with "dry" (unlubricated) rolling contacts are also excluded. Since for automotive applications the high-power density at the rolling contacts usually mandates the use of a coolant, this last limitation of the definition has little practical bearing on automotive applications.

ROLLING CONTACT PHENOMENA

In order to facilitate discussion of the various functional attributes of traction drives and the critical technical issues which arise from their consideration for automotive use, it will also be helpful to comment on several phenomena which occur during transfer of power through rolling contacts. Kinematically speaking, any smooth body of revolution pressing against another smooth body of revolution and causing it to turn is a traction drive. Ball and roller bearings operate essentially in the same mode as traction drives. The difference is that in traction drives, useful power is intentionally transmitted from the driving member to the driven member; while in a bearing, the power transmitted from the driving member to the driven member, i.e., from the inner race to the rollers, is used up internally to drive the cage and rollers (or balls) at the proper kinematic speed.

In order to improve the durability of the rolling contacts, it is necessary to lubricate

*Numbers in brackets designate references at the end of the paper.

and cool them. The rheological behavior of the fluid film which exists between two elastic bodies which are rolling upon each other under high normal and shear forces is quite complex and produces a number of equally complex elastohydrodynamic phenomena which have far-reaching effects on traction drive performance.

ELASTOHYDRODYNAMIC FLUID FILM -Although it is outside the scope of this overview to give a detailed, quantitative account of the various elastohydrodynamic relationships which apply to traction drive contacts, it is nevertheless necessary to make some qualitative observations. In a properly designed traction drive, the power transfer from the driving roller to the driven roller is solely through shearing of the fluid film and not, as is often erroneously assumed, through body-to-body friction. This happens because the contact between the rolling bodies, although frequently referred to as a "point contact," is in reality a finite area whose orthographic projection is usually an ellipse. The fluid, which is trapped between the rollers, does not have enough time to flow out and is subjected to a high compressive stress (usually a maximum Hertz stress of 100,000 to 500,000 psi). These stresses increase the instantaneous viscosity of the fluid by several orders of magnitude, and it is the shear strength of this semi-solidified lubricant which determines the coefficient of traction at the contact and ultimately the torque which can be transmitted through the drive. The film thickness and the stress distribution within the contact can be calculated using elastohydrodynamic theory. Although in the last two decades great advances have been made in the theory, analytical attempts to predict the film conditions for a specific application often require making many assumptions, thus reducing accuracy below that which is needed for the design of a practical drive. Experimental determination of the coefficient of traction is still probably the most reliable and frequently used method.

There are ample indications that there are three different regions of operation which exist within a lubricated rolling contact: (1) a "hydrodynamic region," where the stress is low and the film thickness is large; (2) an "elastohydrodynamic region," where the film thickness is relatively unaffected by load; and (3) a "boundary lubrication region," where the fluid film breaks down and metal-to-metal contact prevails [5].

Figure 2, taken from Reference 6, shows an example of the behavior of the fluid film and of the coefficient of traction in a typical traction drive contact operating at a constant speed. At very light loads, the coefficient of traction is nearly zero since only a very small shear force can be transmitted through a "thick" hydrodynamic film whose viscosity is essentially the same as it was under ambient conditions. In the elastohydrodynamic region, increasing compressive stress drastically raises film viscosity (to nearly solid); thus, the coefficient of traction rises. In the boundary lubrication

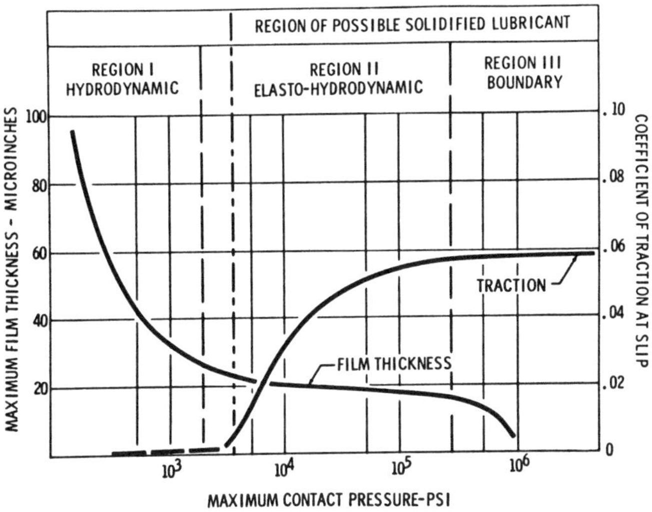

Figure 2 - Different regions of rolling element lubrication at some constant speed.

region, the film breaks down, partial metal-to-metal contact occurs, and the coefficient of traction reaches a high, more or less constant value. Continued operation in the boundary region will result in damage to the rolling surfaces. Typically, traction drives are designed operate in the elastohydrodynamic region.

TRACTION AND CREEP - In any traction drive, a measure of tractive (sic, torque) capacity of the contact is the coefficient of traction. It can be defined as the ratio of the tangential force to the normal force while the contact is rolling. The limiting value of the coefficient of traction is the coefficient of traction at slip, which is observed when the rolling contact is on the verge of gross skidding. This happens when the value of the tangential force is approaching the limiting value of the shear strength of the fluid film at a given normal force.

The coefficient of traction is greatly affected by the type of lubricant used. Typically, the coefficient of traction at slip ranges from .04 to .09, depending on the type of lubricant and on the specific operating conditions. As is illustrated in Figure 3, taken from Reference 7, there is a certain amount of "creep" as the traction coefficient increases. As the torque transmitted through the contact is increased, the shear forces acting at the contact also increase, resulting in tangential displacements within the semi-solidified fluid film. The creep is expressed in percent reduction in speed of the driven rolling body and therefore represents a power loss. It is further evident in Figure 3 that creep is strongly dependent on the level of traction and that it increases very rapidly at a critical traction value.

SPINNING AND ROLLING MOTIONS - On the basis of contact kinematics, traction drives can be

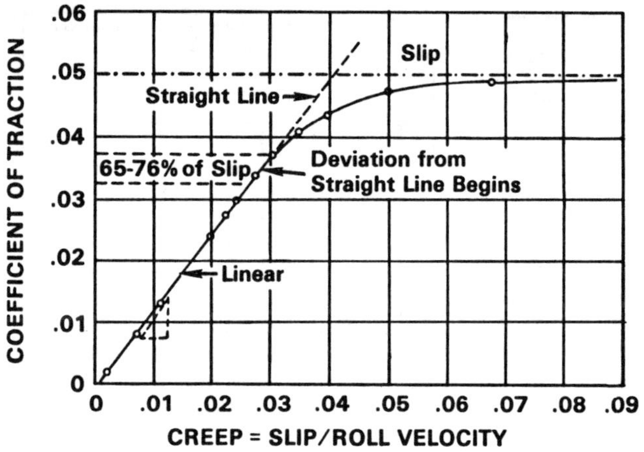

Figure 3 - Typical dependence of traction on creep.

divided into two types: (1) those having contacts with pure rolling motion and (2) those where, in addition to rolling, the contacts are also subjected to a superimposed spin motion, which is a rotary motion about some axis perpendicular to the contact plane. As will be shown later, that spinning at the contact has very far-reaching effects on tractive capacity and power losses within the rolling contacts.

Figure 4 shows a greatly exaggerated and simplified velocity diagram for a "spinning-rolling" contact under two conditions: (a) no transmitted torque, shown on the left, and (b) transmitted torque, shown on the right [7].

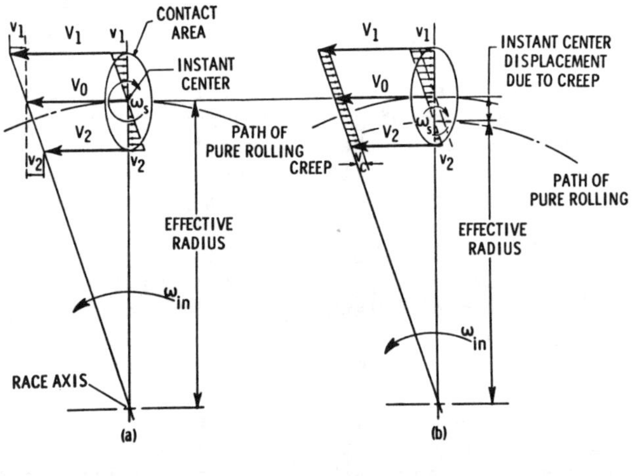

Figure 4 - Velocity diagram of a "spinning-rolling" contact in a traction drive.

(A) Velocity diagram of a spinning-rolling contact not transmitting torque.

(B) Velocity diagram of a spinning-rolling contact transmitting torque.

This type of contact area is representative of a ball thrust bearing or a toric transmission as it is seen on the race when the principal plane of the race is parallel to the paper. In Diagram (a), due to the finite length of the elliptical contact area, the velocity vectors V_1, V_0, and V_2 at various locations on the contact are not of the same magnitude since they lie at different distances from the geometric axis of race rotation. This results in the spinning velocity vectors v_1 and v_2, shown on the left, to be of opposite sense. Their instant center of rotation is exactly coincident with the geometric center of the contact area and lies on the curvilinear contact path of pure rolling. The resulting contact rotation is called "spinning" and is indicated by ω_s.

When tangential force is applied to this contact, shear forces are generated which result in tangential elastohydrodynamic displacements between the two bodies and, consequently, some creep. The creep velocity, indicated in Diagram (b) on the right as vector v_c, can be visualized crudely as a small, constant vector opposite in sense to the rolling vector. It is obvious now that spinning vectors v_1 and v_2 at extremes of contact will not be the same anymore; thus, the instant center of spinning must displace itself downward and consequently reduce ω_s. If sufficiently large tangential force is applied, the instant center can move further downward and even outside of the contact area as is the case at slip. When the instant center reaches the center of the race, rolling ceases completely and the contact is in gross skidding.

It must be emphasized here that the depicted velocity vectors represent only macro effect across the overall contact area. The actual micro displacements are not at all simply distributed throughout the contact patch. This has been quite amply indicated by numerous experimental data and analytical solutions of the elastohydrodynamic problem. For the traction drive designer, the more practical "macro" displacements are of primary interest.

The contact between two cylinders having parallel axes or two cones with coincident apexes are typical representatives of pure rolling contacts. Figure 5 depicts two plain cylinders under load [7]. The size and static pressure distribution at these contacts is of the Hertzian type. Again, on a macro scale due to the absence of spinning, the velocity vectors V_1, V_2, and V_0 at all points of the rolling contact are the same (Diagram A). Superposition of the creep vector v_c in Diagram B does not alter the ratio of V_1/V_2, and consequently no displacement of the instant center of rotation of the contacts takes place. Thus, there is no spinning in this contact without creep, and none is introduced due to creep.

The effect of spin on traction is well illustrated in Figure 6. Since the spinning motion in a rolling contact can be expected to amplify the hydrodynamic effects and the shear strains, one can anticipate such a contact to exhibit lower traction, higher creep, and

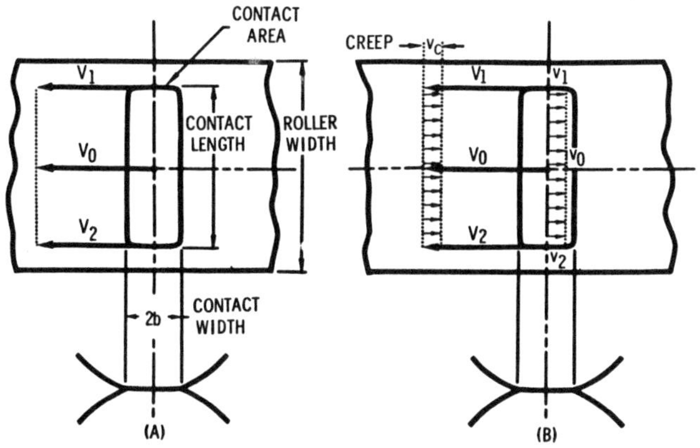

Figure 5 - Velocity diagram of a "pure rolling" contact in a traction drive.

(A) Simplified velocity diagram of a rolling contact not transmitting torque.
(B) Simplified velocity diagram of a rolling contact transmitting torque.

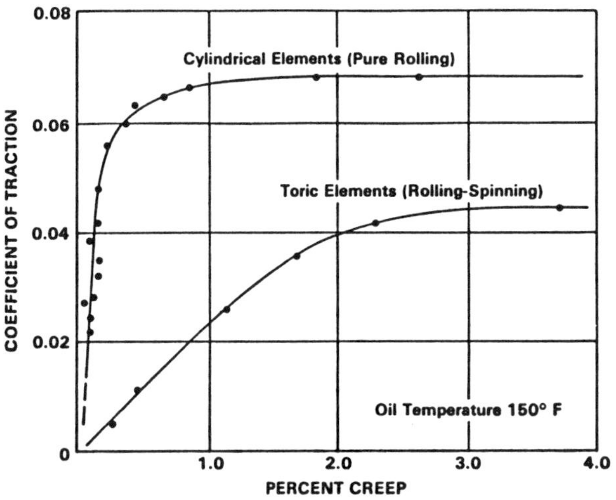

Figure 6 - Typical comparison of contact traction and creep between "pure rolling" and "rolling-spinning" contacts operating at the same Hertz stress and velocity (data from ASLE Transactions, Vol. 12, 1969, pp. 151-161).

greater drop in traction with speed. Higher creep can also be expected in a spinning elliptical contact due to the fact that the maximum Hertz stress occurs at a small region near the center rather than a line; thus, the region of maximum tangential tractive capacity is considerably narrower than that of a cylinder or cone contact, resulting in higher tangential strains. At low speed, however, where hydrodynamic effects become insignificant, there should be little difference between the traction of cylinders and that of the toric elements. Experimental data shown in Figure 6 support this reasoning [7].

EFFICIENCY AND DRIVE RATIO - The efficiency with which power is transmitted through the rolling contact is so important when considering the performance of a drive. Whereas with gears the overall efficiency is equal to the torque efficiency, with traction drives and some fluid drives one must distinguish between speed efficiency and torque efficiency. A numerical definition of these two efficiencies is given below.

$$\eta_t = \frac{\text{actual output torque}}{\text{ideal output torque}} \quad (1)$$

$$\eta_s = \frac{\text{actual output speed}}{\text{ideal output speed}} \quad (2)$$

$$C = \frac{\text{ideal output speed - actual output speed}}{\text{ideal output speed}} \quad (3)$$

where the ideal torque and speed are calculated on the basis of kinematic ratio. Then

$$\eta_s = 1 - C \quad (4)$$

$$\eta = \eta_t \times \eta_s \quad (5)$$

where η = overall efficiency (expressed decimally), η_t = torque efficiency (expressed decimally), η_s = speed efficiency (expressed decimally), and C = creep (expressed decimally).

In general, the torque efficiency is a measure of spinning losses (including some insignificant losses due to rolling resistance), and the speed efficiency is a measure of contact creep due to tangential force. Since the overall efficiency is the product of the speed and torque efficiencies, a loss in either type of efficiency can be detrimental to the performance of a drive.

Figure 7 shows the typical behavior of the three efficiencies for a typical traction drive CVT [8]. Since the speed efficiency drops with torque while the torque efficiency rises, and since the overall efficiency is the product of the two efficiencies, a point of maximum overall efficiency obviously exists. It should be noted in Figure 7 that typically the overall efficiency curve of a traction drive is quite flat over most of the torque region.

The speed ratio of a traction drive is defined as the ratio of the output speed divided by the input speed. At zero torque, the speed ratio can be also defined as a ratio of input radius divided by the output radius of the rolling bodies. However, this ratio will not be correct at torques other than zero because of creep. For the same reason, unlike in gears, one cannot express torque ratio as a reciprocal of the speed ratio.

DURABILITY AND SIMILITUDE RELATIONSHIPS - Indirectly, the coefficient of traction also affects the fatigue life of a traction drive. This happens because the life of a rolling contact element is inversely proportional to some power of the normal load, whereas the torque capacity is only directly proportional to the load. Thus, increasing the drive's torque

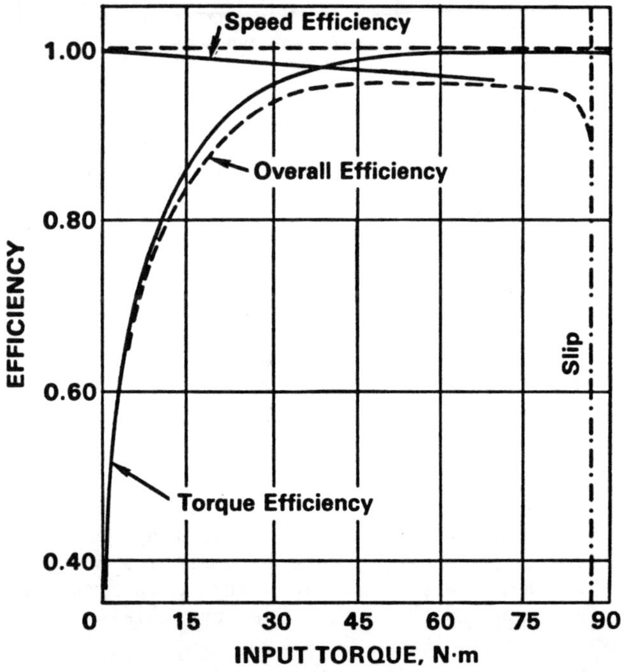

Figure 7 - Typical behavior of speed, torque, and overall efficiencies as a function of input torque.

capacity by increasing the coefficient of traction is far more desirable than by increasing the normal load. From a rather thorough study, it was found that naphthenic base oils and certain synthetic fluids have the highest tractive properties [8,9].

Other variables affecting the coefficient of traction are the rolling speed, the lubricant temperature, the surface finish quality, the Hertzian contact pressure, the degree of contact spinning, and the geometry of the rolling bodies. All these variables, together with the properties of the material used in the manufacturing of the rolling components, interact to produce a drive with a predictable power capacity, size, fatigue, durability, and reliability.

The relationships governing the rolling contact durability in a traction drive closely resemble those used for rolling contact bearings. The relationship between the several parameters mentioned previously and their model effect on life prediction are:

$$L_1/L_2 = (T_2/T_1)^c \times (R_1/R_2)^d \times (N_{p1}/N_{p2})^e \times (S_2/S_1)^f \qquad (6)$$

where L = life; T = torque; R = characteristic dimension; N = speed; N_p = number of rollers; S = Hertz stress; c,d,e,f = experimentally or analytically determined exponents; and 1,2 = subscripts designating two different sets of conditions.

The durability of the traction drive can be brought to any level by properly adjusting the previously listed parameters. The reliability of an assembly consisting of many rolling elements can be computed by the use of Miner's cumulative damage theory and from the known Weibull distribution. Considerable success has been achieved in obtaining experimental data and then applying analytical techniques that provide rather accurate life predictions. A more detailed description of the statistical and mathematical approaches frequently used in these calculations is given in References 8, 10, and 11.

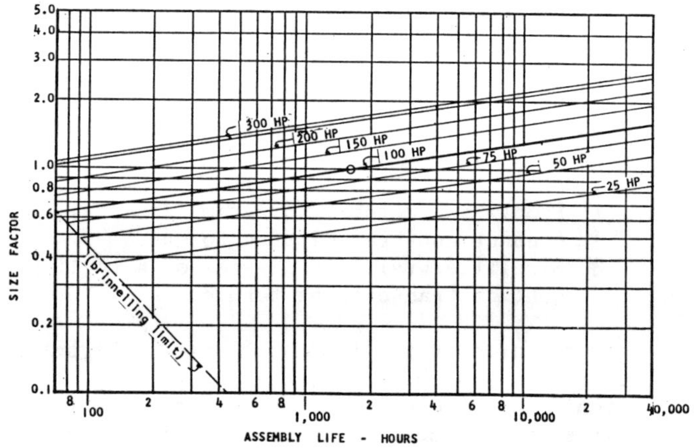

Figure 8 - B_{10} life of a typical traction drive as a function of size and power at a constant input speed of 3600 rpm.

An example of size effect on the B_{10} life* of a typical traction drive unit operating at different power levels is shown in Figure 8. Here, the 100-hp drive with a B_{10} life of 1500 hours was arbitrarily given a size factor of 1. The dashed line is the expected brinelling limit and defines the smallest permissible unit for given power levels [10]. A relationship between life in hours and reliability expressed as a percent survival of the units for any level of transmitted power is shown in Figure 9. At the design point indicated, 90 percent of the units survived at least 1500 hours while transmitting 10 hp at a 3600-rpm input speed [10]. Even at the brinelling limit of 330 hp, the B_{10} life of the unit is a statistically predictable 42 hours. For higher reliability, B_1 life** may be considered. For example, a drive has a B_1 life of 275 hours while transmitting 100 hp. These results apply only to the life of the power transmitting rolling contacts; other components, such as bearings, shafts, or springs, may have other lives and have to be considered separately.

---

*B_{10} life is defined as hours of operation which will be exceeded without failure by at least 90 percent of units transmitting given power.

**B_1 life is defined as hours of operation which will be exceeded without failure by at least 99 percent of units transmitting given power.

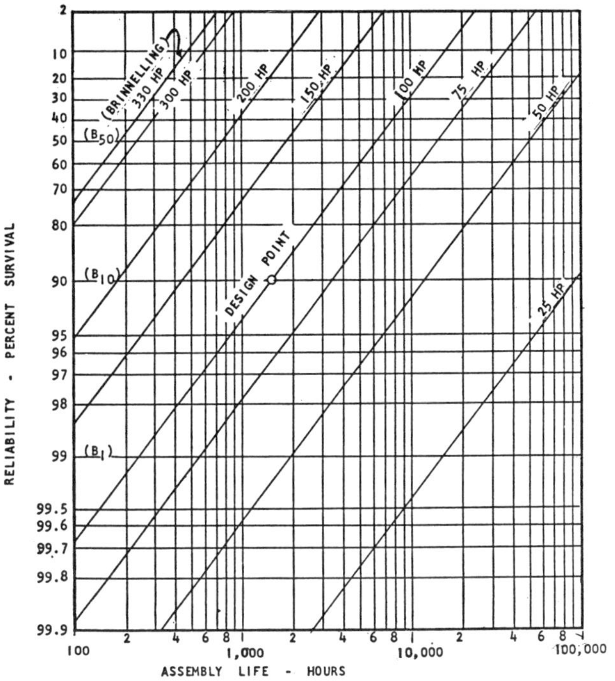

Figure 9 - Reliability of a typical traction drive operating at different power levels at 3600 rpm input, size factor = 1.0.

To determine an optimum drive configuration, it is necessary to vary each parameter individually between permissible physical limits to observe the effects of the change on life and size and then to repeat the calculation for all other parameters until the optimum combination is found. Such procedures have been programmed for a digital computer [10].

TORIC CVT

An effective way to illustrate the various operational characteristics of a generic traction drive CVT is to use a specific drive as an example. A logical choice for the purpose of this discussion is the toric CVT since, as was mentioned before, it was a primary subject of extensive research and development effort in order to adapt it to automotive transmissions.

BASIC TORIC GEOMETRY - Let us first review briefly how a toric drive operates and what are its useful attributes. A typical single unit toric drive is shown in Figure 10, taken from Reference 7. In such an arrangement, one toric race serves as input and the other as output. Three rollers are equally spaced at 120 degrees within the toroidal cavity. The outside diameters of the rollers are equal to the transverse diameter of the torus, while the centers of the rollers are located on its pitch diameter. The input torque is equal to the product of the normal component of force P, the coefficient of traction, and the input radius. The output

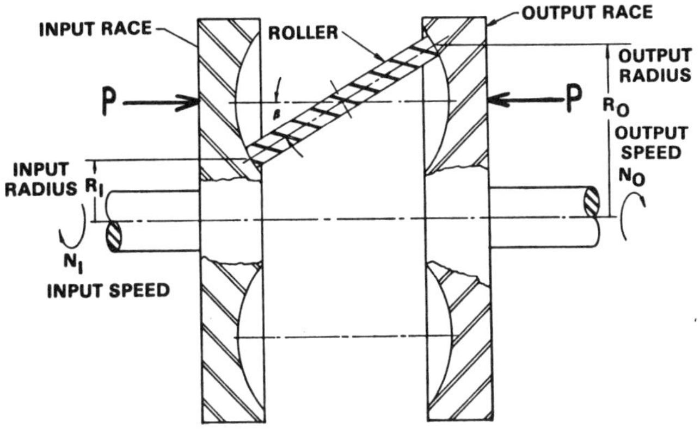

Figure 10 - Typical single unit toric drive.

contact has the same normal load and nominally has the same coefficient of traction, but its torque radius is larger; thus, the roller position shown in Figure 10 results in a torque multiplication or speed reduction ratio.

The ratio is changed by simultaneously changing the "tilt" position of all three rollers. Even under high torque, this is done, almost effortlessly, by "inclining" the rollers about an axis going through the two contacts and thus "steering" the rollers along a spiral path on the races to a new ratio position. When the rollers are in a horizontal position ($\beta = 0$), the ratio is 1:1; when the rollers are at an angle which is symmetric to that shown, an overdrive ratio results. When the drive is not rotating, the ratio can be changed only after the contact normal load (sic, torque) is removed.

The rolling surfaces are cooled with oil jets, while the power transfer from input to output is accomplished through the elastohydrodynamic fluid film existing at the roller-race contacts. Unlike in most other traction drives, in the toric drive, the ratio control forces are almost completely decoupled from the torque forces and only negligible power is required for ratio control.

DOUBLE TORIC CONFIGURATION - All past automotive toric CVT effort centered on the "double unit toric" kinematic arrangement shown in Figure 11. Two toric sections are coaxially arranged and transmit the torque in parallel [7]. Input torque is applied via an "input drum" to the center race and the output torque is then collected on the inner-most shaft which also serves as the transmission output. Since there are three rollers in each section, one roller carries only one-sixth of the total power flowing through the CVT. This reduces the size of the components, but unfortunately it also creates very stringent constraints on roller position and ratio control.

The axial force necessary for torque transmittal is generated automatically by a mechanical "torque-loading" ball/cam device shown on the right. The axial force generated in this way is always proportional to the instantaneous torque at the output contacts and is in part

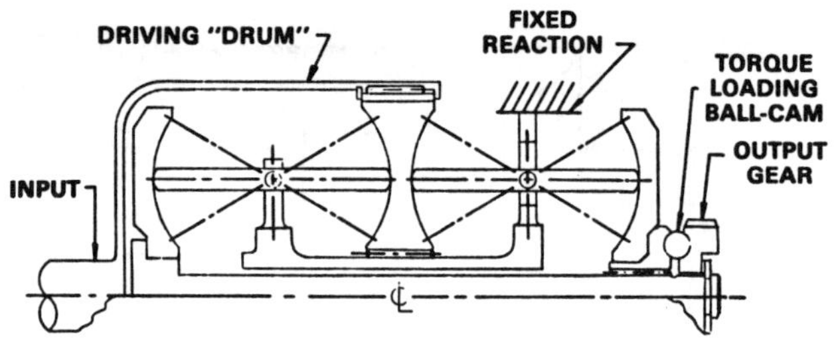

Figure 11 - Typical double unit toric transmission.

responsible for the high efficiency of toric drives. The ball/cam torque loader is, however, subject to complex dynamic phenomena, which can be troublesome under transient automotive conditions.

SOME EXPERIMENTAL OBSERVATIONS - An important characteristic of any CVT is its ability to accomplish ratio change rapidly and with low control effort. This capability of a toric drive can be best illustrated by plotting the total angle of spiral in race revolutions needed to traverse the full ratio range for various roller inclination or "steering angles" (Figure 12). For example, at two degrees of inclination and at 3000 rpm, the total ratio range of 6.25 can be traversed in less than five revolutions or in less than one-tenth of a second. This illustrates the great capability for fast response of these units and also points out the potential for serious stability problems if synchronization and dimensional position accuracy of all rollers is not maintained throughout the life of the transmission.

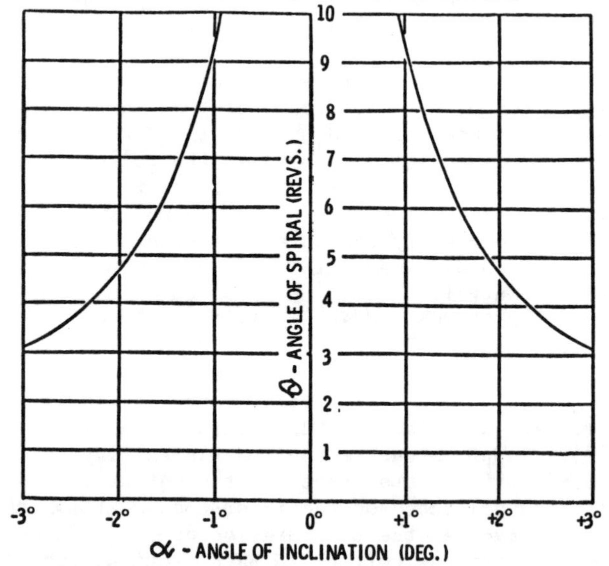

Figure 12 - Total angle of spiral in race revolutions necessary to traverse full ratio range (-30° ≤ B ≤ +30°) for various inclinations

Roller synchronization also plays an important role in overall efficiency, as shown in Figure 13. This figure shows the efficiency of the toric unit only and not that of the whole transmission. If the difference in ratio between the sections is small, then the efficiency can be very high over a wide range of loads. However, even a small error in synchronization results in a significant drop in efficiency. For example, a 2.47-degree error more than doubles transmission losses, as is manifested by the efficiency drop from nearly 97 percent to 93 percent. This CVT was designed to operate along the 80 percent slip line shown by solid dots in Figure 13.

Among the various elastohydrodynamic phenomena which play an important role in determining the power capacity and durability of traction drives, thermal effects take a leading

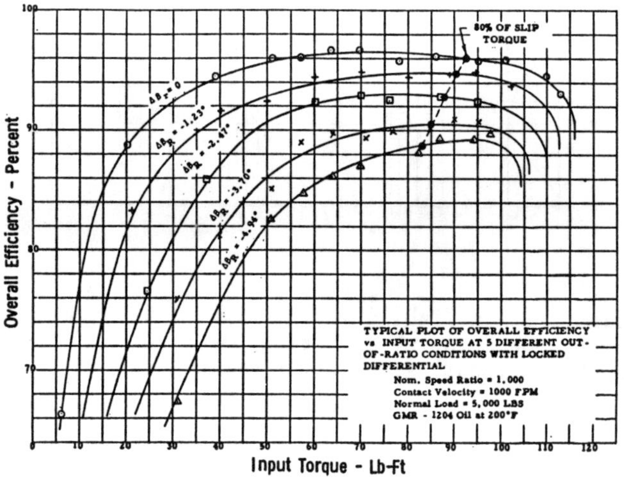

Figure 13 - Effect of ratio error on overall efficiency in a toric CVT.

place. For instance, the classical analysis would lead one to believe that in order to transmit higher torque through a contact, all one needs to do is to increase the normal load on the contact. However, this, for the most part, appears to be true only in the moderate load area where, for instance, industrial drives typically operate. In automotive applications, where the transmission must frequently handle sudden bursts of high torque, the existence of an apparent paradox was discovered [8]. Due to what is believed to be the thermal effects within the elastohydrodynamic film, the load effect on torque capacity reverses itself and increasing the normal load beyond a certain level actually reduces traction and consequently the torque capacity of the rolling contacts. Support for this discovery was later provided by the elastohydrodynamic theory [13]. This points to the need for careful scrutiny and understanding of the thermal and other contact phenomena before high-capacity traction drives can be properly designed. There are also several other rolling contact phenomena for which full understanding is lacking, such as, for instance, optimization of the Hertzian contact geometry with respect to traction, interaction between steering and tractive forces during ratio transients, etc.

Because the power in a traction drive CVT is transmitted through smooth and continuous action of the tractive forces between round, lubricated, rolling components machined to high precision, these drives tend to exhibit very quiet operation, as is evident from Figure 15, taken from Reference 10. The steady-state power flow through the elastohydrodynamic contact interface results in the absence of periodic vibration disturbances and also adds an effective viscous damping component into the driveline dynamics. This is unlike gears where the tooth mesh frequency excitations may result in severe torsional and axial vibrations of the driveline. Even a theoretically perfect gear set, with several teeth in simultaneous contact, will generate torsional oscillations as part of the torque is transferred from one tooth to the next.

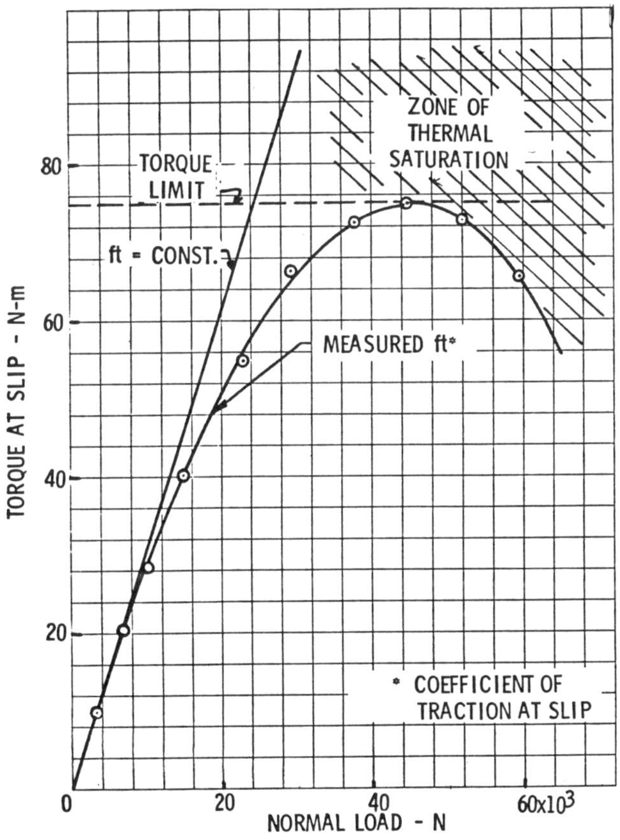

Figure 14 - Thermal effect on traction in a toric CVT.

Having defined a generic traction drive and reviewed the various rolling contact phenomena which occur during power transfer through such a drive, and further, having described a toric automotive CVT concept and some of its operational characteristics, let us now focus our attention on their various attributes.

ADVANTAGES AND LIMITATIONS OF TRACTION DRIVES

Traction drive CVTs can be configured using a variety of rolling body shapes and kinematic arrangements, as is apparent from Figure 1. Consequently, different advantages and limitations may apply to different drives. However, as was mentioned earlier, automotive traction drive CVTs were usually similar to the "toric" geometry; therefore, certain common characteristics can be readily identified. Nevertheless, the reader is cautioned that the degree to which a specific drive exhibits certain attributes, may they be positive or negative, has to be judged by analyzing each concept individually.

ADVANTAGES - Most of the advantages offered by traction drive CVTs over other types of CVTs stem from the basic mechanism of transmitting power through traction forces existing within lubricated rolling contacts. Let us now summarize these advantages.

The wide ratio range* over which the mechanical drive ratio can be varied in a continuous manner certainly is an important attribute of a CVT. Whereas competitive CVTs, such as variable-sheave drives, have a ratio range of about 5, the ratio range of the toric drive can be as high as 9. However, the ratio range of past automotive designs was limited to between 6 and 7.

The potential for high mechanical efficiency is another important characteristic of traction drives. Rolling motion is intrinsically very efficient, with losses approaching

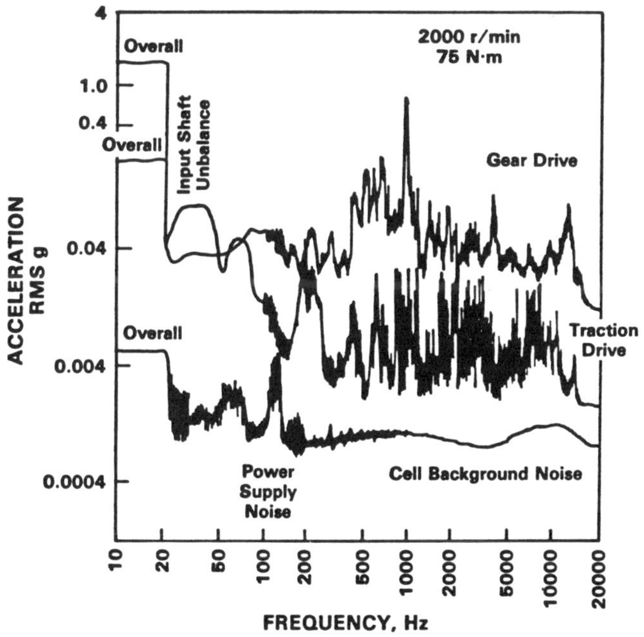

Figure 15 - Frequency spectrum comparison of torsional vibrations between traction drive and gear drive running at 2000 r/min and 75 N•m of torque (data from AIAA Paper No. 67-429).

---

*Ratio range is defined as a quotient of the highest and the lowest ratio exhibited by a given unit.

zero. However, in most CVTs, "creep," "spin," and dimensional imperfections increase the rolling losses; even for the best of drives, the efficiency is less than 95 percent, as is evident in Figure 13.

The ability to have a high rate of ratio change is also a desired characteristic of traction drive CVTs. Because the changing of the ratio is accomplished by steering rather than skidding the rollers to a desired contact path, it can be accomplished very quickly, as is shown in Figure 12. This feature is of particular interest in systems where asynchronous shifts must be performed (i.e., hydrid and split-torque transmissions).

Changing the drive ratio under load can be accomplished not only rapidly but also with relatively small control forces in all traction drives that use "steering of rollers" for ratio control. This is possible because the contact forces needed for torque transfer can lie in a plane which is perpendicular to the plane in which the ratio control forces operate and are therefore relatively decoupled from each other.

The potential for balanced and self-contained internal forces is another important feature. Traction drive units can be arranged in such a way that it becomes unnecessary to support the rolling contact normal forces by separate bearings. This is usually done by having one traction unit serve as a bearing for another. The double-unit toric transmission shown in Figure 11 is a good example of such an arrangement. Efficiency, weight, and noise are greatly enhanced in such an arrangement.

A rather high degree of torsional damping is another useful characteristic of traction drives. The dependence of traction on creep (Figure 3) results in considerable torsional damping, which has a potential for reducing torsional vibration which might be present in the driveline.

Traction drives operate more quietly when compared to chains, gears, hydrostatic drives, and other intermittent action devices. Because of the continuous power flow through the rolling contact in a traction drive, not only is the overall noise level lower, but also the noise signature is white and therefore less objectionable (Figure 15).

Ease of packaging is also important when considering automotive CVTs. Traction drives in general, and toric drives in particular, offer packaging advantages because they are concentric and also because they can support themselves within a housing with a minimum of bearings. However, if long durability is required, the increased weight and size may offset other packaging gains.

The manufacturing of the principal rolling bodies for a traction drive consists essentially of turning and grinding of smooth bodies of revolution. This offers obvious advantages over machining of gear teeth, sprockets, and chains. This advantage may be somewhat offset by the higher dimensional accuracy usually required in traction drives.

Although not all traction drive CVTs exhibit all of the above positive attributes, the toric drive seems to incorporate all of them.

LIMITATIONS - Paradoxically, the same mechanism of transmitting power through traction of rolling contacts, which is the source of all the positive attributes of traction drives, is also the source of their principal limitations.

The weight and size of traction drive CVTs tend to exceed that of other CVTs (i.e., variable-diameter sheaves, hydrostatics). This is due to the high required rigidity of component parts which must carry the high contact loads needed for torque transmittal. Additionally, as was shown in Figure 8, the fatigue life of the drive is related to the size of the rollers, while the weight of the rollers is a cube function of their linear dimension factor. Thus, the longer the required life, the higher the weight becomes.

The relatively large quantity of clean, bearing type, high-quality steel typically required for automotive traction drives is another potential limitation. Manufacturing of these steels tends to be expensive (usually Carbon Vacuum Deoxiding Process), and their heat treatment is complicated (sic, expensive and prone to production errors).

The high dimensional precision to which the drive rollers and the races have to be fabricated exceeds that of typical automotive components. Consistent production of such precise components in automotive volumes can be a major concern.

Although the normal forces acting on the traction contacts are internally self-contained, nevertheless, they are large (up to 30,000 lbs) and can create fretting and bending fatigue problems. Frequently, an adequate solution to this difficulty is not readily obtained within the limited space of an automotive transmission.

Skidding or gross slip of the traction contact can have a catastrophic effect on the drive. Precautions are taken to prevent this from happening through the use of a torque-actuated preloading mechanism. However, should slip occur even once in the life of a drive (under certain dynamic conditions or in the case of a malfunction), failure of the contacts will result.

In order to reduce the size of a drive to that suitable for automotive use, special lubricants are usually used which have higher traction properties than automatic transmission fluids. These fluids are more expensive, and their properties are rather sensitive to contamination and service abuse. The high cost has a negative impact on the economic feasibility of the drive and the deterioration of fluid raises a concern for changes in torque capacity with time.

Sensitivity to wear and other service-related factors is another limitation of traction drive CVTs. These drives tend to be highly "engineered" precision devices. They offer more in terms of performance, but they also require tighter overall system tolerances, correct maintenance, and more technically demanding service.

It is apparent from this discussion that the outstanding positive attributes of traction drive CVTs which tantalized automotive transmission designers of the early 1930s are tempered by several negative attributes which caused the abandonment of these early production attempts. But rapid technological advances which followed World War II held out promise of solving some of the early problems and stimulated a new wave of interest in traction drive CVTs which continues until the present.

RECENT ACTIVITIES ON TRACTION DRIVE CVTs

In the early 1960s, technological advances in servo-control systems, electronics, materials, manufacturing, and analytical techniques (including the arrival of digital computers) have convinced many that another look was warranted at the traction drive CVT concept.

A new major research and development program was launched at the General Motors Research Laboratories to develop a toric transmission for passenger cars [14]. As a result of this program, lower-cost steels were developed that showed up to a 300-percent increase in durability; high traction fluids were discovered, resulting in an increase in torque capacity by nearly 30 percent; and novel, improved kinematic arrangements were developed which solved many of the earlier ratio-control problems [14]. New analytical tools enabled preparation of more effective designs with better functional characteristics. Several different-size passenger cars and one city bus were equipped with prototype toric transmissions and were successfully tested. A cross section of a representative toric transmission of that period is shown in Figure 16. The unit was slightly larger and heavier than a conventional transmission and featured a torque converter which was located on the output side of the CVT in order to reduce the torque in the CVT [15,16]. However, at the end of 1967, the toric transmission project at GM was terminated. Concerns about reliability and cost were not satisfactorily resolved to justify release for production.

As a spin-off from this CVT effort, several fixed-ratio traction drives were developed by GM engineers. These drives were found to run 16 times more quietly than gears and were therefore of interest for underwater and other applications where quiet operation is of importance. Although several prototype drives were built and evaluated successfully in underwater applications, no volume production followed [10].

When in 1967 General Motors terminated its activity on traction drives, others continued, some until the present time. Most notable was Perbury, Inc., in England, now promoted by the British Technology Group (BTG). Perbury engineers equipped several passenger cars with experimental toric drives of their proprietary design. Their latest design is said to utilize

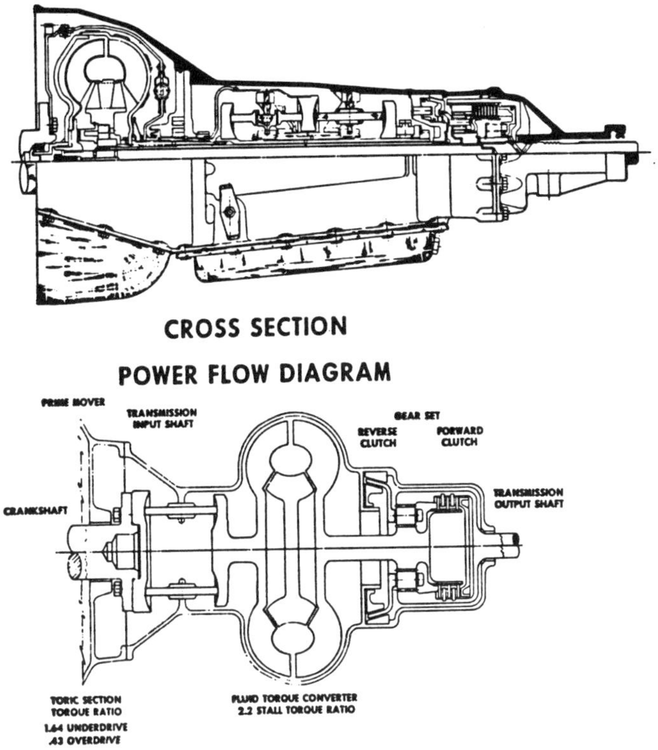

Figure 16 - Example of a toric traction drive CVT for passenger cars (GMR circa 1962).

microcomputers for roller position and ratio control. A variety of proprietary kinematic structures and a multitude of customized ratio control strategies are also used [17]. The Perbury/BTG unit appears today to be the most advanced toric drive. An example of a Perbury transmission is shown in Figure 17, taken from Reference 18. A development of a truck and bus CVT based on the Perbury concept is also being pursued by British Leyland [19].

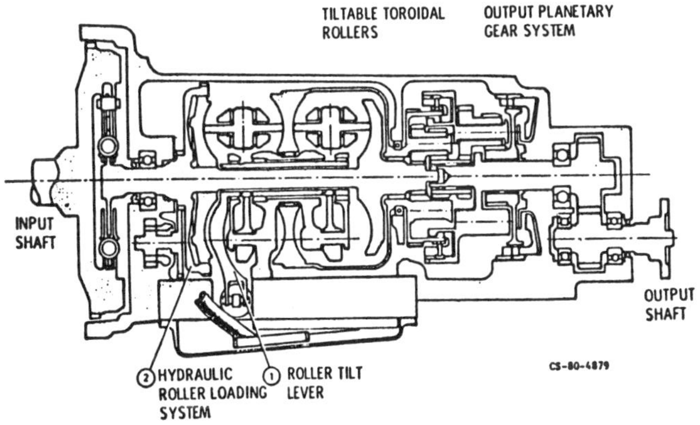

Figure 17 - Perbury/BTG toric traction drive CVT for a passenger car.

An interesting spin-off application of the toric drive was developed by AVCO Lycoming Division in the form of a CVT accessory drive for aircraft. Several thousand of these units were manufactured in the United States and in England, ranging in power capacity up to 60 KVA. This development is being mentioned here because the power capacity of this unit approached that of a CVT for a small passenger car. After several years of production, the Lycoming unit is no longer being manufactured, citing service and operational difficulties as the reason.

Tracor Corporation and the Wright Aeronautics Division have separately demonstrated unique traction drive CVT concepts in several vehicles, but they appear to have abandoned the effort in the late 1970s.

Airesearch Corporation, under contract with the U.S. Department of Energy, developed a design for a toric CVT for use in a hybrid passenger car. Traction Propulsion, Inc., also participated in the project until 1984, when the Department of Energy terminated the contract before the design was fully implemented. A cross section of a transmission representative of that effort is shown in Figure 18 (Reference 20).

Vadetec Corporation demonstrated several designs of an ingenious nutating cone-type traction drive CVT designed for automobiles and tractors [21], Figure 19. The exact status of this project is not fully known, but recent comments in the press indicate that the tractor CVT development is continuing at another company under license from Vadetec, while at Vadetec a new kinematic arrangement, more suitable for passenger cars, is being developed [22].

In the last few years, frequent reports of considerable activity on toroidal CVTs have been forthcoming from Japan. Most of the work seems to be centered at the University of Tokyo Institute of Industrial Science and the University of Yokohama [23]. Also, progress on high traction fluid research has been reported by the research laboratory of Mitsubishi Oil Company [24]. It is not clear from the Japanese reports whether their CVT experiments have progressed from the laboratory to vehicles.

The preceding discussion of recent traction drive activities touched briefly on only the more extensive attempts at automotive CVT applications. Literally hundreds of other traction drive CVT ideas were also invented and/or reported in the literature, but most did not proceed beyond "paper" designs.

CRITICAL TECHNICAL ISSUES

After reviewing the concept, attributes, and history of automotive traction drive CVTs, two obvious questions arise: (1) why is there such a recurring interest in them? and (2) why is there no apparent market success? The answer to the first question is clear. Traction drive CVTs have many attractive attributes which can be very useful in automotive powertrains.

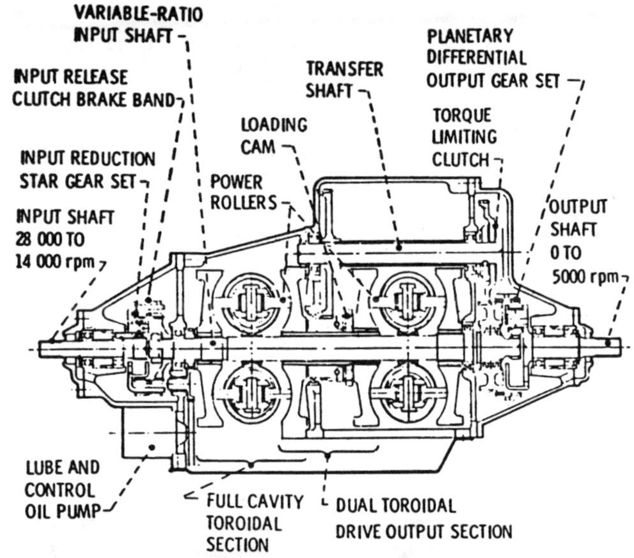

Figure 18 - Toric CVT for DOE hybrid passenger car.

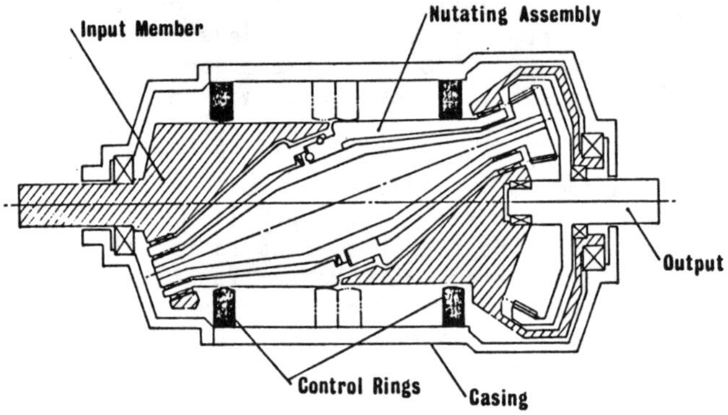

Figure 19 - Nutating traction drive CVT (Vadetec circa 1982).

The answer to the second question -- why no apparent market success? -- is more complex. Undoubtedly, there were various "business" reasons, which differed from company to company. But there also exists a number of critical technical issues associated with traction drives which so far have eluded satisfactory solution. Although in recent years considerable advances have been made in traction drive technology, further improvements are still needed in several areas in order to make them suitable for automotive applications. Let us briefly review some of them.

(1) Difficulty with ratio control in traction drive CVTs seems to be greater than in other CVTs. Problems with roller position synchronization, sensitivity to roller steering angle errors, and complex system dynamics when combined with geometric variability resulting from production tolerances and wear raises serious concerns regarding the capability of the units to provide consistent and reliable operation over the lifetime of the car.

(2) Intolerance to slip, which manifests itself in premature or catastrophic failure of the principal power-transmitting components, namely the rolling surfaces, does not appear to lend itself readily to an immediate reliable solution. Various mechanical, torque-responsive, preloading schemes have been tried, but most of them seem to suffer from dynamic limitations and do not safely cover all transient vehicular torque requirements. The danger of destroying the transmission, even if it slips only once in the lifetime of the car, appears to be too high a risk to take.

(3) Improvements in traction and efficiency through optimization of the rolling contact phenomena are needed. This would reduce the size and losses of traction drives in comparison to conventional automatic transmissions. A better understanding of thermal phenomena occurring at the contact and how they affect performance of the drive would also improve the design.

(4) The need for new materials (steels, lubricants) and the manufacturability of several key components raises concerns regarding cost and durability.

Solutions to these technical difficulties are awaiting ingenious approaches which must be superior to those tried in the past. Recent trends toward the use of microprocessor-based control systems on-board automobiles and in manufacturing, when combined with new advances in control theory and modeling, may provide the answers needed for making traction drive CVTs an automotive success. Published technical literature indicates that most of the designs currently under development make use of these new technologies. However, their history is too recent to enable accurate predictions for success or failure in the future. The degree to which these new approaches will accomplish the task of removing the various technical obstacles remains to be seen.

SUMMARY

The unique performance characteristics of traction drive CVTs have tantalized automotive transmission engineers from the earliest days of automobiles. These "gears without teeth," in which ratios can be changed in a continuous manner rapidly, quietly, and almost effortlessly through traction of metallic rolling bodies, had much to offer the automotive driveline. The kinematic arrangement which was the principal focus in the development of traction drive CVTs was the "toric" drive because of its superior power-to-weight ratio and concentric shape.

Although industrial traction drive CVTs can claim a modest market success, it is not so with automotive applications. Several attempts to develop and mass-produce them failed. The primary reason appears to be a number of critical technical issues which must be resolved before the drives can become viable automotive transmissions. Paradoxically, the obstacles to success stem from the same fundamental power transfer mechanism that is responsible for their desirable attributes -- the power transfer through traction forces within rolling contacts. The critical technical issues which must be resolved are: the difficulties with ratio control, intolerance to slip, the not fully understood elastohydrodynamic phenomena within the rolling contacts, and the manufacturability of certain key components.

Only extensive research and development into these critical areas can provide solutions which could lead to a technical and commercial success for traction drive CVTs. Although the risk is high, so is the potential payoff. There are significant, new, technological developments in controls, analytical techniques, and manufacturing methods which just may succeed in making the traction drive CVT a viable automotive transmission in the near future.

REFERENCES

1. Loewenthal, S. H., "Advances in Traction Drive Technology," NASA Technical Memorandum No. 83397, September 1983, p. 15.
2. Carson, R. W., "Focus on Traction Drives: 100 Years of Traction Drives," Power Transmission Design, Vol. 17, No. 9, May 5, 1975, pp. 84-88.
3. Cahn-Speyer, P., "Mechanical Infinitely Variable Speed Drives," The Engineer's Digest, February 1957, pp. 20-21.
4. Sloan, A. P., Jr., My Years with General Motors, edited by J. McDonald and K. Stevens, Doubleday, 1963, p. 472.
5. Sternlicht, B., Lewis, P., and Flynn, P., "Theory of Lubrication and Failure in Rolling Contacts," Society of Automotive Engineers, Paper No. 60-WA-286.
6. Hewko, L. O., "Contact Traction and Creep of Lubricated Cylindrical Rolling Elements at Very High Surface Speeds," ASLE Transactions, Vol. 12, No. 2, April 1969, pp. 151-161.
7. Hewko, L. O., "Traction Drives and Their Potential Role in Energy Conservation," General Motors Research Laboratories, Research Publication GMR-3218, February 1980.
8. Hewko, L. O., Rounds, F. G., Jr., and Scott, R. L., "Tractive Capacity and Efficiency of Rolling Contacts," Proceedings of the Symposium on Rolling Cotnact Phenomena, edited by J. B. Bidwell, Elsevier, Amsterdam, 1962, pp. 157-185.
9. Hewko, L. O., and Kacmarsky, W. M., "Effect of a High Traction Fluid on Skidding in a High-Speed Roller Bearing," ASLE Paper No. 70-LUB-E, p. 2.
10. Hewko, L. O., "Roller Traction Drive Unit for Extremely Quiet Power Transmissions," AIAA Journal of Hydronautics, 1968, pp. 160-167.
11. Johnson, L. G., The Statistical Treatment of Fatigue Experiments, Elsevier, Amsterdam, 1964.

12. Heilich, F. W., and Shube, E. E., Traction Drives -- Selection and Aplication, Marcel Dekker, Inc., New York, 1983, pp. 144-205.
13. Tevaarwerk, J. L., "Thermal Traction Contact Performance Evaluation," NASA Report No. CR-168173, May 1985.
14. Bidwell, J. B., editor, Proceedings of the General Motors Research Laboratories Symposium on Rolling Contact Phenomena, Elsevier, 1962.
15. Chironis, N. P., "Toric Transmission Smoothes Vehicles Power," Product Engineering, June 2, 1969, pp. 72-73.
16. "GM Progress for Power: A General Motors Report on Vehicular Power Systems," Warren, MI, May 1969, pp. 4-6.
17. Stubbs, P. W. R., "The Development of a Perbury Traction Transmission for Motor Car Applications," American Society of Mechanical Engineers, Paper No. 80-C2/DET-59, August 1980.
18. Loewenthal, S. H., Anderson, N. E., and Rohn, D. A., "Advances in Traction Drive Technology," NASA Technical Memorandum, No. 83397, 1983.
19. Greenwood, C. J., "The Design, Construction, and Operation of a Commercial Vehicle Continuously Variable Transmission," Driveline '84, Institution of Mechanical Engineers, March 1984, pp. 89-99.
20. Parker, R. J., and Loewenthal, S. H., "Design Studies of Continuously Variable Transmission for Electric Vehicles," NASA Technical Memorandum, No. TM-81642, February 1981.
21. Kemper, Y., and Elfes, L., "A Continuously Variable Traction Drive for Heavy-Duty Agricultural and Industrial Applications," Society of Automotive Engineers, Paper No. 810948, September 1981.
22. Kemper, Y., and Tewaarwerk, J. L., "A New Approach to CVT Traction Drives in Automotive Applications," Society of Automotive Engineers, Paper No. 861351, September 1986.
23. Tanaka, H., and Ishihara, T., "Electro-Hydraulic Digital Control of Cone-Roller Toroidal Traction Drive Automatic Power Transmissions," Journal of Dynamic Systems, Vol. 106, December 1984.
24. Muraki, M., and Kumura, Y., "Traction Characteristics of Synthetic Hydrocarbon Oils," Japanese Journal of Lubrication, Vol. 30, No. 10, 1985, pp. 65-72.

BIBLIOGRAPHY

The literature in the Bibliography references additional information on Continuously Variable Transmissions for passenger vehicles. The CVT concept has been addressed by many organizations. In order to broaden the usefulness of this publication, listed below are publications, including abstracts, from some of these other sources.

Copies of SAE papers are available either in original or photocopy form from SAE. For ordering information contact the Publications Division, SAE, 400 Commonwealth Drive, Warrendale, Pennsylvania, 15096, USA.

For your convenience, this Bibliography is divided into five sections:

- Efficiency and Performance
- System Control and Design
- Metal Belt or Chain Drives
- Non-Metallic Belt Drives
- Traction Drives

EFFICIENCY AND PERFORMANCE

Evaluation of Split-Path Extended Range Continuously-Variable Transmissions for Automotive Applications, N. H. Beachley, C. Anscomb, C. R. Burrows, Journal of the Franklin Institute, Vol. 317, No. 4, April, 1984, p. 235-262

One potentially important application of continuously variable transmissions (CVT's) is the improvement in fuel economy of internal combustion engine vehicles. Split-path configurations can be used to extend the ratio coverage, providing even a geared neutral and reverse capability with some combinations of parameters. The increased coverage is achieved, however, only at the expense of lower overall CVT efficiency and a larger required variable unit capacity. All possible single-regime designs which use a single differential to extend the ratio coverage of the basic variable unit are considered.

Performance of a Prototype Continuously Variable-Ratio Traction Drive, E. G. Trachman, Y. Kemper, Lubrication Engineering, Vol. 39, No. 3, March, 1983, p. 163-168

This paper discusses the development of a novel mechanical continuously variable-ratio transmission which has the power density required for highway vehicles. The concept, design, and advantages of the new CVT are presented along with the transmission's performance. The overall efficiency of the transmission is high. It was measured to be 94 percent at rated power.

Analytical Study of Vehicle Performance and Fuel Economy Potential of Continuously Variable Transmission, A. Numazawa, S. Okawa, Hydraulic Pneumatic Mechanical Power, Vol. 27, No. 313, January, 1981, p. 23-27

The paper describes an analytical study conducted to show the vehicle performance and fuel economy potential of a CVT. A new method of determining the characteristics of engine and CVT gear ratio spread is described, satisfying the fuel economy and the required vehicle performance target by computer simulation. A numerical example is investigated.

Fiat Research Center Hybrid Vehicle Prototype, L. Morello, R. Piccolo, L. Ippolito, SAE No. 790014, February, 1979

The scope of the engine-electric hybrid propulsion systems studies that have been developed by Fiat Research Center is the energy consumption evaluation in comparison with conventional traction drives. After a general study of the hybrid propulsion systems mathematical models were developed for the evaluation of performance and energy balance. The construction of a prototype with parallel hybrid propulsion system has been completed to validate the computer simulation models. Computer and on-road test results are reported. Criteria used to select the parameters to the optimization of the system are discussed.

Evaluation of Advanced Automobile Transmissions, R. W. Hull, C. W. Coon, Jr., Dept. of Energy, Washington, D.C., DOE/CS/54231-T1, 1981

Two advanced design transmissions, one now in production (the Ford integral overdrive) and the other in prototype state of development (Orshansky continuously variable), were evaluated to determine the increase in fuel economy achievable through the use of a transmission that would provide an improved engine speed/load relationship. The EPA Federal Test Procedure was used for the fuel consumption and emissions measurements. Ford Fairmont test cars, powered by a 302 CID engine, showed 12.8 percent and 17.6 percent increase in mileage for the Ford integral overdrive (FIOD) and the Orshansky continuously variable transmission (CVT) respectively when compared with the baseline C-4 transmission. While achieving the fuel economy gains, emissions increased slightly with the FIOD transmission and were lower (measured in a different car with modified engine) for the CVT equipped car. When tested using light truck inertia loadings, the FIOD continued to show 7 to 8 percent fuel economy improvements with no trend toward higher exhaust emissions. A production lead time for the CVT was estimated to be 4 to 4½ years. Capital investment costs were estimated to be $261 million for the CVT and the incremental cost increase to the consumer was estimated to be in a range of $46 to $62 over the cost of the C-4 transmission.

Results of Vehicle-Performance and Fuel Consumption Tests of the Research Car Uni-Car with Electronic Engine-Transmission-Control, J. Helling, U. Rohs, E. Stall, Society of Automotive Engineers (Australasia) Report No. SAEA 82074, 1982

The driving efficiency of motor vehicles during vehicle operation is not only determined by the quality of the engine but also by the cooperation between engine and transmission. The reason for this is that the engine power is produced with very different efficiencies contingent on torque and rotational speed. On the other hand, there is only one operation point for each power requirements, at which power is produced with least expense of fuel. It is not possible to achieve the greatest fuel efficiency with a manually operated transmission and fuel supply control. On the contrary, the highest fuel efficiency is reached by using continuously variable transmission (CVT) with an 'intelligent'' automatic control. This concept of drive management is installed in the research car Uni-Car developed by a German university study group by order of the Ministry for Research and Technology. Theoretical research and results of vehicle performance and fuel consumption tests using this CVT with the automatic control are reported.

Design and Development of a Continuously Variable Ratio Transmission for an Automotive Vehicle. Phase IV—Quarterly Progress Report, Dept. of Energy, Washington, D.C., 1978

Progress in the design and development of a continuously variable ratio transmission for an automotive vehicle is reported. The major automotive hydromechanical transmission development problem continues to be the reduction of hydrostatic noise and the project plan, therefore, concentrated on the new hydrostatic module. The potential for achieving acceptably low noise levels in the second generation hydromechanical transmission is to be assessed by comparing the noise levels of the hydrostatic modules for the first and second generation transmissions. A set of twelve test points was selected comprising of road load steady state and wide-open-throttle acceleration at 10, 20, 30, 40, 50 and 60 mph. The module operating conditions for the two transmissions at each of these twelve points were calculated. Baseline noise data was measured on the first generation module. The results are given testing of co-axial hydrostatic module for second generation hydromechanical transmission will be emphasized.

SYSTEM CONTROL AND DESIGN

CVT: Driver's Impressions are Crucial, A. Hope, Automotive Engineer (London), Vol. 9, No. 1, February/March, 1984, p. 47-49

A discussion is presented of the technicalities of a CVT system already adopted by one OEM and possibly to be taken up by another. An investigation is presented of systems in detail: electronic control under development by Ford and the Van Doorne hardware productionised by Fiat.

CVT Control Systems: Some Design Considerations, Automotive Engineering, Vol. 92, No. 10, October, 1984, p. 63-69

Continuously variable transmission have improved greatly in reliability, durability, and efficiency. Vehicle system control must now be reexamined. Several general design factors and goals are outlined.

Leyland Constantly Variable Transmission, K. Garrett, Automotive Engineer (London), Vol. 8, No. 3, June/July, 1983, p. 28-29

Commercial vehicles equipped with a new constantly variable transmission (CVT) by Leyland will be in the hands of operators, for user trials, by the end of 1984, and production is planned to begin in 1986. Moreover, the same CVT, together with a flywheel energy storage system for regenerative braking, will be in a Leyland National bus by the end of 1984 and this is expected to be on user trials by 1986. With this CVT it will be possible to design the engine specifically for maximum efficiency and to match the transmission to it for obtaining the required performance and driveability characteristics. Such a concept shows prospects for improving fuel economy by about 22 percent, while the addition of flywheel regenerative braking should bring this figure up to 30 or perhaps even 40 percent. This transmission comprises a mechanism called the variator and a planetary gear set. Control is exercised by a microprocessor. The variator input assembly comprises a pair of discs on a common shaft, and the output is a third disc interposed coaxially between them but rotating in the opposite direction on a roller bearing between it and the shaft. Three small discs, or rollers, transmit the drive from each input disc to the central output disc, so there are six of these rollers in all. Tilting them varies the ratio of the drive: when the contact between the rollers and the input disc is at a small radius and that between them and the output disc at a larger radius, the output ratio is a reduction and vice versa.

Getting Rid of the Gearbox, Engineering (London), Vol. 223, No. 4, April, 1983, p. 285-287

The article discusses a practicable, stepless, continuously-variable-speed drive developed by Leyland vehicles. The Leyland continuously variable transmission (CVT) includes a variator unit that consists of a driven rotor comprising two discs on a common shaft which in turn drive a third disc located between them and through which their shaft passes but does not contact. Each driving disc transmits power to that side of the driven disc facing it, via rollers disposed radially at intervals of 120 degrees. These rollers, in turn, contact the surfaces of curved toroidal trackways in the faces of the discs and by means of a hydraulic actuator are able to alter in synchronism the angle of tilt of their axes of rotation. This enables them to establish a stepless change of speed ratio between the inlet and output discs. Among the subjects covered in the article are input/output speed ratios, low and high regimes, microcomputer control, and possible regenerative braking applications.

Assessment of Flywheel Energy Storage in Electric Vehicles, C. R. Burrows, G. Price, F. G. Perry, SAE No. 800885, 1980

The analysis and design of a hybrid drive for a 7.5 ton urban delivery van are described. The drive consists of a flywheel to supply the power for vehicle acceleration and a battery pack and compound-wound dc motor to supply the steady-state power requirements. The motor output is automatically regulated and the continuously variable transmission (CVT) is controlled to match the flywheel to the vehicle demand. Various types of CVT are considered and their relative merits are compared. Results of a digital simulation are presented comparing the hybrid drive with an equivalent electric drive when the vehicle follows a standard duty cycle. The simulation is used to study the effect of varying the capacity of the flywheel and the benefits of regenerative braking are evaluated.

Differential Split Power Transmissions for a Single Shaft Passenger Car Gas Turbine Engine, D. L. Carriere, American Society of Mechanical Engineers No. 80-GT-19, 1980

Descriptions, schematics, equations, and efficiency curves for a unique single shaft gas turbine engine transmission, consisting of a power splitting differential reduction gear combined with either a variable stator torque converter variable ratio traction drive as a speed variator and with a current production automatic four-speed gear box as a ratio expander, are contained in this paper.

METAL BELT OR CHAIN DRIVES

Design Study of Steel V-Belt CVT for Electric Vehicles, J. C. Swain, T. A. Klausing, J. P. Wilcox, National Aeronautics and Space Administration, NASA-CR-159845, DOE/NAA/0116-80/1

A continuously variable transmission (CVT) design layout was completed. The intended application was for coupling the flywheel to the driveline of a flywheel battery hybrid electric vehicle. The requirements were that the CVT accommodate flywheel speeds from 14000 to 28000 rpm and driveline speeds of 850 to 5000 rpm without slipping. Below 850 rpm a slipping clutch was used between the CVT and the driveline. The CVT was required to accommodate 330 ft-lb maximum torque and 100 hp maximum transient. The weighted average power was 22 hp, the maximum allowable full range shift time was 2 seconds and the required lift was 2600 hours. The resulting design utilized two steel V-belts in series to accommodate the required wide speed ratio. The size of the CVT, including the slipping clutch, was 20.6 inches long, 9.8 inches high and 13.8 inches wide. The estimated weight was 155 lbs. An overall potential efficiency of 95 percent was projected for the average power condition.

NOT METALLIC BELT DRIVES

"Rubber Band" Cars are Coming Back, Mechanical Engineering, Vol. 106, No. 12, December, 1984, p. 34-38

Continuously variable transmissions, or CVT's, can boost fuel economy by 30 percent compared with conventional automatic transmissions. Invented in 1896, CVT's are now back in favor once again, with ongoing research at Fiat, Ford, General Motors, Subaru, and others. The Dutch Daffodil (or Daf) introduced a CVT using rubber belts in 1959 and produced it for many years; Volvo has produced limited-run CVT cars since 1974. A CVT, for automobile applications, is usually a variable-pitch belt drive with pulleys. The variable-pitch belt drive provides any reduction ratio continuously over a specified range. The adjustable ratio is obtained by use of adjustable-pitch, conical sheaves. When the sheaves move laterally on their shafts, the sheave separation determines the depth at which the belt rides inside the sheave. In effect, this separation creates different sheave diameters which results in different ratios. With the CVT, the speed ratio can be selected independently of transmitted torque. What this means is that with a CVT, a car's engine can always be operated at its most fuel-efficient speed. Control systems for both the CVT and the engine are also described as well as CVT's using other means than the belt-sheave system.

TRACTION DRIVES

Advances in Traction Drive Technology, S. H. Loewenthal, N. E. Anderson, D. A. Rohn, NASA Tech. Memo 83397, 1983

Traction drives are among the simplest of all speed changing mechanisms. Although they have been in industrial use for more than 50 years, their operating characteristics and performance capabilities are not widely known. This paper briefly traces their technical evolution from early uses as main transmission in automobiles at the turn of the century to modern, high-powered traction drives capable of transmitting hundreds of horsepower. Recent advances in technology are described which enable today's traction drive to be a serious candidate for off-highway vehicles and helicopter applications.

Multiroller Traction Drive Speed Reducer—Evaluation for Automotive Gas Turbine Engine, D. A. Rohn, N. E. Anderson, S. H. Loewenthal, NASA Tech. Memo 2027, 1982

Tests were conducted on a nominal 14:1 fixed-ratio Nasvytis multiroller traction drive retrofitted as the speed reducer in an automotive gas turbine engine. Power turbine speeds of 45000 rpm and a drive output of 102 kW (137 hp) were reached. The drive operated under both variable roller loading (proportional to torque) and fixed roller loading (automatic loading mechanism locked). The drive operated smoothly and efficiently as the engine speed reducer. Engine specific fuel consumption with the traction speed reducer was comparable to that with the original helical gearset.

Transaxle Design for a Traction Continuously Variable Transmission for Automobiles, C. E. Kraus, SAE No. 800102, 1980

A continuously variable, wide ratio range, traction transmission (CVT) is discussed as a transaxle design for automotive use. A configuration positively driving both wheels and eliminating a separate differential promises lower manufacturing costs, better vehicle performance and high fuel mileage. The over 90% efficient and very quiet CVT depends on a traction design able to take a very high shock loading with a long maintenance free life. Traction requirements and solutions are discussed. Applications to internal combustion engine, electric and hybrid vehicles explain the need for a CVT with a wide ratio range.

Perbury Transmission, F. G. de B. Perry, Journal of Engineering for Power, ASME Transactions, Vol. 102, No. 4, October, 1980, p. 937-942

The Perbury transmission is a form of toroidal traction drive. This paper presents a brief overall view of its principles, background, present status of development, applications and future development areas.

Automotive CVT, J. H. Kraus, Mechanical Engineering, Vol. 98, No. 10, October, 1976, p. 38-43

A continuously variable transmission, CVT, can optimize the automotive propulsion system by allowing the engine to be operated in the most efficient manner regardless of road speed. For maximum acceleration the CVT allows the engine to run at peak power continuously rather than fluctuating with road speed, and the overall power rating fo the engine can be reduced for equivalent performance. The smaller engine increases the engine load factor for normal driving, further improving the engine and propulsion system efficiency. The development of the cone roller toroidal drive (CRTD) beginning in the late 1940's, along with modern bearing steels, synthetic traction lubricants, and modern computer-based analysis of the traction contact have led to a practical, efficient, quiet, wide ratio range transmission for automotive use. The CRTD is a "traction" drive, not a "friction" drive. In traction drives, friction is as undesirable as in any modern drive. With lubricated traction, where the use of the lubricant both guards against friction and helps cool the traction elements, the actual power transmission between the contacting elements is through the resistance to shear and slip of the lubricant film under very high contact pressure. Starting with the selected 12:1 ratio range cone roller toroidal drive, a complete automotive transmission can be designed. This transmission has a ratio range that varies from 1.54:1 in overdrive to 6.48:1 in reduction. An automatic clutch is used to start the car rather than the conventional hydraulic torque converter to allow normal vehicle operation with the engine below converter lockup speed of about 1600 rpm. This clutch must bring the car to about 3 1-2 mph (5.5 km/h) with an 800 rpm idling engine. The clutch also provides a torque limit through the drive train. Efficiency is generally in the range between 90 and 95 percent over the entire ratio and load range.

New and Better Traction Drives are Here, R. W. Carson, Machine Design, Vol. 46, No. 10, April, 1974, p. 148-155

The state of the art of variable-speed drives using friction gears is reported. The possibility of transmitting more power by this means has resulted from two recent developments: Cleaner steels having high fatigue strength and with impurities eliminated that could cause surface flaking under repeated high local stresses and a new lubricant that offers a 50% increase in traction coefficient, high heat resistance, increased metal fatigue life, and compatibility with metals, seals, and paint. Seven basic types of friction drive are described: ring cone with planetary gearing, the Beier disk, the Koop Variator (ball-and-disk); free-ball;' spool (between spherical disks); toroidal (tiltable rollers between input and output halves of toroidal disks). A toroidal drive is described which is under development for use in a 220 hp automobile. Most of the commercially available units have ratings in the range of 5 to 20 hp.

Lubricants for Traction Drives, R. L. Green, F. L. Langenfeld, Machine Design, Vol. 46, No. 11, May, 1974, p. 108-113

Various lubricants suitable for use with variable-speed drives using friction gears are described, with particular emphasis on a Monsanto product called Santotrac. In the friction-drive application, the lubricant operates in the elastohydrodynamic modes; it must not only minimize wear of the contacting surfaces but also transmit torque. For this latter function, the traction coefficient of the lubricant becomes important. Traction coefficient is the ratio of tangential force transmitted to the normal force on the traction elements. Santotrac, described as a synthetic cycloaliphatic hydrocarbon, has a traction coefficient of 9.5%, which is roughly double that of the extreme-pressure lubricant SAE 90. In addition, Santotrac is claimed to have good compatibility with metal, oil seals, and paint, and to approximately double the fatigue life of rolling elements. A tabulation is given of the significant characteristics of the various lubricants available for friction drives, and these characteristics are individually discussed. An experimental Pinto automobile using a continuously variable friction drive and Santotrac lubricant is described briefly, for which it is claimed that, compared to a standard automatic transmission, mileage was improved 30 to 50% and acceleration times were cut about 40%.

INDEX

232 page booklet. Printed in U.S.A.